An Introduction to
Programming
with
Mathematica®

SECOND EDITION

An Introduction to

Programming
with
Mathematica®

SECOND EDITION

Includes diskette

Richard J. Gaylord | Samuel N. Kamin | Paul R. Wellin

Richard J. Gaylord
Dept. of Materials Science
University of Illinois
Urbana–Champaign
Urbana, IL 61801 USA

Samuel N. Kamin
Dept. of Computer Science
University of Illinois
Urbana–Champaign
Urbana, IL 61801 USA

Paul R. Wellin
Dept. of Mathematics
California State University
Sonoma
Rohnert Park, CA 94928 USA

Published by TELOS, The Electronic Library of Science, Santa Clara, CA.
Publisher: Allan M. Wylde
Publishing Associates: Kate McNally Young, Keisha Sherbecoe
Product Manager: Carol Wilson
TELOS Production Manager: Jan V. Benes
Electronic Production Advisor: Joe Kaiping
Cover and Page Designer: André Kuzniarek

Cataloging in Publication Data is available from the Library of Congress.

Printed on acid-free paper.

Photocomposed pages prepared from the authors' TEX files.
Printed and bound by Hamilton Printing Co., Rensselaer, NY.
Printed in the United States of America.

9 8 7 6 5 4 3 2 1

ISBN 0-387-94434-6 Springer-Verlag New York Berlin Heidelberg

TELOS, The Electronic Library of Science, is an imprint of Springer-Verlag New York with publishing facilities in Santa Clara, California. Its publishing program encompasses the natural and physical sciences, computer science, economics, mathematics, and engineering. All TELOS publications have a computational orientation to them, as TELOS' primary publishing strategy is to wed the traditional print medium with the emerging new electronic media in order to provide the reader with a truly interactive multimedia information environment. To achieve this, every TELOS publication delivered on paper has an associated electronic component. This can take the form of book/diskette combinations, book/CD-ROM packages, books delivered via networks, electronic journals, newsletters, plus a multitude of other exciting possibilities. Since TELOS is not committed to any one technology, any delivery medium can be considered.

The range of TELOS publications extends from research level reference works through textbook materials for the higher education audience, practical handbooks for working professionals, as well as more broadly accessible science, computer science, and high technology trade publications. Many TELOS publications are interdisciplinary in nature, and most are targeted for the individual buyer, which dictates that TELOS publications be priced accordingly.

Of the numerous definitions of the Greek word "telos," the one most representative of our publishing philosophy is "to turn," or "turning point." We perceive the establishment of the TELOS publishing program to be a significant step towards attaining a new plateau of high quality information packaging and dissemination in the interactive learning environment of the future. TELOS welcomes you to join us in the exploration and development of this frontier as a reader and user, an author, editor, consultant, strategic partner, or in whatever other capacity might be appropriate.

TELOS, The Electronic Library of Science
Springer-Verlag Publishers
3600 Pruneridge Avenue, Suite 200
Santa Clara, CA 95051

THE
ELECTRONIC
LIBRARY
OF
SCIENCE

TELOS Diskettes

Unless otherwise designated, computer diskettes packaged with TELOS publications are 3.5" high-density DOS-formatted diskettes. They may be read by any IBM-compatible computer running DOS or Windows. They may also be read by computers running NEXTSTEP, by most UNIX machines, and by Macintosh computers using a file exchange utility.

In those cases where the diskettes require the availability of specific software programs in order to run them, or to take full advantage of their capabilities, then the specific requirements regarding these software packages will be indicated.

TELOS CD-ROM Discs

For buyers of TELOS publications containing CD-ROM discs, or in those cases where the product is a stand-alone CD-ROM, it it always indicated on which specific platform, or platforms, the disc is designed to run. For example, Macintosh only; Windows only; cross-platform, and so forth.

TELOSpub.com (Online)

Interact with TELOS online via the Internet by setting your World-Wide-Web browser to the URL: *http://www.telospub.com*.

The TELOS Web site features new product information and updates, an on-line catalog and ordering, samples from our publications, information about TELOS, data-files related to and enhancements of our products, and a broad selection of other unique features. Presented in hypertext format with rich graphics, it's your best way to discover what's new at TELOS.

TELOS also maintains these additional Internet resources:

gopher://gopher.telospub.com
ftp://ftp.telospub.com

For up-to-date information regarding TELOS online services, send the one-line e-mail message:
send info to: info@TELOSpub.com.

*To Carole, for a quarter-century of forbearance;
and to Darwin, for constant companionship
during the writing of this book.*
Richard

*To my mother, for her faith in me; and in loving
memory of my mother-in-law.*
Sam

*To Sheri, Sam, and Oona for keeping me going;
and to Bob, who nudged me down the road.*
Paul

Preface

Technical Computing for the Rest of Us

Computers have wrought a fundamental change in the nature of research, and in education in science and engineering. Experimentalists routinely use computers to collect and analyze data while theoreticians use computers to manipulate equations numerically and symbolically. For both, computer simulation studies have become an indispensable investigative tool.

In response, technical education is changing to incorporate the use of computers into the curriculum both as a topic and as a medium for the presentation of technical material.

The use of computers in all of these areas has been accelerating rapidly as computer hardware development has progressed to the point where nearly every person can afford his or her own self-contained computing environment. It even seems likely that the struggle by universities and colleges to obtain funds to establish computer laboratories of desktop computers will be obviated by students bringing their own powerful notebook computers to the classroom.

Yet, even as the hardware hurdle has been overcome, another obstacle has held back this "democratic" spread of technical computing power. That hindrance has been the paucity of software that is both powerful and user-friendly. While there are software packages for carrying out various kinds of technical computing, there are few that provide a fully integrated technical computing environment, including a programming language suitable for individuals who are not full-time programmers, but who need to create programs to carry out their work. *Mathematica* provides such an environment. It includes:

1. Built-in mathematical and graphical capabilities, that are both powerful and flexible.

2. A programming language that can be used to extend its capabilities virtually without limit. That language is interactive, has the capability to perform both numeric and symbolic manipulation, makes broad use of pattern matching, and supports the functional style of programming favored by many computer scientists (while incorporating constructs for more conventional programming styles).

3. Extensive on-line help facilities, including the Function Browser, a new feature in Version 2.2 that makes it easy to learn about built-in functions and get their syntax correct.

4. The ability to connect *Mathematica* to other computing environments and other languages.

5. An interface that allows text and graphics to appear together in documents.

In this book, we focus on the *Mathematica* programming language. While there are many books, including the reference manual by Stephen Wolfram [Wolfram 1991], that discuss various aspects of *Mathematica*, there has been a need for a text explaining how to use the underlying programming language so that *Mathematica's* capabilities can be fully utilized. This is the first book that explains *Mathematica's* programming language for the beginning programmer, emphasizing the programming styles that are most efficient and idiomatic in *Mathematica*: functional and transformational (or rule-based).

The Audience for this Book

This book was written for two distinct groups of readers:

- *Mathematica* users who have no prior programming experience. We start from the very beginning, explaining how to build *nested function calls*, create *anonymous* functions, and use *higher-order* functions. Recursive and iterative programming are also explained.

    We include in the category of beginners *Mathematica* users who have programmed in conventional languages such as Fortran, C, BASIC and Pascal. In many instances, these programmers will write *Mathematica* programs in a procedural style, producing code that looks like "Fortran in *Mathematica*." By understanding how *Mathematica* programs can be constructed using the functional and rule-based styles of programming, much simpler and more efficient code can be written.

- Those who want to learn to program, and would like a friendly and useful language to start with. Most low-level languages, like C and Fortran, require too many lines of complicated code in order to do something interesting while many high-level languages, like LISP, use an unnatural syntax and are difficult to master. *Mathematica* has a more natural syntax and provides very high-level built-in operations with which one can do a lot of interesting things right away.

Changes in the Second Edition

There are two major changes for this edition—exercises and solutions, and a new chapter on applications. In addition there are hundreds of more minor changes involving reorganization and cosmetics.

Dozens of exercises have been added to the previous edition. Some of these new exercises are of an elementary level, and others are more advanced. In addition, we have added solutions to many exercises with occasional commentary to guide the reader. The book contains solutions to over half of the exercises. Most of the remaining solutions are found on the diskette included with the book.

We have added a new chapter on applications (Chapter 11) and moved the chapter on contexts and packages to Chapter 12. The applications in Chapter 11 are widely different and give a good sense of how *Mathematica* can be used to solve a variety of problems of significant computational complexity. Chapter 12 (Contexts and Packages) has been expanded with an additional package that includes a discussion of options and defaults.

In several places throughout the book, there are references to new features that will be present in the next version of *Mathematica* after Version 2.2. We made note of those differences from Version 2.2 and earlier that are significant for the presentation.

Readers of the first edition found a handful of mistakes, which have been corrected in this edition. We welcome further communication from our readers.

How to Use This Book

There are several ways in which this book should be useful:

- As a primary text in an introductory course on programming. Perhaps the most obvious use would be in an introductory computer science course for students in science, engineering, or mathematics, who would solve technical problems of interest to them while at the same time learning to program. The *Mathematica* programming language supports many programming styles that can be transferred to other programming languages; the programming skills learned using *Mathematica* can be transferred to conventional programming languages.

- As a supplemental text in a course in which *Mathematica* is being used as a tool for studying another technical subject. In these kinds of courses, the principles of *Mathematica* programming can be introduced as needed during the course, although we have found that a structured introduction to *Mathematica* programming at the beginning of the semester works best.

- As a self-study book, particularly for *Mathematica* users who need to use its programming capabilities more fully or who are interested in understanding how the *Mathematica* programming language works.

Conventions Used in This Book

All input and output in this book appear in a different font from the regular text. This is true for examples that appear in the middle of text, such as Expand[(a + b)^4], as well as displayed *Mathematica* code. So, for example, lines of input (what you type at your computer) appear as

```
In[1]:= 3 + 5
```

whereas all output (what *Mathematica* prints on your computer screen) appears in a slightly lighter font than the input.

```
Out[1]= 8
```

You do not have to type the prompts In[1]:= or Out[1]=; *Mathematica* will do that for you.

In addition to being placed on the floppy diskette, all of the programs that are defined in this book are indexed in two different ways—under the program's name and also under the heading **Programs**. So, for example, the function runEncode defined on page 178 is listed in the index under runEncode and also under **Programs, runEncode**.

What is in This Book

We should perhaps start with what is *not* in this book, namely, a complete list and explanation of the hundreds of operations built into *Mathematica*. For that list, the indispensable reference is [Wolfram 1991].

This book is about teaching you how to program. We assume that you have never programmed a computer (though you may already be a *Mathematica* user), and we take you step by step through the various programming techniques you will want to use in *Mathematica*.

The chapter structure is as follows:

Chapter 1: Preliminaries. This chapter contains some of the fundamentals of *Mathematica*, such as how to use the *Mathematica* interface and how to enter expressions. Experienced *Mathematica* users should feel free to skim this material.

Chapter 2: Overview of *Mathematica*. For the reader who has no knowledge of *Mathematica* at all, we've included a chapter showing some of its built-in capabilities, including graphics. Like Chapter 1, it can be skimmed lightly if you are already a *Mathematica* user.

Chapter 3: Lists. The heart of the book begins here, with a discussion of this most important of data types. Aside from numbers, no other data type is more useful in programming. (For example, our bowling program consists mainly of computations on lists.) We will describe the use of built-in functions to manipulate lists.

Chapter 4: Functions. Here we continue the discussion of Chapter 3, introducing the nesting of function calls and higher-order functions. The combination of built-in functions introduced in this and the previous chapters can solve many programming problems.

Chapter 5: Evaluation. Before going further into programming, we find that we need to explain exactly how *Mathematica* carries out your computations. This permits us to introduce the technique of rule-based programming.

Chapter 6: Conditionals. A vital feature of computers is their ability to make decisions to take different actions depending upon properties of their inputs. This chapter shows how the *Mathematica* programmer can include such decision-making in his programs.

Chapter 7: Recursion. This method of programming—in which a function is defined in terms of itself—is heavily used in *Mathematica*.

Chapter 8: Iteration. This is another programming method, in which a result is obtained by repeating certain steps a specified number of times. This programming style is more characteristic of conventional languages like C than it is of *Mathematica*, but is still important.

Chapter 9: Numerics. *Mathematica* contains a variety of types of numbers. This chapter discusses their similarities and differences and points out some of the issues that must be confronted when computing with both imprecise numbers and exact numbers.

Chapter 10: Graphics Programming. This chapter introduces the basic concepts of *Mathematica* graphics. It uses some of the techniques of the previous chapters to create graphics-based programs and to solve problems that are inherently graphical in nature.

Chapter 11: Applications. *Mathematica* is used to solve larger-scale programming problems. Included are *Mathematica* implementations of the Game of Life using functional and rule-based programming, random walks, and finally the creation of a mini picture-description language.

Chapter 12: Contexts and Packages. The last chapter introduces the method of organizing libraries of *Mathematica* functions into convenient units called *packages*.

The Diskette

The material in this book has been made available in electronic form. *Mathematica* notebooks containing most of the examples, exercises, and solutions are included on a floppy diskette on the inside back cover of this book. The file names correspond directly to the chapter structure. So for example, the notebook containing the material from Chapter 1 is called `Chap1.ma`.

The diskette also contains files with *Mathematica* programs that can be loaded into your *Mathematica* sessions as needed. These files (normally called *packages*), have a filename extension ending with `.m`. These can be used in your *Mathematica* session by first making sure the files are in a directory that *Mathematica* can find (SetDirectory["*PathToFiles*"]), and then reading them in (<<file.m).

The diskette is a 3.5", 1.44 megabyte high-density disk that is formatted in a DOS-compatible format. It should be readable by any IBM-compatible computer, as well as by any Unix machine, NeXT computer, or Macintosh computer. The files are available in several operating system formats on the diskette: DOS, Macintosh, and Unix archives. This should prevent file transfer problems. The disk contains a ReadMe.txt file with further instructions on its usage.

In addition to the diskette, these materials are available from Wolfram Research's *MathSource*, an electronic distribution service of *Mathematica* materials. To obtain information about *MathSource*, send an email message to mathsource@wri.com containing the one-line message help intro. If you just wish to obtain the materials from this book, then send the message: find 0204938. *MathSource* will then send you a file indicating the materials from this book that are available for you to download. The intro file describes the *MathSource* service in greater detail, including instructions on how to query and obtain any of the materials it provides.

The *Mathematica* notebooks and packages for this book can also be accessed via the World-Wide Web by setting your Web browser to http://www.telospub.com. The materials on this site will be updated periodically, as needed.

A Final Note to The Reader

While the basic aspects of *Mathematica* programming are discussed in this book, there are a great many more things that can be said about *Mathematica* and the *Mathematica* programming language. The most comprehensive reference source is the manual that comes with each copy of the software: *Mathematica, A System for Doing Mathematics by Computer* [Wolfram 1991].

MathSource contains hundreds of programming examples and notebooks from all areas of science and engineering. These materials can be obtained by sending electronic mail to mathsource@wri.com. A description of how to access the materials on *MathSource* is contained on the floppy diskette included with this book.

For those with access to the Internet, a newsgroup exists that is devoted to *Mathematica*: comp.soft-sys.math.mathematica. In addition, a mailing list is also available that essentially mirrors the newsgroup. To subscribe to the mailing list, send a message to mathgroup-request@christensen.cybernetics.net asking to be added to the list. Furthermore, common questions are answered in a FAQ (frequently-asked questions) at http://www.wri.com/techsupport/.

Mathematica in Education and Research is a quarterly journal that is published by *TELOS*/Springer-Verlag that includes articles and notes about the use of *Mathematica* in the classroom. It includes regular columns on programming and also a student column. To subscribe, send a request to *Mathematica in Education and Research*, *TELOS*/Springer-Verlag, 333 Meadowlands Parkway, Secaucus, NJ 07094 USA, or send an email request to `MathInEd@telospub.com`.

The Mathematica Journal is a quarterly journal that publishes articles about all aspects of *Mathematica*. For subscription information, contact Miller Freeman, Inc., 600 Harrison Street, San Francisco, CA; 415-905-2334.

Finally, there are now over 100 books that are about *Mathematica* or that use *Mathematica* to teach a subject. A partial listing can be found in the bibliography at the end of this text. For a complete up-to-date publications list, contact Wolfram Research.

Colophon

This book was produced using LaTeX from original TeX source files and *Mathematica* notebooks. *Mathematica* Version 2.2 was used throughout, except in several places where the new features of Version 3.0 are explicitly discussed. The notebooks were converted to LaTeX files using the **nb2tex** converter available from *MathSource*. All graphics were produced in *Mathematica* and imported to the TeX documents by means of the options available in the `epsf.sty` style file. PostScript files were generated with **dvips** and cut into individual page PostScript files before being output to film.

Acknowledgments

We have been fortunate in having invaluable help in this project from two organizations: *TELOS* and Wolfram Research, Inc.

Our publisher, Allan Wylde, immediately recognized the need for this book when we proposed it to him and he strongly encouraged and supported us during the project's nine month gestation. We thank him for his insight and willingness to tackle a non-traditional project of this sort. Kate McNally-Young, the publishing associate, Keisha Sherbacoe, the publishing assistant, and Jan Benes, the production editor provided assistance throughout the project.

Many individuals at Wolfram Research contributed in various ways to the project, including technical assistance, providing beta copies of new versions of *Mathematica*, manuscript reviewing, and cheerleading our efforts. In particular, we would like to thank Stephen Wolfram, Prem Chawla, Joe Kaiping, Andre Kuzniarek, Ben Friedman, Mark Moline, Amy Young, and Joe Grohens.

The reviewers of the manuscript provided very helpful criticism and encouragement: We thank Claude Anderson, Bill Davis, Allan Hayes, Ralph Johnson, Jerry Keiper, John Novak, Jose Rial, Michael Schaferkotter, Bruce Smith, and Doug Stein.

We also thank the reviewers of the second edition: Claude Anderson, Paul C. Hoffman, Martin Maxey, Stephanie Rieser, and Winthrop W. Smith.

Richard J. Gaylord would like to thank Shawn Sheridan who was his guru for both *Mathematica* and the Macintosh.

Samuel Kamin would like to acknowledge the Computer Science Department of the University of Illinois for its excellent working environment and the support of his colleagues there. Above all, he would like to thank Judy and Rebecca for their love and patience.

Paul Wellin would like to thank his wife Sheri for her support, understanding, and sense of humor throughout this project.

Finally, we would like to hear of any errors that you, the reader, may find in the book or in the electronic materials so that we can fix them for future editions.

Richard J. Gaylord
(gaylord@ux1.cso.uiuc.edu)

Samuel N. Kamin
(kamin@cs.uiuc.edu)

Paul R. Wellin
(wellin@groucho.sonoma.edu)

August 1995

Contents

1 Preliminaries

Mathematica is a very large and seemingly complex system. It contains hundreds of functions for performing various tasks in science, mathematics, and engineering. It has its own language with well-defined rules and syntax. In this preliminary chapter, we will give some of the basic information that users of *Mathematica* need to know, such as how to start *Mathematica*, how to get out of it, how to enter simple inputs and get answers, etc. In addition, we will give a hint of the internal structure of this system in order to begin the process of understanding what this book is all about—the *Mathematica* programming language.

1.1 Introduction

Mathematica has been aptly described as a sophisticated calculator. With it, you can enter complicated formulas and get their values.

```
In[1]:= Sin[.86] - Log[1.23](1 + .08/12)^12
Out[1]= 0.533646
```

You can store values in memory.

```
In[2]:= rent = 350
Out[2]= 350
```

```
In[3]:= food = 175
Out[3]= 175
```

```
In[4]:= heat = 83
Out[4]= 83
```

```
In[5]:= rent + food + heat
Out[5]= 608
```

You can factor polynomials, find roots of functions, calculate definite and indefinite integrals, plot functions, and much, much more.

```
In[6]:= Factor[x^2 - 1]
```

$Out[6]= (-1 + x) (1 + x)$

```
In[7]:= Solve[x^3 - 2x + 1 == 0, x]
```

$Out[7]= \{\{x \to 1\}, \{x \to \dfrac{-1 - Sqrt[5]}{2}\}, \{x \to \dfrac{-1 + Sqrt[5]}{2}\}\}$

```
In[8]:= Integrate[x^2.5 - 5.0911x, {x, 0, 20}]
```

$Out[8]= 9203.81$

```
In[9]:= Integrate[1/(1 - x^3), x]
```

$$Out[9]= \dfrac{ArcTan[\dfrac{1 + 2\,x}{Sqrt[3]}]}{Sqrt[3]} - \dfrac{Log[1 - x]}{3} + \dfrac{Log[1 + x + x^2]}{6}$$

```
In[10]:= Plot[Sin[x + Sqrt[2] Sin[x^2]], {x, -Pi, Pi}]
```

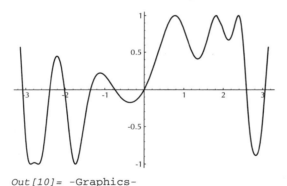

$Out[10]= -Graphics-$

Mathematica has a convenient user interface, in which you can easily refer to the result of the previous calculation using the symbol %.

```
In[11]:= 2^64
```

$Out[11]= 18446744073709551616$

```
In[12]:= %
```

$Out[12]= 18446744073709551616$

You can also refer to the result of any earlier calculation using its Out[i] label or, equivalently, %i.

```
In[13]:= Out[1]

Out[13]= 0.533646


In[14]:= %6

Out[14]= (-1 + x) (1 + x)
```

With a copy of the reference manual by Stephen Wolfram [Wolfram 1991] or one of the abbreviated reference manuals ([Blachman 1991] or [Wolfram 1992]) listing the hundreds of operations included in *Mathematica*, it would seem you can compute just about anything you might want.

But that impression is mistaken. There are simply more kinds of calculations than could possibly be included in a single program. Whether you're interested in computing bowling scores or finding the mean square distance of a random walk on a torus, *Mathematica* doesn't have a built-in function to do it. What it *does* have—and what really makes it the amazingly useful tool it is—is the capability for users to define their own functions. This is called *programming*, and it is what this book is all about.

To give you an idea of what you can do, here is a program to compute the score of a game of bowling, given a list of the number of pins scored by each ball.

```
In[15]:= BowlingScore[pins_] := Module[{score},
            score[{x_, y_, z_}] := x + y + z;
            score[{10, y_, z_, r___}] := 10+y+z+score[{y, z, r}];
            score[{x_, y_, z_, r___}] := x+y+z+score[{z, r}] /;
                                         x + y == 10;
            score[{x_, y_, r___}] := x+y+score[{r}] /; x+y < 10;
            score[If[pins[[-2]] + pins[[-1]] >= 10, pins,
                Append[pins, 0]]]]
```

Here is the computation for a "perfect" game—12 strikes in a row.

```
In[16]:= BowlingScore[{10, 10, 10, 10, 10, 10, 10, 10, 10, 10, 10, 10}]

Out[16]= 300
```

You're not supposed to understand this now—that's why we wrote this book! What you should understand is that there are many useful calculations that *Mathematica* doesn't know how to do, so to make full use of its many capabilities, you will sometimes need to program. The purpose of this book is to show you how.

Another purpose is to teach you the basic principles of programming. These principles—combining built-in functions, using conditionals, recursion, and iteration—are applicable (with great differences in detail, to be sure) to all other programming languages.

1.2 Using *Mathematica*

First things first. Before you can do any serious work, you'll need to know how to get a *Mathematica* session started, how to stop it, and how to get out of trouble when you get into it. These procedures depend somewhat on the system you're using. You should read the system-specific information that came with your copy of *Mathematica*, and you may need to consult a local *Mathematica* guru if our advice here is not applicable to your system.

1.2.1 | Getting Into and Out of *Mathematica*

There are two basic types of interfaces to *Mathematica*: the "textual" (or "command line") interface and the "notebook" (or "graphical") interface. These are discussed in more detail in Section 1.4, but for some of what we're discussing here, it helps to know which one you're using. If you look at Sections 1.4.1 and 1.4.2, you'll be able to tell as soon as you've started your *Mathematica* session which interface you have. Probably, if you're using a Macintosh, a PC running Windows, or most Unix workstations, you have the notebook interface; if you're using a PC not running Windows, or a VAX workstation, you probably have the textual interface.

Starting *Mathematica*

How you start *Mathematica* on your computer will depend upon whether you have the notebook front end, or are using the command line to enter commands:

> *Notebook interface:* Select the *Mathematica* icon (shown on page 21), and double-click.
> *Textual interface:* Enter the command **math** at a shell prompt.

Entering input

Input is always typed exactly as it appears in this book. To get *Mathematica* to evaluate an expression that you have entered:

> *Notebook interface:* After typing your input, press Shift-Return or Enter.
> *Textual interface:* Press Enter.

Note that in the textual interface, every line is evaluated unless it is not a complete expression (this is illustrated on page 9). In the notebook interface, none of the lines you type are evaluated until you press Shift-Enter; then they all are.

Ending a *Mathematica* session

Notebook interface: Select **Quit** from the **File** menu.

Textual interface: Type Quit at the prompt.

Getting out of trouble

From time to time, you will type an input which will cause *Mathematica* to misbehave in some way, perhaps by just going silent for a long time (if, for example, you've inadvertently asked it to do something very difficult), or perhaps by printing out screen after screen of information that you can't use. In this case, you can try to "interrupt" the calculation. How you do this depends on the machine you are working on.

> *Macintosh or NeXT:* Type Command-. (the Command key and the period) and then type a.
>
> *PC with Windows:* Type Alt-. (the Alt key and the period)
>
> *PC without Windows:* Type Ctrl-Break
>
> *Unix:* Type Ctrl-C and then type a and then Return

With the notebook interface, these attempts to stop the computation will sometimes fail. If after waiting a reasonable amount of time (say, a few minutes), *Mathematica* still seems to be stuck, you will have to "kill the kernel."[1] This is accomplished by selecting **Quit/Disconnect Kernel** from the **Action** menu. In versions of *Mathematica* earlier than Version 2.2, *Mathematica* will then have to be restarted. Starting with Version 2.2, the kernel itself can be restarted without killing the Front End by first selecting **Kernels and Tasks...** under the **Action** menu and then selecting New Kernel in the dialog window.

1.2.2 | Getting Help

If you are aware of the name of a function but are unsure of what it does, the easiest way to find out about it is to enter ?function. For example, if you want to find out about the function ParametricPlot, you should type

---

[1] Before attempting to kill the kernel, try to convince yourself that the computation is really in a loop from which it will not return and that it is not just an intensive computation that requires a lot of time.

```
In[1]:= ?ParametricPlot
```

```
ParametricPlot[{fx, fy}, {t, tmin, tmax}] produces a
   parametric plot with x and y coordinates fx and fy
   generated as a function of t. ParametricPlot[{{fx,
   fy}, {gx, gy}, ...}, {t, tmin, tmax}] plots several
   parametric curves.
```

Also, if you were not sure of the spelling of a command (Integrate, for example), you could type the following to display all built-in functions that start with Int:

```
In[2]:= ?Int*
```

```
Integer                    InterpolationOrder
IntegerDigits              Interrupt
IntegerQ                   Intersection
Integrate                  Interval
InterpolatingFunction      IntervalIntersection
InterpolatingPolynomial    IntervalMemberQ
Interpolation              IntervalUnion
```

Users with the notebook interface can select the menu item **Complete Selection** under the **Action** menu. This will give a list similar to that above. They can also use command completion and templates to find out more about functions for which they have only partial knowledge.

Beginning with Version 2.2, notebook versions of *Mathematica* contain a useful addition to the help system called the *Function Browser*. This is explained in Section 1.6.1.

1.2.3 | The Syntax of Inputs

When you first start *Mathematica*, you will notice that it enters the In and Out prompts for you. You do not type these prompts. Notebook users will see these prompts *after* they evaluate their input, whereas command line users will have a prompt displayed by *Mathematica* waiting for their input:

```
In[1]:= 39/13
```

```
Out[1]= 3
```

Your inputs to *Mathematica* will consist of *expressions*, numerical and otherwise. Numerical expressions are typed as in ordinary mathematics, as shown earlier. However, since you can't enter a fraction on multiple lines, you have to use a slash (/) for division, as we did above. Similarly, use the caret (^) for exponentiation:

```
In[2]:= 2^5
Out[2]= 32
```

Mathematica allows you to indicate multiplication by juxtaposing two expressions, as in mathematics, but it also gives you the option of using the asterisk (*) for that purpose, as is traditional in computer programming:

```
In[3]:= 2 5
Out[3]= 10

In[4]:= 2*5
Out[4]= 10
```

Mathematica also gives operations the same *precedence* as in mathematics. In particular, multiplication and division have a higher precedence than addition and subtraction, so that 3 + 4 * 5 equals 23 and not 35.

Functions are also written as they are in mathematics books, except that function names are capitalized and their arguments are in square brackets:

```
In[5]:= Factor[x^5 - 1]
                          2    3    4
Out[5]= (-1 + x) (1 + x + x  + x  + x )
```

Almost all of the built-in functions are spelled out in full, as in the above example. (The exceptions to this rule are well known abbreviations such as D for differentiation, Sqrt for square roots, Log for logarithms, Det for the determinant of a matrix, etc.) Spelling out the name of a function in full is quite useful when you are not sure whether a function exists to perform a particular task. For example, if we wanted *Mathematica* to compute the conjugate of a complex number, an educated guess would be:

```
In[6]:= Conjugate[3 + 4 I]
Out[6]= 3 - 4 I
```

Whereas square brackets [and] are used to enclose the arguments to functions, curly braces { and } are used to indicate a *list* or range of values; *Mathematica* has very powerful list-manipulating capabilities that will be explored in detail in Chapter 3. Lists may also appear as additional arguments to functions, such as in `Plot` and `Integrate`.

In[7]:= **Plot[Sin[x + Sqrt[2] Sin[x]], {x, -2Pi, 2Pi}]**

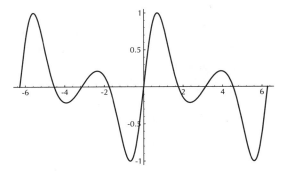

Out[7]= -Graphics-

In[8]:= **Integrate[Cos[x], {x, a, b}]**

Out[8]= -Sin[a] + Sin[b]

In the `Plot` example, the list `{x, -2Pi, 2Pi}` indicates that the function $\sin(x + \sqrt{2}\sin(x))$ is to be plotted over an interval as x takes on values from -2π to 2π. The `Integrate` command above is equivalent to the mathematical expression $\int_a^b \cos(x)\, dx$.

When you end an expression with a semicolon (;), *Mathematica* computes its value but does not print it. This is very helpful when the result of the expression would be very long and you don't need to see it. In the following example, we first create a list of the integers from 1 to 10,000, suppressing their display with the semicolon, and then compute their sum and average.

In[9]:= **x = Range[10000];**

In[10]:= **Apply[Plus, x]**

Out[10]= 50005000

In[11]:= **% / 10000**

$$Out[11]= \frac{10001}{2}$$

An expression can be entered on multiple lines, but only if *Mathematica* can tell that it's not finished after the first line. For example, you can enter 3* on one line and 4 on the next,

```
In[12]:= 3 *

        4
Out[12]= 12
```

but not 3 on the first line and *4 on the second.

```
In[13]:= 3
Out[13]= 3

In[14]:= *4
Syntax::sntxb: Expression cannot begin with "*4".
```

If you use parentheses, you can avoid this problem.

```
In[15]:= (3
           *4)
Out[15]= 12
```

With the notebook interface, you can input as many lines as you like; *Mathematica* will evaluate them all when you enter **Shift-Enter** still obeying the rules stated above for incomplete lines.

Finally, you can enter a *comment*—some words that are not evaluated—by entering the words between (* and *).

```
In[16]:= D[Sin[x],      (* differentiate Sin[x]   *)
             {x, 1}]    (* with respect to x once *)
Out[16]= Cos[x]
```

Postfix Form

As stated above, all functions in *Mathematica* can be written in the form head[arg_1, arg_2, ...]. Occasionally we will find it convenient to use another form for inputting expressions, known as *postfix form*. Instead of writing an expression in the standard form f[*expr*], the postfix form *expr* // f can also be used. This form is quite useful especially when added to the end of an already composed expression. Here we use the function `Factor` in postfix form on an expression we factored previously:

```
In[1]:= x^5 - 1 //Factor
                          2     3     4
Out[1]= (-1 + x) (1 + x + x  + x  + x )
```

This is equivalent to `Factor[x^5 - 1]`.

1.2.4 | Errors

In the course of using and programming in *Mathematica*, you will encounter various sorts of errors, some obvious, some very subtle, some easily rectified, and others not. We've already mentioned (page 5) that it is possible to send *Mathematica* into an infinite loop—where it goes on forever doing nothing useful—and that in this case you can interrupt the computation. In this section, we discuss those situations where *Mathematica* does finish the computation, but without giving you the answer you expected.

Perhaps the most frequent error you will make is misspelling the name of a function. Here is an illustration of the kind of thing that will usually happen in this case.

```
In[1]:= Sine[1.5]

General::spell:
    Possible spelling error: new symbol name "Sine"
       is similar  to existing symbols {Line, Sin, Sinh}.

Out[1]= Sine[1.5]
```

Whenever you type a name that is *close* to an existing name, *Mathematica* will print a warning message like the one above. You may often use such names intentionally, in which case these messages can be annoying. In that case, it is best to turn off the warnings.

```
In[2]:= Off[General::spell, General::spell1]
```

Now, *Mathematica* will not report that this function name might be misspelled, but since it can't find a definition associated with `Sine`, it returns your input unevaluated. (You can turn these spell warnings back on by evaluating `On[General::spell, General::spell1]`.)

```
In[3]:= Sine[1.5]
```

```
Out[3]= Sine[1.5]
```

Having your original expression returned unevaluated—as if this were perfectly normal—is a problem you'll often run into. Aside from misspelling a function name, or simply using a function that doesn't exist, another case where this occurs is when you give the wrong number of arguments to a function, especially to a user-defined function. For example, the BowlingScore function takes a single list argument; if we accidentally leave out the list braces, then we are actually giving BowlingScore twelve arguments.

```
In[4]:= BowlingScore[10, 10, 10, 10, 10, 10, 10, 10, 10, 10, 10, 10]
```

```
Out[4]= BowlingScore[10, 10, 10, 10, 10, 10, 10, 10, 10, 10, 10, 10]
```

Of course, some kinds of inputs cause genuine error messages. Syntax errors are an example, as seen earlier on page 9. The built-in functions are designed to usually warn you of such errors in input. In the first example below, we have supplied the Log function with an incorrect number of arguments (it expects one or two arguments only). In the second example, FactorInteger operates on integers only and so the real number argument causes the error condition.

```
In[5]:= Log[2, 16, 3]

        Log::argt: Log called with 3 arguments; 1 or 2
            arguments are expected.

Out[5]= Log[2, 16, 3]
```

```
In[6]:= FactorInteger[12.5]

        FactorInteger::facn:
            First argument 12.5 in FactorInteger[12.5]
                is not an exact number.

Out[6]= FactorInteger[12.5]
```

There are situations where *Mathematica* will return a correct-looking answer which is in fact wrong. For example, the following expression should return a root of the function $x - \sin(x)$ using Newton's method, starting with an initial guess of $x_0 = 0.2$.

```
In[7]:= FindRoot[x - Sin[x], {x, 0.2}]

Out[7]= {x -> 0.0116915}
```

This answer is quite inaccurate. The function $x - \sin(x)$ clearly has a root at $x = 0$, but it appears as if FindRoot is unable to find it with an initial guess that is quite close to the desired root. As it turns out, this function is particularly "pathological" in that it converges to its roots very, very slowly. *Mathematica*'s functions are set up in such a way so that their default behavior does the right thing in as broad a manner as possible. These functions can sometimes miss certain peculiarities. In these types of situations, a careful analysis of the problem you are working on is often necessary. Sometimes *Mathematica* will provide special tools or options to built-in functions to deal with special cases. Other times, you will have to extend *Mathematica*'s capabilities by writing programs that effectively deal with your domain of problems.

1.3 The *Mathematica* Language

In this section, we cover the material that we feel you should be comfortable with before proceeding. Some of the concepts are mathematical in nature, but most concern features of *Mathematica* that will be used throughout this book. Spend some time and become familiar with these constructs, but don't spend *too* much time on them. It is better to know that they exist and where to find them. When you have need for them, you can come back to this section and study it more carefully.

1.3.1 | Internal Forms of Expressions

When doing a simple arithmetic operation such as $3 + 4 * 5$, you are usually not concerned with exactly how a system such as *Mathematica* actually performs the additions or multiplications. Yet you will find it extremely useful to be able to see the internal representation of such expressions as this will allow you to manipulate them in a consistent and powerful manner.

Internally, *Mathematica* groups the objects that it operates on into different types: integers are distinct from real numbers; lists are distinct from numbers. One of the reasons that it is useful to identify these different *data types* is that specialized algorithms can be used on certain classes of objects that will help to speed the computations involved.

The Head function can be used to identify types of objects. For simple numbers, it will report whether the number is an integer, a rational number, a real number, or a complex number.

```
In[1]:= {Head[7], Head[1/7], Head[7.0], Head[7 + 2 I]}

Out[1]= {Integer, Rational, Real, Complex}
```

In fact, every *Mathematica* expression has a Head that gives some information about the type of expression.

```
In[2]:= Head[a + b]

Out[2]= Plus
```

```
In[3]:= Head[{1, 2, 3, 4, 5}]

Out[3]= List
```

Mathematica's internal representation of any expression can be displayed using the FullForm function. For example, to see how the previous two expressions are stored, we enter

```
In[4]:= FullForm[a + b]

Out[4]= Plus[a, b]
```

```
In[5]:= FullForm[{1, 2, 3, 4, 5}]

Out[5]= List[1, 2, 3, 4, 5]
```

One of the things to notice is that the binary operation a + b is really represented internally in functional form using the built-in function Plus. The Plus function just sums its arguments.

```
In[6]:= Plus[1, 2, 3, 4, 5]

Out[6]= 15
```

In more complicated expressions, the internal form may be a bit harder to decipher.

```
In[7]:= FullForm[a x^2 + b x + c]

Out[7]= Plus[c, Times[b, x], Times[a, Power[x, 2]]]
```

In any case, the important point to remember is that every function can be spelled out and used with functional syntax. The standard input forms, in which the operators appear between their operands, are just easier ways of writing such expressions.

1.3.2 | Predicates and Boolean Operations

Predicates

In addition to being able to determine the type of data you are working with, you are often presented with the problem of deciding whether your data or expression passes certain criteria. A *predicate* is a function that returns a value of true or false depending upon whether its argument passes a test. For example, the predicate PrimeQ tests for the primality of its arguments.

```
In[1]:= PrimeQ[144]

Out[1]= False


In[2]:= PrimeQ[2^31 - 1]

Out[2]= True
```

Other predicates are available for testing numbers to see whether they are even, odd, integral, and so on.

```
In[3]:= OddQ[21]

Out[3]= True


In[4]:= EvenQ[21]

Out[4]= False


In[5]:= IntegerQ[5/9]

Out[5]= False
```

Relational and Logical Operators

Relational operators are used to compare two expressions and return a value of True or False. The relational operators in *Mathematica* are Equal (==), Unequal (!=), Greater (>), Less (<), GreaterEqual (>=),, and LessEqual (<=). They can be used to compare numbers or arbitrary expressions.

```
In[1]:= 7 < 5

Out[1]= False
```

```
In[2]:= Equal[3, 7 - 4, 6/2]

Out[2]= True


In[3]:= x^2 - 1 == (x^4 -1) / (x^2 + 1) //Simplify

Out[3]= True
```

Note that the relational operators have lower precedence than arithmetic operators. So the second example above is interpreted as 3 == (7 – 4) and not as (3 == 7) – 4. Table 1.1 lists the relational operators and their various input forms. The reader is referred to [Wolfram 1991] for a complete discussion of the relational operators.

Standard form	Functional form	Meaning
x == y	Equal[x, y]	equal
x != y	Unequal[x, y]	unequal
x > y	Greater[x, y]	greater than
x < y	Less[x, y]	less than
x >= y	GreaterEqual[x, y]	greater than or equal
x <= y	LessEqual[x, y]	less than or equal

Table 1.1: Relational operators.

The logical operators (sometimes known as Boolean operators) determine the truth of an expression based on Boolean arithmetic. For example, the conjunction of two true statements is always true.

```
In[4]:= 4 < 5 && 8 > 1

Out[4]= True
```

The Boolean operation "And" is represented in *Mathematica* by And, with shorthand notation &&. Here is a table that gives all the possible values for the And operator.

```
In[5]:= TableForm[
          {{True&&True, True&&False}, {False&&True, False&&False}},
          TableHeadings -> {{T, F}, {T, F}}]

Out[5]//TableForm=      T       F

                T    True    False

                F    False   False
```

The logical "Or" operator || is true when either of its arguments is true.

```
In[6]:= 4 == 3 || 3 == 6/2
```

Out[6]= True

```
In[7]:= 0 == 0 || 3 == 6/2
```

Out[7]= True

Note the difference between this boolean "Or" and the common notion of "or." A phrase such as, "It is cold or it is hot," uses the word "or" in an *exclusive* sense; that is, it excludes the possibility that it is *both* cold and hot. The logical Or is inclusive in the sense that if A and B are both true, then A || B is also true.

```
In[8]:= True || True
```

Out[8]= True

Mathematica also contains an operator for the exclusive or, Xor.

```
In[9]:= Xor[True, True]
```

Out[9]= False

```
In[10]:= Xor[True, False]
```

Out[10]= True

Table 1.2 shows the logical operators and their input forms.

Standard form	Functional form	Meaning
!x	Not[x]	not
x != y	Unequal[x, y]	unequal
x && y	And[x, y]	and
x || y	Or[x, y]	or
(x || y) && ! (x && y)	Xor[x, y]	exclusive or

Table 1.2: Logical operators.

1.3.3 | Evaluation of Expressions

Trace and TracePrint

When writing programs, things invariably go wrong. A useful function for tracking down errors is the Trace function, which generates a list of all the steps used in evaluating an expression. As a simple example, we `Trace` the evaluation of $3 * (4 + 5)$:

```
In[1]:= Trace[3 * (4 + 5)]
Out[1]= {{4 + 5, 9}, 3 9, 27}
```

It should be plain from this example that *Mathematica* follows the same routine as that learned in grade school. First the addition inside the parentheses $(4 + 5)$ is performed to produce 9, and then the multiplication $(3 \; 9)$ is performed to get the final result of 27.

`TracePrint` is similar to `Trace`, but its output is sometimes clearer. This function displays output in a tree-like form enabling you to see the various levels at which each subexpression is evaluated.

```
In[2]:= TracePrint[3 * (4 + 5)]
Out[2]=  3 (4 + 5)
           Times
           3
           4 + 5
           Plus
           4
           5
           9
         3 9
         27
```

Timing

You will often find it useful to be able to time the running of a computation. This is done by using the `Timing` function. Here we factor a large integer, 1000!, and compute the CPU time necessary to perform the evaluation.[2]

```
In[1]:= Timing[ FactorInteger[1000!]; ]
Out[1]= {3.31667 Second, Null}
```

Since the output of this call to the `FactorInteger` function would run several pages long, we suppress it by placing a semicolon after the `FactorInteger` call.[3] As seen earlier, a semicolon causes the printing of the value of an expression to be suppressed; here, the word `Null` is printed where the list of factors would have been.

---

[2]Recall that $n! = n(n-1)(n-2) \cdots 3 \cdot 2 \cdot 1$

[3]We could, of course, store the result in a variable for later reference.

There are a few points to keep in mind when using the Timing function. First, the amount of time that is reported by Timing is only the time that the CPU spends on the calculation. It does not include any time needed to format and display the result.

Second, the time it takes for an evaluation to be completed will vary depending upon the computer you are using as well as on the current state of *Mathematica*. Whenever possible, *Mathematica* tries to keep internal tables of certain calculations that will speed evaluation times of similar computations. For example, we can calculate the 200th Bernoulli number as follows.

```
In[2]:= Timing[ BernoulliB[200] ]

Out[2]= {6.03333 Second, -(4983840494283334147649286321403
        99662108495887457206674968055822617263669621523687
        56886580230221099913260141269761327939105865452714
        53405158400992904780263503828028843717123593379842
        741228611598002800191101978885558936711151 /
        1366530)}
```

Now, when we recompute this number, *Mathematica* consults internal tables that it constructed during the previous computation and the time is greatly reduced.

```
In[3]:= Timing[ BernoulliB[200] ]

Out[3]= {0. Second, -(498384049428333414764928632140399
        66210849588745720667496805582261726366962152368756
        88658602302210999132601412697613279391058654527145
        34051584009929047802635038280288437171235933798427
        4122861159800280019110197888555893671151 / 1366530)
        }
```

As mentioned above, the Timing function can be useful in examining the running time of algorithms. For example, it is widely known that factoring integers is a very difficult process in general and that the time that it takes to factor an integer increases almost exponentially with the size of the integer. We can get a glimpse of this by using the Timing function to time how long it takes to factor each of the integers 100!, 200!, 300!, ..., 1400!, 1500!.

```
In[4]:= timings = Table[Timing[FactorInteger[(100j)!]][[1]],
                    {j, 1, 15}]
```

```
Out[4]= {0.05 Second, 0.116667 Second, 0.2 Second,
         0.35 Second, 0.7 Second, 0.933333 Second,
         1.3 Second, 1.73333 Second, 2.45 Second,
         2.85 Second, 3.7 Second, 4.41667 Second, 5.5 Second,
         6.46667 Second, 7.25 Second}
```

A plot of the data clearly shows the exponential growth.

```
In[5]:= ListPlot[timings/Second, PlotStyle -> PointSize[.02]]
```

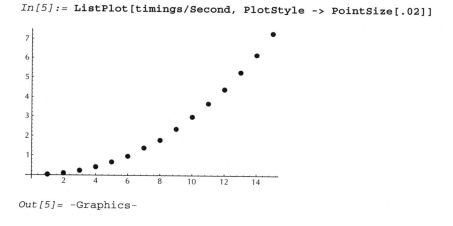

```
Out[5]= -Graphics-
```

A ListPlot of the logarithm of these numbers would appear linear, confirming the exponential nature of the original data.

1.3.4 | Attributes

All functions in *Mathematica* have certain properties, called *attributes*. These attributes can make a function commutative or associative, or they may give the function the ability to be applied to a list. The Attributes function can be used to view the attributes of a function.

```
In[1]:= Attributes[Plus]
```

```
Out[1]= {Flat, Listable, OneIdentity, Orderless, Protected}
```

The Flat attribute indicates that this function (Plus) is associative. That is, given three elements to add, it doesn't matter which two are added first. In mathematics, this is known as *associativity* and is written as $a + (b + c) = (a + b) + c$ for arbitrary numbers *a*, *b*, and *c*. In *Mathematica* this could be indicated by saying that the two expressions `Plus[a, Plus[b, c]]` and `Plus[Plus[a, b], c]` are equivalent to the flattened form `Plus[a, b, c]`. When *Mathematica* knows that a function has the attribute `Flat`, it writes it in flattened form.

```
In[2]:= Plus[Plus[a, b], c]

Out[2]= a + b + c
```

The `Orderless` attribute indicates that the function is commutative; that is, *a* + *b* = *b* + *a* for arbitrary expressions *a* and *b*. This allows *Mathematica* to write such an expression in an order that is useful for computation. It does this by sorting the elements into a *canonical order*. For expressions consisting of letters and words, this ordering is clearly alphabetic.

```
In[3]:= s + u + p + e + r + b

Out[3]= b + e + p + r + s + u
```

Sometimes a canonical order is readily apparent.

```
In[4]:= x^3 + x^5 + x^4 + x^2 + 1 + x
```
$$Out[4]= 1 + x + x^2 + x^3 + x^4 + x^5$$

Other times, it is not so apparent.

```
In[5]:= x^3 y^2 + y^7 x^5 + y x^4 + y^9 x^2 + 1 + x
```
$$Out[5]= 1 + x + x^4 y + x^3 y^2 + x^5 y^7 + x^2 y^9$$

When a symbol has the attribute `Protected`, the user is prevented from modifying the function in any significant way. Thus, all built-in operations have this attribute.

Functions with the attribute `OneIdentity` have the property that repeated application of the function to the same argument will have no effect. For example, the expression `Plus[Plus[a, b]]` is equivalent to `Plus[a, b]`, hence only one addition is performed.

```
In[6]:= FullForm[Plus[Plus[a + b]]]

Out[6]//FullForm= Plus[a, b]
```

The other `Attribute` for the `Plus` function, (`Listable`), along with several others, will be discussed later. Consult the manual [Wolfram 1991, p. 272] for a complete list of the `Attributes` that a symbol can have.

Although it is unusual to want to alter the attributes of a built-in function, it is fairly common to change the default attributes of a user-defined function. The `SetAttributes` function is used to change the attributes of a function. We will see a use of `SetAttributes` in Chapter 8.

1.4 The *Mathematica* Interface

1.4.1 | The Notebook Front End

On many systems, such as Macintosh, NeXT, X Windows, and Windows, a graphical user interface, called the *notebook front end*, is available. To start *Mathematica* on such a system, you will have to find and then double-click on the *Mathematica* icon, which will look something like this:

The computer will then load parts of *Mathematica* into its memory and soon a blank window will appear on the screen. This window is the visual interface to a *Mathematica* notebook and it has many features that are useful to the user.

Notebooks allow you to write text, perform computations, and create graphics all in one document. Notebooks also have many of the features of common word-processors, so those familiar with word-processing will find the notebook interface easy to learn. In addition, the notebook provides features for outlining material which you may find useful for giving talks and demonstrations.

When a blank notebook first appears on the screen (either from just starting *Mathematica* or from selecting **New** in the **Notebook** menu), you can start typing immediately. For example, if you type `N[Pi]` and then press the Enter key,[4] *Mathematica* will evaluate the result and print the decimal approximation to π on the screen.

---

[4]On DOS machines, press the Insert key or Shift-Return to evaluate an expression. If you are using other platforms, check your *Mathematica* User's Guide.

Notice that when you evaluate an expression in a notebook, *Mathematica* adds input and output prompts. In the example notebook above, these are denoted In[1]:= and Out[1]=. These prompts can be thought of as markers (or labels) that you can refer to during your *Mathematica* session.

You should also note that when you started typing, *Mathematica* placed a *bracket* on the far right side of the window that enclosed the *cell* that you were working in. These *cell brackets* are helpful for organizational purposes within the notebook. Double-clicking on cell brackets will open any collapsed cells, or close any open cells. For example, in the following notebook (which is included on the diskette at the back of this book), notice that the cell brackets in section titles contain rectangles of varying widths. The widths of these rectangles symbolize the amount of material available within a closed (or collapsed) cell.

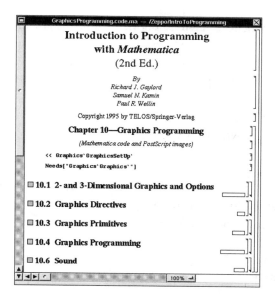

Double-clicking on the cell bracket containing the **10.2 Graphics Directives** cell will open the cell to display its contents:

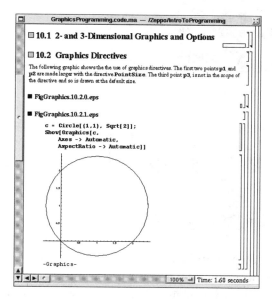

Using cell brackets in this manner allows you to organize your work in an orderly manner, as well as to outline material. For a complete description of cell brackets and many other Interface features, you should consult the documentation that came with your version of *Mathematica*.

New input can be entered whenever there is a horizontal line that runs across the width of the notebook. If one is not present where you wish to place an input cell, move the cursor up and down until it changes to a horizontal bar and then click the mouse once. A horizontal line should now appear across the width of the window. You can immediately start typing and an input cell will be created.

For information on other features such as saving, printing, and editing notebooks, consult the manuals that came with your version of *Mathematica*.

The Function Browser

Beginning with Version 2.2, the Macintosh and NeXT notebook versions of *Mathematica* contain a very useful addition to the help system called the *Function Browser*.[5] The Function Browser allows you to search for functions easily and it provides extensive documentation and templates.

To start the Function Browser, select **Open Function Browser** under the **Info** or **Windows** menu item. You should quickly see something like the following:

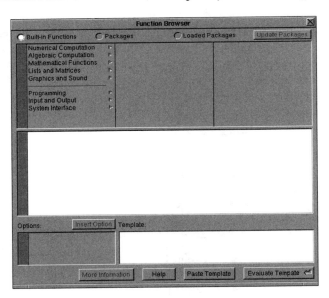

Notice the three buttons at the top of the Function Browser window. Clicking on the **Packages** button will give you access to all of the packages that come with each implementation of *Mathematica*. Similarly, clicking on the **Loaded Packages** button will give you access to all packages that you have loaded during the current session.

---

[5] *Mathematica* on X window systems includes a feature called *MathBook*, which is very similar to the Function Browser. In the version of *Mathematica* after Ver. 2.2, the Function Browser will be renamed the Help Browser and will be included on Windows systems as well.

Suppose you were looking for information about three-dimensional parametric graphics. First select **Graphics and Sound** on the left, then **3D Plots** and finally **ParametricPlot3D**. The Function Browser should look like this:

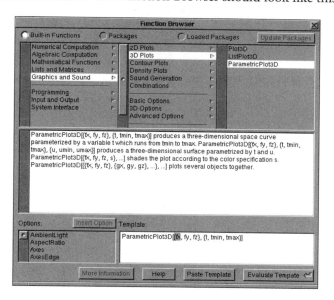

Notice that in the main window, the Function Browser has displayed information about the `ParametricPlot3D` function. This is identical to the usage message you would get if you entered `?ParametricPlot3D`.

In the lower right corner, the Function Browser has created a template which would allow you to paste a correct function call into your current notebook.

In a small window in the lower left corner is a list of all available `Options` for the `ParametricPlot3D` function. Selecting any one of these options will cause the Function Browser to list a descriptive message about the option in the message window. In addition, you can instruct the Function Browser to paste any of these options into the Template field. For example, selecting the option `AmbientLight` and then **Insert Option** will display information about the `AmbientLight` option and then paste it into the Template field.

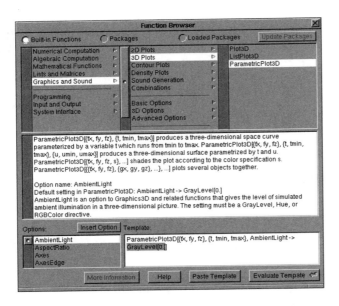

Many additional features are available in the Function Browser and you are advised to consult your documentation for a complete list and description.

1.4.2 | The Command Line Interface

Using *Mathematica* on a computer that does not have a notebook interface is not a disadvantage if the user can just remember a few additional commands. To start up *Mathematica* from the command line, open a shell and type the word **math**. You will be greeted with something similar to the following (output will vary depending upon the computer system you are using).

```
localhost> math
Mathematica 2.2 for SPARC
Copyright 1988-94 Wolfram Research, Inc.
 - Terminal graphics initialized -

In[1]:=
```

Input can now be typed and evaluated by pressing the Return key after inputting your expression at the command line. For example, to compute the first 35 digits of π, type N[Pi, 35] at the In[1]:= prompt and then press the Return key.

```
localhost> math
Mathematica 2.2 for SPARC
Copyright 1988-94 Wolfram Research, Inc.
 - Terminal graphics initialized -

In[1]:= N[Pi, 35]

Out[1]= 3.1415926535897932384626433832795 0288

In[2]:=
```

To terminate the *Mathematica* session, type Quit on the command line, and then press the Return key.

Functions are available for saving sessions and graphic images to files. In addition, you may want to edit files containing *Mathematica* inputs separately, using a text editor. Such files conventionally end with '.m'; they are read into *Mathematica* by entering '<<*filename*.m'. Since the details of this often vary from one system to another, you should consult your user documentation for full explanations.

2 A Brief Overview of *Mathematica*

Mathematica is a powerful tool for computing, programming, data analysis, knowledge representation, and visualization of information. Although our focus in this text will be on the programming capabilities of the *Mathematica* language, these other features should not be viewed as unrelated. You often need to write programs that carry out complex computations and represent data in a graphical manner to help you visualize a problem you are working on. In this chapter we introduce the elementary operations available in *Mathematica*. We will give a hint of the richness of the *Mathematica* programming language.

2.1 Numerical and Symbolic Computations

Mathematica differs from calculators and simple computer programs in its ability to calculate exact results and to compute to an arbitrary degree of precision.

```
In[1]:= 1/15 + 1/35 + 1/63
```

$$Out[1]= \frac{1}{9}$$

```
In[2]:= 1111111^2
```

```
Out[2]= 1234567654321
```

```
In[3]:= 9348 * 437 - 923874^3 + 378/2346 (1283 - 3764)
```

$$Out[3]= -(\frac{30832943920360709 6571}{391})$$

```
In[4]:= 2^500
```

```
Out[4]= 32733906078961141870013189696827599152216642\
        04604306478948329136809613379640467455488\
        32700923259041571508866841275600710092172\
        56545885939305332852758 9376
```

Since this last integer contains over 150 digits, the backslash symbol (\) is used as a continuation marker to indicate that the output wraps around to the next line. To convert this number to scientific notation use the N function. Wrapping the built-in function N around any expression will convert it to an approximate value.

```
In[5]:= N[2^500]
```
$$Out[5]= 3.27339\ 10^{150}$$

To calculate $\sqrt{75}$, use the built-in function Sqrt.

```
In[6]:= Sqrt[75]
```
```
Out[6]= 5 Sqrt[3]
```

Mathematica simplifies the result, but does not give a numerical approximation. Since $\sqrt{75}$ is an irrational number, a decimal result would only be approximate. You could, of course, explicitly ask for the approximation.

```
In[7]:= N[%]
```
```
Out[7]= 8.66025
```

Mathematical constants such as e, π, i, and γ are all built-in:

```
In[8]:= N[E]
```
```
Out[8]= 2.71828
```

```
In[9]:= N[Pi]
```
```
Out[9]= 3.14159
```

```
In[10]:= Sqrt[-9]
```
```
Out[10]= 3 I
```

```
In[11]:= N[EulerGamma]
```
```
Out[11]= 0.577216
```

Notice that *Mathematica* displays six significant digits when N is used (even though it stores more than six digits internally). You can use N to get higher precision than the default. For example, to calculate π to 35-digit precision, you have to indicate an optional argument to N.

```
In[12]:= N[Pi, 35]
```

```
Out[12]= 3.1415926535897932384626433832795028
```

All of the standard mathematical functions are available. For example, the natural logarithm is given by Log:

```
In[13]:= Log[E]
```

```
Out[13]= 1
```

Logarithms in other bases can also be obtained. For example, the logarithm of 1024 in base 2 (*i.e.*, $\log_2 1024$) is written in a form that closely approximates the standard mathematical notation.

```
In[14]:= Log[2, 1024]
```

```
Out[14]= 10
```

The trigonometric functions (Sin, Cos, Tan, Sec, etc.) and their inverses (ArcSin, ArcCos, etc.) can operate on any type of number or expression.

```
In[15]:= Cos[Pi/3]
```

$$Out[15]= \frac{1}{2}$$

```
In[16]:= ArcTan[1]
```

$$Out[16]= \frac{Pi}{4}$$

```
In[17]:= Expand[Sin[x]^2, Trig -> True]
```

$$Out[17]= \frac{1}{2} - \frac{Cos[2\ x]}{2}$$

Expand is used to give a symbolic expansion of an expression, and the Trig -> True option tells *Mathematica* to use built-in rules for simplifying terms containing trigonometric functions.

Numerous function are available for solving equations. For equations that can be solved by simple algebraic techniques, the Solve function is used. For example, this solves the quadratic equation $ax^2 + bx + c = 0$ for x.

In[18]:= **Solve[a x^2 + b x + c == 0, x]**

$$Out[18]= \{\{x \rightarrow \frac{-b - \text{Sqrt}[b^2 - 4\ a\ c]}{2\ a}\}, \{x \rightarrow \frac{-b + \text{Sqrt}[b^2 - 4\ a\ c]}{2\ a}\}\}$$

Of course, some equations cannot be solved by simple algebraic manipulation.

In[19]:= **Solve[Cos[x] == x, x]**

```
Solve::tdep:
    The equations appear to involve transcendental
        functions of the variables in an essentially
        non-algebraic way.
```

Out[19]= Solve[Cos[x] == x, x]

For equations such as these, numerical techniques must be employed. Find-Root implements Newton's method of root finding (see Sections 8.1 and 9.4 for a discussion of this method). It requires an initial guess of the approximate location of the root. In the following example, we give it an initial guess of $x_0 = 1.0$.

In[20]:= **FindRoot[Cos[x] == x, {x, 1.0}]**

Out[20]= {x -> 0.739085}

Indefinite integration on symbolic expressions can be performed using the Integrate function. This computes $\int 1/(1 - x^3)\ dx$:

In[21]:= **Integrate[1/(1 - x^3), x]**

$$Out[21]= \frac{\text{ArcTan}[\frac{1 + 2\ x}{\text{Sqrt}[3]}]}{\text{Sqrt}[3]} - \frac{\text{Log}[1 - x]}{3} + \frac{\text{Log}[1 + x + x^2]}{6}$$

You can confirm that the derivative of this expression is what you started with by using the differentiation function D (although in this case, some simplification is necessary to obtain the original form).

In[22]:= **D[%, x]**

$$Out[22]= \frac{1}{3\ (1 - x)} + \frac{1 + 2\ x}{6\ (1 + x + x^2)} + \frac{2}{3\ (1 + \frac{(1 + 2\ x)^2}{3})}$$

In[23]:= **Simplify[%]**

$$Out[23]= \frac{1}{1 - x^3}$$

Mathematica has numerous ways of representing differentiation. As shown in the above example, you can mimic regular mathematical notation with the D function. If you wanted to compute the second derivative of log x, you could do it as follows:

In[24]:= **D[Log[x], {x, 2}]**

$$Out[24]= -x^{-2}$$

Here is the chain rule of first semester calculus:

In[25]:= **D[f[g[x]], x]**

Out[25]= f'[g[x]] g'[x]

The familiar "prime" notation can also be used to compute derivatives.

In[26]:= **Log'[x]**

$$Out[26]= \frac{1}{x}$$

In fact, the prime notation f'[x] is shorthand for Derivative[1][f][x].

In[27]:= **FullForm[f'[x]]**

Out[27]//FullForm= Derivative[1][f][x]

In addition to indefinite integration, you can also compute definite integrals. To compute $\int_0^{\pi} sin(x)\, dx$ for example, enter:

In[28]:= **Integrate[Sin[x], {x, 0, Pi}]**

Out[28]= 2

Some functions do not possess simple anti-derivatives.

In[29]:= **Integrate[E^Cos[x], {x, 0, Pi}]**

Out[29]= Integrate[E^Cos[x], {x, 0, Pi}]

This result indicates that *Mathematica* was unable to integrate this function. Fortunately, numerical integration routines are available to obtain an approximation to such integrals. The NIntegrate function uses an adaptive algorithm to perform numerical integration.

In[30]:= **NIntegrate[E^Cos[x], {x, 0, Pi}]**

Out[30]= 3.97746

Double integrals are easily computed. The following integrates with respect to y first; that is, it computes $\int_0^1 \int_0^x \sqrt{x^2 + y^2} \, dy \, dx$.

In[31]:= **Integrate[Sqrt[x^2 + y^2], {x, 0, 1}, {y, 0, x}]**

Out[31]= $\dfrac{\text{Sqrt}[2] + \text{Log}[1 + \text{Sqrt}[2]]}{6}$

Differential equations are solved using the built-in DSolve function. To solve the equation $y''(x) = ay(x)$ for the function $y(x)$ in the independent variable x, write,

In[32]:= **DSolve[y"[x] == a y[x], y[x], x]**

Out[32]= $\left\{\left\{y[x] \rightarrow \dfrac{C[1]}{E^{\text{Sqrt}[a] \, x}} + E^{\text{Sqrt}[a] \, x} \, C[2]\right\}\right\}$

By specifying two initial conditions, you can find both constants of integration.

In[33]:= **DSolve[{y"[x] == a y[x], y'[0]==1, y[0]==0}, y[x], x]**

Out[33]= $\left\{\left\{y[x] \rightarrow \dfrac{-1}{2 \, \text{Sqrt}[a] \, E^{\text{Sqrt}[a] \, x}} + \dfrac{E^{\text{Sqrt}[a] \, x}}{2 \, \text{Sqrt}[a]}\right\}\right\}$

Functions are also available for solving differential equations by numerical techniques (NDSolve in particular).

2.2 Functions

Mathematica contains an extraordinary range of functions for doing the computations of mathematics and science. When these built-in functions are not sufficient for a particular task, there are a wide range of packages that contain hundreds of additional functions. These packages come with every implementation of *Mathematica* and significantly extend its capabilities. In this section we will first look at a very small set of the built-in functions in order to get an idea of the syntax and functionality, and then conclude with a discussion of *Mathematica's* packages.

2.2.1 | Functions of Number Theory

The function Mod[*k*, *n*] gives the remainder upon dividing *k* by *n*. For example, 28 divided by 5 leaves a remainder of 3.

```
In[1]:= Mod[28, 5]
Out[1]= 3
```

If you were interested in the divisors of 28 you should use the built-in Divisors function.

```
In[2]:= Divisors[28]
Out[2]= {1, 2, 4, 7, 14, 28}
```

The Euler totient function $\phi(n)$ gives the number of positive integers less than *n* that are relatively prime to *n* (contain no common factors).

```
In[3]:= EulerPhi[28]
Out[3]= 12
```

Almost all students of mathematics at some point study the binomial theorem whose discovery is due to Isaac Newton in 1676:

$$(a+b)^n = \sum_k \binom{n}{k} a^k b^{n-k}$$

where $\binom{n}{k}$ is the number of ways to choose a *k*-element subset from an *n*-element set. Students in pre-calculus use this formula to find binomial products.

```
In[4]:= Expand[(a + b)^2]
Out[4]= a^2 + 2 a b + b^2
```

```
In[5]:= Expand[(a + b)^3]
Out[5]= a^3 + 3 a^2 b + 3 a b^2 + b^3
```

In[6]:= **Expand[(a + b)^4]**

Out[6]= $a^4 + 4 a^3 b + 6 a^2 b^2 + 4 a b^3 + b^4$

The coefficients of these expansions crop up so often in mathematics that they have been given a special name: *the binomial coefficients*. They can be displayed explicitly in *Mathematica*. The coefficient of the $(k + 1)$ term in the nth row is given by Binomial[n, k]. For example, to find the coefficient of the ab^3 term, enter

In[7]:= **Binomial[4, 3]**

Out[7]= 4

Binomial[n, k] is *Mathematica*'s notation for the binomial coefficient $\binom{n}{k}$. We can display the coefficients of $(a + b)^4$ as follows:

In[8]:= **Table[Binomial[4, j], {j, 0, 4}]**

Out[8]= {1, 4, 6, 4, 1}

Table[*expr*, {*i, imin, imax*}] creates a list by evaluating *expr* as *i* takes on values from *imin* to *imax*. For example, the command Table[i^2, {i, 1, 5}] could be used to list the squares of the first five integers.

Pascal's Triangle, named after Blaise Pascal (1623–1662), lists the binomial coefficients in an organized fashion. Here are its first 9 rows (using TableForm to format the rows in a more readable form):

In[9]:= **TableForm[Table[Binomial[n, k], {n, 0, 8}, {k, 0, n}]]**

Out[9]//TableForm=

1								
1	1							
1	2	1						
1	3	3	1					
1	4	6	4	1				
1	5	10	10	5	1			
1	6	15	20	15	6	1		
1	7	21	35	35	21	7	1	
1	8	28	56	70	56	28	8	1

Thousands of identities related to Pascal's Triangle and the binomial coefficients have been discovered over the years. For example, the third column of Pascal's Triangle given by $\binom{n}{2}$ contains the "triangular numbers."

```
In[10]:= Table[Binomial[n, 2], {n, 1, 10}]
```

$Out[10]= \{0, 1, 3, 6, 10, 15, 21, 28, 36, 45\}$

At the age of ten, Carl Friedrich Gauss (1777–1855) surprised his teacher by quickly computing the sum $1 + 2 + 3 + \cdots + 97 + 98 + 99 + 100$. He evidently knew that the general term for the sum of the first *n* integers involved a binomial expression.

```
In[11]:= Binomial[n + 1, 2]
```

$$Out[11]= \frac{n \, (1 + n)}{2}$$

2.2.2 | Functions of Linear Algebra

Scientists and engineers have developed a vast array of tools for working with data sets of one or more dimensions. The development and application of functions and routines for operating on vectors and matrices falls under the domain of linear algebra. Some of the techniques are symbolic and others are purely numerical. We will look at some of the algorithms and numerical issues in linear algebra in detail in Chapters 7–9. In this section, we will give a brief introduction to the built-in capability that *Mathematica* provides in this area.

A vector is represented as a list of elements.

```
In[1]:= vec = {0, 1, 3, 1}
```

$Out[1]= \{0, 1, 3, 1\}$

You can multiply a vector by a scalar,

```
In[2]:= 5 vec
```

$Out[2]= \{0, 5, 15, 5\}$

and add two vectors.

```
In[3]:= vec + {1, 2, 3, 4}
```

$Out[3]= \{1, 3, 6, 5\}$

Matrices are represented as lists of lists (nested lists). Functions are available for displaying them in a more traditional form.

```
In[4]:= mat = {{a, b}, {c, d}};
```

```
In[5]:= MatrixForm[mat]
```

$$Out[5]//MatrixForm= \begin{matrix} a & b \\ c & d \end{matrix}$$

You can multiply a matrix by a scalar, and multiply two or more matrices (of appropriate rank).

```
In[6]:= 3 mat
```

```
Out[6]= {{3 a, 3 b}, {3 c, 3 d}}
```

```
In[7]:= MatrixForm[mat . mat]
```

$$Out[7]//MatrixForm= \begin{matrix} a^2 + b\ c & a\ b + b\ d \\ a\ c + c\ d & b\ c + d^2 \end{matrix}$$

When working with approximate numbers, some care should be taken in interpreting results. Here is a 2×2 matrix consisting of floating point numbers.

```
In[8]:= matf = {{.324, 1.3}, {.01, 5.8}};
```

Here is the inverse of this matrix.

```
In[9]:= invmatf = Inverse[matf]
```

```
Out[9]= {{3.10792, -0.696603}, {-0.00535848, 0.173615}}
```

Multiplying the original matrix by its inverse should give the identity matrix.

```
In[10]:= matf . invmatf
```

$$Out[10]= \{\{1., 1.12622\ 10^{-17}\}, \{8.77797\ 10^{-17}, 1.\}\}$$

The small numerical errors that were introduced are a result of using approximate arithmetic and is discussed in some detail in Chapter 9.

The off-diagonal terms can be trimmed using Chop.

```
In[11]:= Chop[%]
```

```
Out[11]= {{1., 0}, {0, 1.}}
```

When solving linear systems of equations, you often can reduce the system to a single matrix equation and solve the problem using techniques of linear algebra that are optimized for such problems. For example, suppose you are trying to solve the following system of equations:

$$2x + 3y = 8$$
$$x - 5y = 1$$

This can easily be recast as the matrix equation:

$$\begin{pmatrix} 2 & 3 \\ 1 & -5 \end{pmatrix} \cdot \begin{pmatrix} x \\ y \end{pmatrix} = \begin{pmatrix} 8 \\ 1 \end{pmatrix}$$

If we let $M = \begin{pmatrix} 2 & 3 \\ 1 & -5 \end{pmatrix}$, $X = \begin{pmatrix} x \\ y \end{pmatrix}$, and $B = \begin{pmatrix} 8 \\ 1 \end{pmatrix}$, then, the solution to $M \cdot X = B$ is given by $X = M^{-1} \cdot B$. Here is an implementation of this.

```
In[12]:= M = {{2, 3}, {-1, 5}};
```

```
In[13]:= X = {x, y};
```

```
In[14]:= B = {8, 1};
```

```
In[15]:= Inverse[M] . B
```

$$Out[15]= \{\frac{37}{13}, \frac{10}{13}\}$$

A quick check verifies the result.

```
In[16]:= M . % == B
```

```
Out[16]= True
```

A more efficient way to solve such systems is to use the built-in linear algebra routines directly.

```
In[17]:= LinearSolve[M, B]
```

$$Out[17]= \{\frac{37}{13}, \frac{10}{13}\}$$

2.2.3 | Random Number Generators

When performing statistical or numerical tests, it is often useful to be able to generate sequences of random numbers. If you suspect that two expressions are equivalent, you could replace common variables by a random number and compare the results.

Mathematica contains a function Random that can generate random numbers in any specified range. For example, to generate a random real number in the range 0 to 5, you would enter:

```
In[1]:= Random[Real, {0, 5}]

Out[1]= 2.280354
```

You can omit the range of numbers and *Mathematica* will then use the default range of 0 to 1:

```
In[2]:= Random[Real]

Out[2]= 0.460106
```

In fact, the default number type is Real, so you could omit that as well and *Mathematica* will compute a random real number in the range from 0 to 1.

```
In[3]:= Random[]

Out[3]= 0.822251
```

To produce a random integer in the range from 0 to 100, you enter,

```
In[4]:= Random[Integer, {0, 100}]

Out[4]= 59
```

One thing to notice about the Random function is that it gives a different result each time you call it. This has to do with the random-number generator that *Mathematica* uses. When you start up *Mathematica*, it makes note of the time of day on your computer and uses that as the "seed" value for its random number generator. When you call the function Random, the algorithm generates a new number as requested and the state of the random number generator is advanced. There are times when you might want to make sure that within a session, you always get the same sequence of random numbers. Or you might be working on a simulated experiment and would

like to repeat a sequence of numbers. To do this, you must seed the generator yourself
by explicitly issuing a `SeedRandom` call.

```
In[5]:= SeedRandom[128]
```

```
In[6]:= {Random[], Random[], Random[]}
Out[6]= {0.355565, 0.486779, 0.00573919}
```

Reseeding the generator will cause the same sequence of numbers to be generated.

```
In[7]:= SeedRandom[128]
```

```
In[8]:= {Random[], Random[], Random[]}
Out[8]= {0.355565, 0.486779, 0.00573919}
```

The sequence of numbers that `Random` generates are uniformly distributed over
the range you specify and appear to be random. In fact, they are not really random as
they are generated from an algorithm that starts with a seed value. Start with the same
seed, and you will get the same sequence. Although it is not the purpose of this book
to give a detailed discussion of such matters, we do discuss random number generators
and statistical tests for determining the "randomness" of such sequences in Section 9.2.

2.2.4 | Packages

As mentioned earlier, even though *Mathematica* contains hundreds of built-in
functions that are available to you when you start each session, you will periodically
find yourself wanting additional features. With each implementation of *Mathematica*,
a set of packages is included that extends the functionality of the system. In addition,
you can write your own packages to customize your set-up.

The packages are contained in a directory (or folder) on your system where
Mathematica can find them. Since this is different on each system, you should consult
your documentation for information on where to find these files.

Each package contains functions that are specialized for a particular task. Once
loaded, they will perform just like the built-in functions; that is, they will have usage
messages and will print out error messages when you use them incorrectly.

For example, suppose we needed information about some chemical elements.
This is not contained in the standard version of *Mathematica* that is loaded when
you start up, but there is a package `ChemicalElements.m` that is located in the

Miscellaneous sub-directory of the Packages directory. We load this package by issuing the following command:

In[1]:= **Needs["Miscellaneous`ChemicalElements`"]**

Once this package is loaded, all of the functions defined in it are available to use in our current session. For example, the package contains functions to show abbreviations of chemical elements, find atomic weights, heats of vaporization, etc.

In[2]:= **Wolfram**

Out[2]= Tungsten

In[3]:= **Abbreviation[Wolfram]**

Out[3]= W

In[4]:= **AtomicWeight[W]**

Out[4]= 183.85

In[5]:= **HeatOfVaporization[W]**

$$Out[5]= \frac{824.2 \text{ Joule Kilo}}{\text{Mole}}$$

In[6]:= **ElectronConfiguration[W]**

Out[6]= {{2}, {2, 6}, {2, 6, 10}, {2, 6, 10, 14}, {2, 6, 4}, {2}}

Any function defined in a package and then read in to the current session can be explored in the same manner as you would explore the built-in functions of *Mathematica*. So, for example, the function ElectronConfiguration defined in the ChemicalElements.m package contains a usage message that can be accessed in the usual way.

In[7]:= **?ElectronConfiguration**

ElectronConfiguration[element] depicts the electron
    configuration of the specified element in a list of
    the form
    {{1s},{2s,2p},{3s,3p,3d},{4s,4p,4d,4f},{5s,5p,5d,5f}
    ,{6s,6p,6d},{7s}}.

Once a package is read in, information about any of its constituent functions is available in the Function Browser as well (Version 2.2 and later).

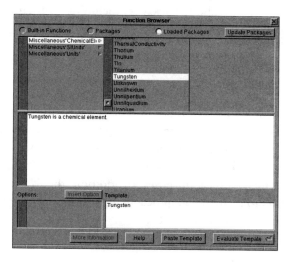

If your work requires that you use the ChemicalElements package on a regular basis, it would be convenient if it were automatically loaded every time you started a new *Mathematica* session. This customization of *Mathematica* can be accomplished by placing the line

```
Needs["Miscellaneous`ChemicalElements`"]
```

in the file init.m. This initialization file is read by *Mathematica* every time it starts up, so any packages that you would like to be loaded on start-up should be placed there.

An alternative way to use functions from a package is to have *Mathematica* just load functions *when* they are needed. You can do this by using DeclarePackage which will automatically evaluate the underlying package when any of the functions defined in this package are used. To accomplish this with the ChemicalElements.m package in a new session of *Mathematica*, you would evaluate:

```
In[8]:= DeclarePackage["Miscellaneous`ChemicalElements`",
            {"AtomicWeight", "ElectronConfiguration"}]

Out[8]= Miscellaneous`ChemicalElements`
```

Now, whenever one of the two functions AtomicWeight and ElectronConfiguration are first called, the package Miscellaneous`ChemicalElements` will be evaluated and all of its function will be available for use. The advantage of this approach is that any necessary packages are not loaded until they are first needed.

We will have much more to say about packages, including a discussion of contexts and writing our own packages, in Chapter 12.

2.3 Graphics

When working with functions or sets of data, it is often desirable to be able to visualize them. *Mathematica* provides a wide range of graphing capabilities. These include two- and three-dimensional plots of functions or data sets, contour and density plots of functions of two variables, bar charts, histograms and pie charts of data sets, and many packages designed for specific graphical purposes. In addition, the *Mathematica* programming language allows you to construct graphical images "from the ground up" using primitive elements, as we will see in Chapter 10.

2.3.1 | Two-Dimensional Plots

Plotting functions of one variable is accomplished with the `Plot` command.

```
In[1]:= Plot[x + Sin[x], {x,-5Pi,5Pi}];
```

```
Out[1]= -Graphics-
```

Although some functions are undefined at certain points, *Mathematica* can deal with these singularities gracefully by sampling additional points near the singularity.

In[2]:= `Plot[Sin[1/x], {x,-1,1}]`

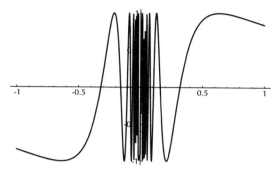

Out[2]= `-Graphics-`

You can combine two or more plots in a single graphic by enclosing them inside curly braces.

In[3]:= `Plot[{Sin[x], Sin[2x]}, {x,0,2Pi}]`

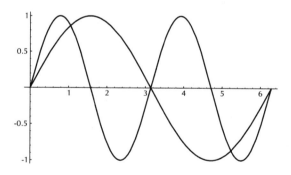

Out[3]= `-Graphics-`

2.3.2 | Parametric Plots

When both the *x* and *y* coordinates of a function depend upon another parameter *t*, we say that the function is represented *parametrically*. Here is a parametrically represented circle:

In[4]:= `circ = ParametricPlot[{Cos[t], Sin[t]}, {t,0,2Pi}]`

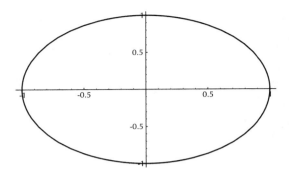

Out[4]= `-Graphics-`

Notice that the ratio of height to width looks unnatural. *Mathematica* tries to fit the plot into a region that is similar to your computer screen and uses a ratio of height to width that is known to be pleasing to the eye. This height to width ratio is known as the `AspectRatio` and has a default value of `1/GoldenRatio`.

In[5]:= `N[1/GoldenRatio]`

Out[5]= `0.618034`

You can get the "true" shape by resetting the `AspectRatio` option.

In[6]:= `Show[circ, AspectRatio -> Automatic]`

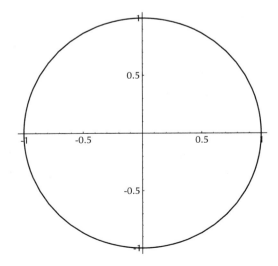

Out[6]= `-Graphics-`

Here is a more interesting parametric plot known as a Lissajous curve.

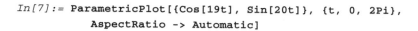

```
In[7]:= ParametricPlot[{Cos[19t], Sin[20t]}, {t, 0, 2Pi},
           AspectRatio -> Automatic]
```

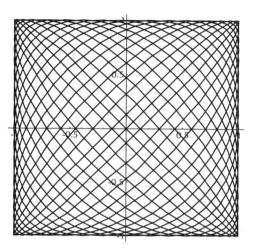

```
Out[7]= -Graphics-
```

2.3.3 | Three-Dimensional Plots

Whereas functions of one variable can be represented as curves in the plane, functions of two variables are visualized as surfaces in space. *Mathematica* has numerous constructs for visualizing these surfaces.

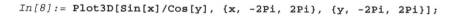

```
In[8]:= Plot3D[Sin[x]/Cos[y], {x, -2Pi, 2Pi}, {y, -2Pi, 2Pi}];
```

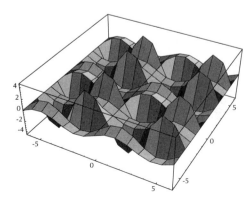

```
Out[8]= -SurfaceGraphics-
```

Notice that this plot is rather choppy. This gives us some insight into how *Mathematica* renders graphics. If you count, you will see that the function `Sin[x]/Cos[y]` is sampled 15 times along both the *x* and *y* axes. The function value is computed at each of these 225 points and then adjacent points are connected by lines and rectangles. This makes for the choppy nature of this plot and could cause some difficulties with functions containing singularities. We can increase the number of points that are sampled and hence improve the smoothness of the surface by adjusting the option `PlotPoints`. The following graphic computes 50 points in both the *x* and *y* directions.

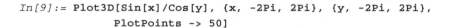

In[9]:= `Plot3D[Sin[x]/Cos[y], {x, -2Pi, 2Pi}, {y, -2Pi, 2Pi},`
`PlotPoints -> 50]`

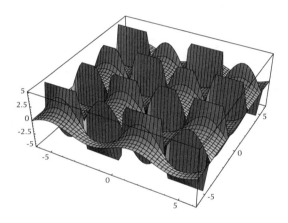

Out[9]= `-SurfaceGraphics-`

Notice that increasing the number of `PlotPoints` does not improve the rendering of these singularities. In the chapter on graphics programming we will discuss some of the subtler issues of graphing functions.

The same surface can be viewed with different tools. `ContourPlot` represents the same function by means of contour lines. Any two points with the same height will appear on the same contour line. Hikers regularly use topographic maps which depict various altitudes by means of different contour lines.

In[10]:= **ContourPlot[Sin[x]/Cos[y], {x, -2Pi, 2Pi}, {y, -2Pi, 2Pi}]**

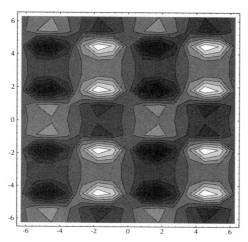

Out[10]= -ContourGraphics-

In this plot, the closer together the lines, the steeper the ascent (or descent) up (or down) the "hill."

The function DensityPlot is also used to represent functions of two variables. In a DensityPlot, the darker the area, the deeper that point is in a trough; lighter areas represent peaks. In the following plot, we have increased the number of points sampled in both the *x* and *y* directions and we have turned off the mesh that is usually placed in a DensityPlot.

In[11]:= **DensityPlot[Sin[x]/Cos[y], {x, -2Pi, 2Pi}, {y, -2Pi, 2Pi},**
PlotPoints -> 150, Mesh -> False]

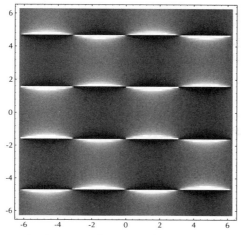

Out[11]= -DensityGraphics-

Just as we plotted parametric curves in the plane, so can we plot three-dimensional surfaces parametrically. Here is a plot of the famous Möbius band, obtained by taking a long sheet, putting a half-twist in it and connecting the two short ends. What is obtained is a surface that contains only one side (excluding the edge). Users of the command-line interface will put each of the inputs on a separate line.

```
In[12]:= x[u_, v_] := (1 + v/2 Cos[u/2]) Cos[u]
         y[u_, v_] := (1 + v/2 Cos[u/2]) Sin[u]
         z[u_, v_] := v/2 Sin[u/2]
```

```
In[13]:= ParametricPlot3D[{x[u, v], y[u, v], z[u, v]},
                          {u, 0, 2Pi}, {v, -1, 1}]
```

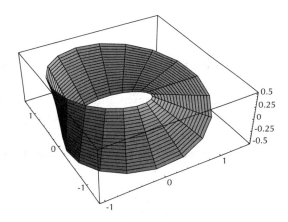

```
Out[13]= -Graphics3D-
```

The Klein bottle is a higher-dimensional version of the Möbius strip. It has no inside nor outside. The code for the following graphic was written by Ari Lehtonen and won the $1000 prize at the 1990 *Mathematica* Graphics Competition.

```
In[14]:= a = 2;
         f = (a + Cos[u/2]Sin[t] - Sin[u/2]Sin[2t]) Cos[u];
         g = (a + Cos[u/2]Sin[t] - Sin[u/2]Sin[2t]) Sin[u];
         h =       Sin[u/2]Sin[t] + Cos[u/2]Sin[2t];
```

```
In[15]:= ParametricPlot3D[{f, g, h}, {t, 0, 2Pi}, {u, 0, 2Pi},
            Boxed -> False, Axes -> False, PlotPoints -> 30]
```

```
Out[15]= -Graphics3D-
```

2.4 Representation of Data

The ability to plot and visualize data is extremely important in all of the sciences—social, natural, and physical. *Mathematica* has capabilities to import and export data from other applications, to plot the data in a variety of forms, and to perform numerical analysis on the data.

In *Mathematica*, data is often represented by means of a *list*. A list is any set of expressions enclosed in braces. Lists can be constructed with the `Table` command, as we have seen before. In addition, many other means of constructing lists are available.

The << function is used to read in information from an external file. In the following example, the file `dataset.m` has been created externally and saved where *Mathematica* can find it. It contains pairs of data points representing body mass *vs.* heat production for 13 different animals. The data is given as (m, r) where m represents the mass of the animal and r the heat production in *kcal* per day.

```
In[1]:= <<dataset.m

Out[1]= {{0.061, 6.951}, {0.403, 28.189}, {0.622, 41.1},
         {2.51, 120.8}, {2.96, 147.9}, {3.33, 182.8},
         {8.2, 368.8}, {28.2, 981.3}, {57.4, 1303.3},
         {72.3, 1512.5}, {340.2, 7100.3}, {711, 10101.1},
         {5000., 29895.}}
```

We can immediately plot the data using the `ListPlot` command.

In[2]:= `ListPlot[<<dataset.m, PlotStyle->PointSize[.02]]`

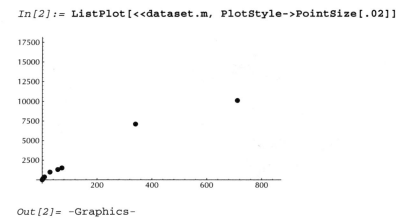

Out[2]= `-Graphics-`

The data is not evenly spread throughout this plot. Information such as this is often best viewed on a logarithmic scale. We can make a log-log plot by first reading in a *Mathematica* package that contains additional functions for logarithmic plots.

In[3]:= `Needs["Graphics`Graphics`"]`

Now we plot the data on log-log "paper."

In[4]:= `LogLogListPlot[<<dataset.m,`
        `PlotStyle->PointSize[.02]]`

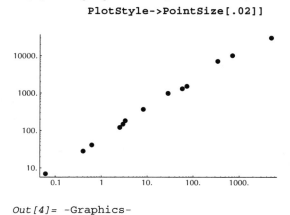

Out[4]= `-Graphics-`

Max Kleiber, an American veterinarian in the 1930s, demonstrated that the metabolic heat rate for a wide range of animals is proportional to $m^{.75}$, where m is the mass. We can check the data against this value by finding the slope of the line through the log-log data. This can be done by first taking the log of each data point.

```
In[5]:= loglogdata = Log[<<dataset.m]

Out[5]= {{-2.79688, 1.93874}, {-0.908819, 3.33893},
         {-0.474815, 3.71601}, {0.920283, 4.79414},
         {1.08519, 4.99654}, {1.20297, 5.20839},
         {2.10413, 5.91025}, {3.33932, 6.88888},
         {4.05004, 7.17265}, {4.28082, 7.32152},
         {5.82953, 8.86789}, {Log[711], 9.2204},
         {8.51719, 10.3054}}
```

We then fit a straight line to this data by performing a linear least squares.

```
In[6]:= Fit[loglogdata, {1, x}, x]

Out[6]= 4.15432 + 0.761477 x
```

We see that the slope of this line is close to Kleiber's expected value of 0.75.

Mathematica contains dozens of functions and packages for performing statistical and numerical analysis of data, including least squares fits, interpolation, Fourier analysis, and numerous graphical capabilities. The interested reader should consult the manual [Wolfram 1991] or [Blachman 1992] for a more thorough listing.

2.5 Programming

Users of every programming language eventually find that the built-in functions are not sufficient for their particular computational needs. They will need to write a program combining the tools offered by the built-in functions with the programming constructs available in the language. *Mathematica* allows for a wide range of programming styles. Our focus will be on its ability to allow the programmer to write "natural code"; that is, code that relies more on the statement of the problem at hand than on various language styles and idiosyncrasies.

In this section, we create two sets of programs that demonstrate some of the programming constructs present in *Mathematica*. We do not expect you to understand all of the details at this point. Hopefully the simplicity of these two programs will be apparent and you will be able to come back to them after having studied a bit more of this text.

2.5.1 | Example—Harmonic Numbers

Suppose you were studying the harmonic numbers H_n, which arise in the study of many computer science algorithms. Here is a mathematical definition for the nth harmonic number,

$$H_n = 1 + \frac{1}{2} + \frac{1}{3} + \frac{1}{4} + \cdots + \frac{1}{n} = \sum_{k=1}^{n} \frac{1}{k}$$

defined for integers $n > 0$. Following the mathematical notation, we can define them in *Mathematica* as follows.

```
In[1]:= harmonic[n_] := Sum[1/k, {k, 1, n}]
```

Notice that our function `harmonic` encloses its argument n in square brackets, as do all *Mathematica* functions. What differs here is the _ character, pronounced "blank." For now, we will just state that the blank is used to denote the argument(s) of the function. (See Chapter 5 for a complete discussion of this topic.)

Let's compute some of the harmonic numbers.

```
In[2]:= harmonic[2]
```

$$Out[2]= \frac{3}{2}$$

Here is the tenth harmonic number together with a numerical approximation.

```
In[3]:= {harmonic[10], N[harmonic[10]]}
```

$$Out[3]= \{\frac{7381}{2520}, \ 2.92897\}$$

Although the harmonic numbers H_n grow very slowly as $n \to \infty$, it is usually proved in a first-year calculus course that they grow without bound. If we try to compute `harmonic[n]` for large n, our computations will quickly become bogged down.

```
In[4]:= N[harmonic[10^4]]
```

```
Out[4]= $Aborted
```

After a few minutes of computation, we interrupted *Mathematica*. [1] The difficulty here was that we were asking the computer to sum up 10,000 different fractions

---

[1] See Section 1.2.1 for details on how to do this on different machines.

and to give us an *exact result*. The least common multiple (LCM) of these fractions has 4349 digits and summing 10,000 fractions, each with a 4349-digit number in the denominator takes a very long time! A more sensible approach will use some of *Mathematica*'s built-in capabilities as well as some mathematical relations.

The closeness of $H_n - \log n$ and γ for large n is generally stated and proven in an analysis course as:

$$\lim_{n \to \infty} H_n - \log n = \gamma$$

The function $\psi(z)$, known as the digamma function, satisfies two formulae that are of interest to us here.

$$\psi(1) = -\gamma$$

and

$$\psi(n) = -\gamma + \sum_{k=1}^{n-1} \frac{1}{k}$$

The ψ function is represented in *Mathematica* as `PolyGamma`, and so with a bit of rearrangement, we can program the harmonic numbers as follows:

```
In[5]:= harmonicSum[n_Integer]:= PolyGamma[n+1] + EulerGamma
```

Here is a list of the first ten harmonic numbers.

```
In[6]:= Table[harmonicSum[n], {n, 10}]
```

$$Out[6]= \{1, \frac{3}{2}, \frac{11}{6}, \frac{25}{12}, \frac{137}{60}, \frac{49}{20}, \frac{363}{140}, \frac{761}{280}, \frac{7129}{2520}, \frac{7381}{2520}\}$$

And here is the time necessary to compute the 10,000[th] harmonic number (we have suppressed the output of this fraction consisting of over 4,300 digits in both the numerator and denominator).

```
In[7]:= Timing[ harmonicSum[10^4]; ]
```

```
Out[7]= {86.35 Second, Null}
```

Although all of the built-in functions in *Mathematica* can be written as user-defined functions, you will find that it is usually best to use the built-in functions whenever possible as they have been optimized for execution speed. When there are no equivalent built-in functions, of course user-defined functions become essential. In fact, creating user-defined functions is the essence of *Mathematica* programming.

2.5.2 | Example—Perfect Numbers

Our second example of a programming problem concerns the search for *perfect numbers.* Perfect numbers are defined as those positive integers that are equal to the sum of their proper divisors (those divisors not including the number itself). The number 6 is the first perfect number.

```
In[1]:= Divisors[6]
Out[1]= {1, 2, 3, 6}

In[2]:= Apply[Plus, {1, 2, 3}]
Out[2]= 6
```

To get a list of the proper divisors, we write a new function `properDivisors` by dropping the last element from the list created by `Divisors`:

```
In[3]:= properDivisors[n_] := Drop[Divisors[n], -1]

In[4]:= properDivisors[6]
Out[4]= {1, 2, 3}

In[5]:= Apply[Plus, properDivisors[6]]
Out[5]= 6
```

Now we would like to create a function to check integers in a specified range for "perfect-ness." The following predicate returns a value of `True` or `False` depending upon whether its argument is perfect or not.

```
In[6]:= perfectQ[n_Integer] := Apply[Plus, properDivisors[n]] == n
```

For example,

```
In[7]:= perfectQ[6]
Out[7]= True

In[8]:= perfectQ[16]
Out[8]= False
```

We are now ready to check a set of integers in a specific range. The function Range[*n*] generates a list of integers in the range 1 through *n*. We only want to select those integers that pass the perfectQ test. The Select[*list*, *crit*] function returns those elements in *list* for which the criterion *crit* is true. Here then are those numbers from 1 through 30 that pass the test for perfectness.

```
In[9]:= Select[Range[30], perfectQ]

Out[9]= {6, 28}
```

Finally, we can write a function that will search for all numbers in the range 1 through *n* that pass the perfectQ test.

```
In[10]:= searchPerfect[n_] := Select[Range[n], perfectQ]
```

To find all perfect numbers less than or equal to 10,000, we enter:

```
In[11]:= searchPerfect[10000]

Out[11]= {6, 28, 496, 8128}
```

In Chapter 4, the reader is encouraged to write a program searchPerfect2[*n, m*] that will return all perfect numbers in a range from *n* to *m*.

It is not known whether there are infinitely many perfect numbers. From the example above, they appear to be very sparse on the number line. Similarly, it is not known whether there are any odd perfect numbers—it is known that *if* there is an odd perfect number, it must be greater than 10^{300}. The interested reader is encouraged to consult [Klee and Wagon 1991] for a more thorough discussion of perfect numbers.

3 | List Manipulation

The list is the fundamental data structure used in *Mathematica* to group objects together. A very extensive set of built-in functions is provided by *Mathematica* to allow us to manipulate lists in a variety of ways ranging from simple operations such as moving list elements around to sophisticated operations such as applying a function to a list. This chapter illustrates some of the many operations that can be performed on lists.

3.1 Introduction

Many computations involve working with a collection of objects. This requires that the objects (also called *data objects*) be gathered together in some way. There are a variety of structures that can be used to store data objects in a computer. The most often used data structure in *Mathematica* is the *list*. This is created using the built-in `List` function.

`List[`arg_1`, `arg_2`, ..., `arg_n`]`

A list can also be written using the standard input form of a sequence of arguments separated by commas and enclosed in braces.

`{`arg_1`, `arg_2`, ..., `arg_n`}`

The arguments of the `List` function (also called the list elements) can be any type of value, including numbers, symbols, functions, character strings, and even other lists:

```
In[1]:= List[2.4, dog, Sin, "read my lips", {5, 3}, Pi, {}]

Out[1]= {2.4, dog, Sin, read my lips, {5, 3}, Pi, {}}
```

Each element has a location within the list that is indicated by its numerical order in the sequence of elements (or in the sequence of arguments to the `List` function). Hence the third element in the list above is Sin. The last element in this list, {}, is a list containing no elements and it is therefore called the *empty list*.

In this chapter, we'll demonstrate the use of built-in *Mathematica* functions to manipulate lists in various ways. In cases where the operation of a function is relatively straightforward, we'll simply demonstrate its use without explanation (the on-line Help system and the *Mathematica* manual [Wolfram 1991] should be consulted for more detailed explanations of all of the built-in functions). The underlying message here is that almost anything you might wish to do to a list can be accomplished using built-in functions. It is important to have as firm a handle on these functions as possible since a key to good, efficient programming in *Mathematica* is to use the built-in functions whenever possible.

3.2 Creating and Measuring Lists

3.2.1 | List Construction

In addition to using the List function to collect data objects, you can also generate lists from scratch by creating the objects and then placing them in a list.

Range[*imin, imax, di*] generates a list of ordered numbers starting from *imin* and going up to, but not exceeding, *imax* in increments of *di*.

```
In[1]:= Range[-4, 7, 3]
Out[1]= {-4, -1, 2, 5}
```

If *di* is not specified, a value of one is used.

```
In[2]:= Range[4, 8]
Out[2]= {4, 5, 6, 7, 8}
```

If neither *imin* nor *di* is specified, then both are given the value of one.

```
In[3]:= Range[4]
Out[3]= {1, 2, 3, 4}
```

It is not necessary for *imin, imax*, or *di* to be integers.

```
In[4]:= Range[1.5, 6.3, .75]
Out[4]= {1.5, 2.25, 3., 3.75, 4.5, 5.25, 6.}
```

Table[*expression*, {*i*, *imin*, *imax*, *di*}] generates a list by evaluating *expression* a number of times:

In[5]:= **Table[3 k, {k, 1, 10, 2}]**

Out[5]= {3, 9, 15, 21, 27}

The first argument, 3 k, is the `expression` that is evaluated to produce the elements in the list in the above example. The second argument to the `Table` function, {*i*, *imin*, *imax*, *di*}, is called the *iterator*. It is a list that specifies the number of times the expression is evaluated and hence the number of elements in the list. The iterator variable may or may not appear in the expression being evaluated. The value *imin* is the value of *i* used in the expression to create the first list element. The value *di* is the incremental increase in the value of *i* used in the expression to create additional list elements. The value *imax* is the maximum value of *i* used in the expression to create the last list element (if incrementing *i* by *di* gives a value greater than *imax*, that value is not used).

In[6]:= **Table[i, {i, 1.5, 6.3, .75}]**

Out[6]= {1.5, 2.25, 3., 3.75, 4.5, 5.25, 6.}

Table[*i*, {*i*, *imin*, *imax*, *di*}] is equivalent to Range[*imin*, *imax*, *di*]. As with the `Range` function, the arguments needed with `Table` can be simplified when the iterator increment is one,

In[7]:= **Table[3 i, {i, 2, 5}]**

Out[7]= {6, 9, 12, 15}

and if both *imin* and *di* are one.

In[8]:= **Table[i^2, {i, 4}]**

Out[8]= {1, 4, 9, 16}

If the iterator variable does not appear in the expression being evaluated, it may be omitted as well. This may or may not result in a list whose elements are all the same. For example,

In[9]:= **Table[darwin, {4}]**

Out[9]= {darwin, darwin, darwin, darwin}

In[10]:= **Table[Random[], {3}]**

Out[10]= {0.0560708, 0.6303, 0.359894}

Table can be used to create a *nested list*; that is, a list containing other lists as elements. This can be done by using more than one iterator.

In[11]:= **Table[i + j, {j, 1, 4}, {i, 1, 3}]**

Out[11]= {{2, 3, 4}, {3, 4, 5}, {4, 5, 6}, {5, 6, 7}}

When there is more than one iterator, their order of appearance is important because the value of the outer iterator is varied for each value of the inner iterator. In the above example, for each value of j (the inner iterator), i was varied from 1 to 3, producing a three-element list for each of the four values of j. If you reverse the iterator order, you will get an entirely different list.

In[12]:= **Table[i + j, {i, 1, 3}, {j,1,4}]**

Out[12]= {{2, 3, 4, 5}, {3, 4, 5, 6}, {4, 5, 6, 7}}

A nested list can be displayed in matrix (or tabular) form.

In[13]:= **Table[i + j, {i, 1, 4}, {j,1,3}] //TableForm**

Out[13]//TableForm=
2	3	4
3	4	5
4	5	6
5	6	7

The value of the outer iterator may depend on the value of the inner iterator, which can result in a nonrectangular list.

In[14]:= **Table[i + j, {i, 1, 3}, {j, 1, i}]**

Out[14]= {{2}, {3, 4}, {4, 5, 6}}

In[15]:= **Table[i + j, {i, 1, 3}, {j, 1, i}] //TableForm**

Out[15]//TableForm=
2		
3	4	
4	5	6

However, the inner iterator may not depend on the outer iterator because, as we have seen, the inner iterator is fixed as the outer one varies.

3.2.2 | Dimensions of Lists

The size of a list can be determined using the `Length` and `Dimensions` functions. For a simple unnested (*linear*) list, the `Length` function tells us how many elements are in the list.

```
In[16]:= Length[{a, b, c, d, e, f}]
Out[16]= 6
```

In a nested list, each inner list is an element of the outer list. Therefore, the `Length` of a nested list indicates the number of inner lists, and not their sizes.

```
In[17]:= Length[{{{1,2}, {3,4}, {5,6}}, {{a,b}, {c,d}, {e,f}}}]
Out[17]= 2
```

To find out more about the inner lists, use the `Dimensions` function. For example,

```
In[18]:= Dimensions[{{{1,2}, {3,4}, {5,6}}, {{a,b}, {c,d}, {e,f}}}]
Out[18]= {2, 3, 2}
```

indicates that there are two inner lists, that each inner list contains three lists, and that the innermost lists each have two elements.

Exercises

1. Generate the list `{{0}, {0, 2}, {0, 2, 4}, {0, 2, 4, 6}, {0, 2, 4, 6, 8}}` in two different ways using the `Table` function.

2. A table containing ten random 1s and 0s can easily be created using `Table[Random[Integer], {10}]`. Create a ten-element list of random 1s, 0s and −1s.

3. Create a ten-element list of random 1s and −1s. This table can be viewed as a list of the steps taken in a random walk along the *x*-axis, where a step can be

Exercises (cont.)

taken in either the positive *x* direction (corresponding to 1) or the negative *x* direction (corresponding to -1) with equal likelihood.

The random walk in one, two, three (and even higher) dimensions is used in science and engineering to represent phenomena that are probabilistic in nature. We will use a variety of random walk models throughout this book to illustrate specific programming points.

4. From a mathematical point of view, a list can be viewed as a vector and a nested list containing inner lists of equal length can be viewed as a matrix (or an array). *Mathematica* has another built-in function Array which creates lists. We can use an undefined function f to see how Array works:

```
In[1]:= Array[f, 5]

Out[1]= {f[1], f[2], f[3], f[4], f[5]}


In[2]:= Array[f, {3, 4}]

Out[2]= {{f[1, 1], f[1, 2], f[1, 3], f[1, 4]},
         {f[2, 1], f[2, 2], f[2, 3], f[2, 4]},
         {f[3, 1], f[3, 2], f[3, 3], f[3, 4]}}
```

Generate both of these lists using the Table function.

5. Predict the dimensions of the list {{{1, a}, {4, d}}, {{2, b}, {3, c}}}. Use the Dimensions function to check your answer.

Solutions

1.
```
In[1]:= Table[j, {i, 0, 8, 2}, {j, 0, i, 2}]

Out[1]= {{0}, {0, 2}, {0, 2, 4}, {0, 2, 4, 6}, {0, 2, 4, 6, 8}}


In[2]:= Table[2j, {i, 0, 4}, {j, 0, i}]

Out[2]= {{0}, {0, 2}, {0, 2, 4}, {0, 2, 4, 6}, {0, 2, 4, 6, 8}}
```

2.
```
In[1]:= Table[Random[Integer, {-1, 1}], {10}]

Out[1]= {0, -1, 0, 0, 1, 0, 0, 1, 1, 1}
```

3. Here are three ways to generate the list.

```
In[1]:= Table[2 Random[Integer] - 1, {10}]

Out[1]= {1, -1, 1, 1, 1, 1, -1, 1, -1, -1}


In[2]:= Table[(-1)^Random[Integer], {10}]

Out[2]= {-1, 1, -1, -1, -1, -1, 1, -1, 1, 1}
```

The following solution will become clearer in the next section.

```
In[3]:= {1, -1}[[Table[Random[Integer, {1, 2}], {10}]]]

Out[3]= {-1, 1, -1, -1, 1, -1, 1, -1, -1, -1}
```

4.

```
In[1]:= Table[f[i], {i, 5}]

Out[1]= {f[1], f[2], f[3], f[4], f[5]}


In[2]:= Table[f[i, j], {i, 3}, {j, 4}]

Out[2]= {{f[1, 1], f[1, 2], f[1, 3], f[1, 4]},
         {f[2, 1], f[2, 2], f[2, 3], f[2, 4]},
         {f[3, 1], f[3, 2], f[3, 3], f[3, 4]}}
```

5.

```
In[1]:= Dimensions[{{{1, a}, {4, d}}, {{2, b}, {3, c}}}]

Out[1]= {2, 2, 2}
```

3.3 Working With the Elements of a List

3.3.1 | Positions in a List

The locations of specific elements in a list can be determined using the Position function.

```
In[1]:= Position[{5, 7, 5, 2, 1, 4}, 5]

Out[1]= {{1}, {3}}
```

This result indicates that the number 5 occurs in the first and third positions in the list. The apparently extra braces are used to avoid confusion with the case when elements are *nested* within a list. For example,

```
In[2]:= Position[{{a, b, c}, {d, e, f}}, f]
Out[2]= {{2, 3}}
```

tells us that f occurs once, in the third position within the second inner list.

3.3.2 | Extracting Elements and Rearranging Lists

Elements can easily be extracted from a specific location in a list.

```
In[3]:= Part[{2, 3, 7, 8, 1, 4}, 3]
Out[3]= 7
```

The Part function has a standard input form (abbreviation).

```
In[4]:= {2, 3, 7, 8, 1, 4}[[3]]
Out[4]= 7

In[5]:= {{1, 2}, {a, b}, {3, 4}, {c, d}, {5, 6}, {e, f}}[[2, 1]]
Out[5]= a
```

If you are interested in the elements from more than one location, you can extract them as follows:

```
In[6]:= {2, 3, 7, 8, 1, 4}[[{2, 1}]]
Out[6]= {3, 2}
```

In addition to being able to extract elements from specific locations in a list, you can extract consecutively placed elements within the list. You can take elements from either the back or the front of a list.

```
In[7]:= Take[{2, 3, 7, 8, 1, 4}, 2]
Out[7]= {2, 3}
```

```
In[8]:= Take[{2, 3, 7, 8, 1, 4}, -2]

Out[8]= {1, 4}
```

If you take consecutive elements from a list other than from the front and the back, you need to remember that the numbering of positions is different front-to-back and back-to-front.

```
In[9]:= Take[{2, 3, 7, 8, 1, 4}, {2, 4}]

Out[9]= {3, 7, 8}

In[10]:= Take[{2, 3, 7, 8, 1, 4}, {-5, -3}]

Out[10]= {3, 7, 8}
```

You can mix both positive and negative indices.

```
In[11]:= Take[{2, 3, 7, 8, 1, 4}, {-5, 4}]

Out[11]= {3, 7, 8}
```

You can discard elements from a list, keeping the rest. Elements can be removed from either end of the list or from consecutive locations.

```
In[12]:= Drop[{2, 3, 7, 8, 1, 4}, 2]

Out[12]= {7, 8, 1, 4}

In[13]:= Drop[{2, 3, 7, 8, 1, 4}, -2]

Out[13]= {2, 3, 7, 8}

In[14]:= Drop[{2, 3, 7, 8, 1, 4}, {3, 5}]

Out[14]= {2, 3, 4}
```

You can remove elements at specific locations as well.

```
In[15]:= Delete[{2, 3, 7, 8, 1, 4}, 2]

Out[15]= {2, 7, 8, 1, 4}

In[16]:= Delete[{2, 3, 7, 8, 1, 4}, {{2}, {5}}]

Out[16]= {2, 7, 8, 4}
```

Certain extractions are used so often that they are given their own functions.

```
In[17]:= First[{2, 3, 7, 8, 1, 4}]

Out[17]= 2

In[18]:= Last[{2, 3, 7, 8, 1, 4}]

Out[18]= 4

In[19]:= Rest[{2, 3, 7, 8, 1, 4}]

Out[19]= {3, 7, 8, 1, 4}
```

There is also a function that picks out the elements in a list that satisfy certain conditions (*i.e.*, that return True when a predicate is applied to them). For example, this finds all of the even numbers in a list.

```
In[20]:= Select[{2, 3, 7, 8, 1, 4}, EvenQ]

Out[20]= {2, 8, 4}
```

The order of the elements in a list can be reversed.

```
In[21]:= Reverse[{2, 7, e, 1, a, 5}]

Out[21]= {5, a, 1, e, 7, 2}
```

The elements can be rearranged into a canonical order (numbers before letters, numbers in increasing order, letters in alphabetical order, etc.).

```
In[22]:= Sort[{2, 7, e, 1, a, 5}]

Out[22]= {1, 2, 2, 5, 7, e}
```

When applied to a nested list, Sort will use the first element of each nested list to determine the order.

```
In[23]:= Sort[{{2, c}, {7, 9}, {e, f, g}, {1, 4.5}, {x, y, z}}]

Out[23]= {{1, 4.5}, {2, c}, {7, 9}, {e, f, g}, {x, y, z}}
```

All of the elements can be rotated a specified number of positions to the right or the left.

In[24]:= **RotateLeft[{2, 7, e, 1, a, 5}, 2]**

Out[24]= {e, 1, a, 5, 2, 7}

In[25]:= **RotateRight[{2, 7, e, 1, a, 5}, 2]**

Out[25]= {a, 5, 2, 7, e, 1}

You can flatten a nested list to various extents. You can remove all of the inner braces, creating a linear list of elements.

In[26]:= **Flatten[{{{3, 1}, {2, 4}}, {{5, 3}, {7, 4}}}]**

Out[26]= {3, 1, 2, 4, 5, 3, 7, 4}

Or, you can limit the degree of leveling, removing only some of the inner lists. For example, two inner lists, each having two ordered pairs, can be turned into a single list of four ordered pairs.

In[27]:= **Flatten[{{{3, 1}, {2, 4}}, {{5, 3}, {7, 4}}}, 1]**

Out[27]= {{3, 1}, {2, 4}, {5, 3}, {7, 4}}

`Partition` rearranges list elements to form a nested list. It may use all of the elements and simply divvy up a list.

In[28]:= **Partition[{2, 3, 7, 8, 1, 4}, 2]**

Out[28]= {{2, 3}, {7, 8}, {1, 4}}

Or, it may only use some of the elements from a list. For example, this takes the elements in the odd-numbered locations:

In[29]:= **Partition[{2, 3, 7, 8, 1, 4}, 1, 2]**

Out[29]= {{2}, {7}, {1}}

You can also create overlapping inner lists, consisting of ordered pairs whose second element is the first element of the next ordered pair.

In[30]:= **Partition[{2, 3, 7, 8, 1, 4}, 2, 1]**

Out[30]= {{2, 3}, {3, 7}, {7, 8}, {8, 1}, {1, 4}}

The Transpose function pairs off the corresponding elements of the inner lists. Its argument is a single list consisting of nested lists.

```
In[31]:= Transpose[{{5, 2, 7, 3}, {4, 6, 8, 4}}]

Out[31]= {{5, 4}, {2, 6}, {7, 8}, {3, 4}}
```

```
In[32]:= Transpose[{{5, 2, 7, 3}, {4, 6, 8, 4}, {6, 5, 3, 1}}]

Out[32]= {{5, 4, 6}, {2, 6, 5}, {7, 8, 3}, {3, 4, 1}}
```

Elements can be added to the front, the back, or to any specified position in a given list.

```
In[33]:= Append[{2, 3, 7, 8, 1, 4}, 5]

Out[33]= {2, 3, 7, 8, 1, 4, 5}
```

```
In[34]:= Prepend[{2, 3, 7, 8, 1, 4}, 5]

Out[34]= {5, 2, 3, 7, 8, 1, 4}
```

```
In[35]:= Insert[{2, 3, 7, 8, 1, 4}, 5, 3]

Out[35]= {2, 3, 5, 7, 8, 1, 4}
```

Elements at specific locations in a list can be replaced with other elements. Here, 3 is replaced by 5 in the second position.

```
In[36]:= ReplacePart[{2, 3, 7, 8, 1, 4}, 5, 2]

Out[36]= {2, 5, 7, 8, 1, 4}
```

Exercises

1. Predict where the 9s are located in the following list:

```
{{2, 1, 10}, {9, 5, 7}, {2, 10, 4}, {10, 1, 9}, {6, 1, 6}}
```

Confirm your prediction using Position.

Exercises (cont.)

2. Given a list of $\{x, y\}$ data points,

$$\{\{x1, y1\}, \{x2, y2\}, \{x3, y3\}, \{x4, y4\}, \{x5, y5\}\}$$

separate the x and y components to get:

$$\{\{x1, x2, x3, x4, x5\}, \{y1, y2, y3, y4, y5\}\}$$

3. Consider a two-dimensional random walk on a square lattice. (A square lattice can be envisioned as a two-dimensional grid—just like the lines on graph paper.) Each step can be in one of four directions: $\{1, 0\}, \{0, 1\}, \{-1, 0\}, \{0, -1\}$, corresponding to steps in the east, north, west and south directions, respectively. Use the list $\{\{1, 0\}, \{0, 1\}, \{-1, 0\}, \{0, -1\}\}$ to create a list of the steps of a 10-step random walk.

4. In three steps, make a list of the elements in even-numbered locations in the list $\{a, b, c, d, e, f, g\}$.

5. Suppose you are given a list S of length n, and a list P containing n different numbers between 1 and n (*i.e.*, P is a *permutation* of Range[n]). Compute the list T such that for all k between 1 and n, $T[[k]] = S[[P[[k]]]]$. For example, if $S = \{a, b, c, d\}$ and $P = \{3, 2, 4, 1\}$, then $T = \{c, b, d, a\}$.

6. Given the lists S and P in the previous exercise, compute the list U such that for all k between 1 and n, $U[[P[[k]]]] = S[[k]]$ (*i.e.*, $S[[i]]$ takes the value from position $P[[i]]$ in U). Thus, for $S = \{a, b, c, d\}$ and $P = \{3, 2, 4, 1\}$, $U = \{d, b, a, c\}$. (Think of it as moving $S[[1]]$ to position $P[[1]]$, $S[[2]]$ to position $P[[2]]$, and so on.) Hint: Start by pairing the elements of P with the elements of S.

Solutions

1.
```
In[1]:= Position[{{2, 1, 10}, {9, 5, 7}, {2, 10, 4},
            {10, 1, 9}, {6, 1, 6}}, 9]

Out[1]= {{2, 1}, {4, 3}}
```

2.
```
In[1]:= Transpose[{{x1,y1}, {x2,y2}, {x3,y3}, {x4,y4}, {x5,y5}}]

Out[1]= {{x1, x2, x3, x4, x5}, {y1, y2, y3, y4, y5}}
```

3. Here is one way to do it:

```
In[1]:= Table[{{1,0}, {-1,0}, {0,1}, {0,-1}}[[Random[Integer, {1,4}]]],
        {10}]

Out[1]= {{0, -1}, {0, -1}, {-1, 0}, {0, 1}, {0, -1}, {0, 1},
         {0, -1}, {0, 1}, {0, 1}, {0, 1}}
```

4. We first drop the first element in the list, then create a nested list of every other element in the remaining list, and finally unnest the resulting list.

```
In[1]:= Rest[{a, b, c, d, e, f, g}]
Out[1]= {b, c, d, e, f, g}

In[2]:= Partition[%, 1, 2]
Out[2]= {{b}, {d}, {f}}

In[3]:= Flatten[%]
Out[3]= {b, d, f}
```

5.
```
In[1]:= {a, b, c, d}[[{3, 2, 4, 1}]]
Out[1]= {c, b, d, a}
```

6.
```
In[1]:= Transpose[{{3, 2, 4, 1}, {a, b, c, d}}]
Out[1]= {{3, a}, {2, b}, {4, c}, {1, d}}

In[2]:= Sort[%]
Out[2]= {{1, d}, {2, b}, {3, a}, {4, c}}

In[3]:= Transpose[%]
Out[3]= {{1, 2, 3, 4}, {d, b, a, c}}

In[4]:= %[[2]]
Out[4]= {d, b, a, c}
```

3.4 Working with Several Lists

A number of the functions described above, such as Transpose, work with several lists if they are inside a nested list structure. We can also work directly with multiple lists.

```
In[1]:= Join[{2, 5, 7, 3}, {d, a, e, j}]
Out[1]= {2, 5, 7, 3, d, a, e, j}

In[2]:= Union[{4, 1, 2}, {5, 1, 2}]
Out[2]= {1, 2, 4, 5}

In[3]:= Union[{4, 1, 2, 5, 1, 2}]
Out[3]= {1, 2, 4, 5}
```

When the Union function is used either on a single list or a number of lists, a list is formed consisting of the original elements in canonical order with all duplicate elements removed. The Complement function gives all those elements in the first list that are not in the other list or lists. Intersection[$list_1$, $list_2$, ...] finds all those elements common to the $list_i$. Complement and Intersection also remove duplicates and sort the elements that remain.

```
In[4]:= Complement[{2, 8, 7, 4, 8, 3}, {3, 5, 4}]
Out[4]= {2, 7, 8}

In[5]:= Intersection[{2, 8, 7, 4, 8, 3}, {3, 5, 4}]
Out[5]= {3, 4}
```

These last three functions, Union, Complement, and Intersection, treat lists somewhat like sets in that there are no duplicates and the order of elements in the lists is not respected.

Exercises

1. How would you perform the same task as `Prepend[{x, y}, z]` using the `Join` function?

2. Starting with the lists {1, 2, 3, 4} and {a, b, c, d}, create the list {2, 4, b, d}.

3. Given two lists, find all those elements that are not common to the two lists. For example, starting with the lists, {a, b, c, d} and {a, b, e, f}, your answer would return the list {c, d, e, f}.

Solutions

1.

```
In[1]:= Join[{z}, {x, y}]

Out[1]= {z, x, y}
```

2.

```
In[1]:= Join[{1, 2, 3, 4}, {a, b, c, d}]

Out[1]= {1, 2, 3, 4, a, b, c, d}

In[2]:= Rest[%]

Out[2]= {2, 3, 4, a, b, c, d}

In[3]:= Partition[%, 1, 2]

Out[3]= {{2}, {4}, {b}, {d}}

In[4]:= Flatten[%]

Out[4]= {2, 4, b, d}
```

3.

```
In[1]:= Complement[Union[{a,b,c,d}, {a,b,e,f}],
            Intersection[{a,b,c,d}, {a,b,e,f}]]

Out[1]= {c, d, e, f}
```

3.5 Higher-Order Functions

There are a number of built-in functions that take other functions as arguments. These are called *higher-order functions* and they are very powerful programming tools.

Map applies a function to each element in a list. This is illustrated using an undefined function *f* and a simple linear list.

```
In[1]:= Map[f, {3, 5, 7, 2, 6}]

Out[1]= {f[3], f[5], f[7], f[2], f[6]}
```

Using the Reverse function with the Map function, you can reverse the order of elements in each list of a nested list.

```
In[2]:= Map[Reverse, {{a, b}, {c, d}, {e, f}}]

Out[2]= {{b, a}, {d, c}, {f, e}}
```

The elements in each of the inner lists in a nested list can be sorted.

```
In[3]:= Map[Sort, {{2, 6, 3, 5}, {7, 4, 1, 3}}]

Out[3]= {{2, 3, 5, 6}, {1, 3, 4, 7}}
```

Using the MapThread function, you can take a function and two or more lists and create a new list in which the corresponding elements of the old lists are paired (or zipped together) as arguments of the function. The function argument must accept as many arguments as there are lists and the lists must be of equal length.

Using an undefined function *g* and a nested list, you can see *g* applied to each of the corresponding elements in the inner lists.

```
In[4]:= MapThread[g, {{a, b, c}, {x, y, z}}]

Out[4]= {g[a, x], g[b, y], g[c, z]}
```

We can raise each element in the first list to the power given by the corresponding element in the second list.

```
In[5]:= MapThread[Power, {{2, 6, 3}, {5, 1, 2}}]

Out[5]= {32, 6, 9}
```

Using the `List` function, the corresponding elements in the three lists are placed in a list structure (note that `Transpose` would do the same thing).

```
In[6]:= MapThread[List, {{5, 3, 2}, {6, 4, 9}, {4, 1, 4}}]

Out[6]= {{5, 6, 4}, {3, 4, 1}, {2, 9, 4}}
```

The `Outer` function applies a function to all of the combinations of the elements in several lists. This is a generalization of the mathematical *outer product.*

```
In[7]:= Outer[f, {a, b}, {2, 3, 4}]

Out[7]= {{f[a, 2], f[a, 3], f[a, 4]},
          {f[b, 2], f[b, 3], f[b, 4]}}
```

Using the `List` function as an argument, you can create lists of ordered pairs that combine the elements of several linear lists.

```
In[8]:= Outer[List, {a, b}, {2, 3, 4}]

Out[8]= {{{a, 2}, {a, 3}, {a, 4}}, {{b, 2}, {b, 3}, {b, 4}}}
```

Many of the built-in functions that take a single argument have the property that when a list is the argument, the function is automatically applied to all of the elements in the list. In other words, these functions are automatically mapped onto the elements of the list. For example,

```
In[9]:= Log[{1.2, 0.4, 3.7}]

Out[9]= {0.182322, -0.916291, 1.30833}
```

This is the same result you get using the `Map` function.

```
In[10]:= Map[Log, {1.2, 0.4, 3.7}]

Out[10]= {0.182322, -0.916291, 1.30833}
```

Many of the built-in functions that take two or more arguments have the property that when multiple lists are the arguments, the function is automatically applied to all of the corresponding elements in the list. In other words, these functions are automatically threaded onto the elements of the list. For example,

```
In[11]:= {4, 6, 3} + {5, 1, 2}
Out[11]= {9, 7, 5}
```

This gives the same result as using the `Plus` function with the `MapThread` function.

```
In[12]:= MapThread[Plus, {{4, 6, 3}, {5, 1, 2}}]
Out[12]= {9, 7, 5}
```

Functions that are either automatically mapped or threaded onto the elements of list arguments are said to be `Listable`. Many of *Mathematica*'s built-in functions have this `Attribute`.

```
In[13]:= Attributes[Log]
Out[13]= {Listable, Protected}
```

You can change the elements of a list (*i.e.*, the arguments of the `List` function) into the arguments of another function as long as the function takes the same number of arguments as there are elements in the list.

```
In[14]:= Apply[f, List[1, 4, 5, 3]]
Out[14]= f[1, 4, 5, 3]
```

We see that the elements of `List` are now the arguments of `f`. You could say that `List` has been replaced by `f`. Here is a specific example:

```
In[15]:= Apply[Plus, {1, 4, 5, 3}]
Out[15]= 13
```

Here, `List[1, 4, 5, 3]` has been changed to `Plus[1, 4, 5, 3]` or, in other words, `List` has been replaced by `Plus`. This list conversion can be applied to an entire list.

```
In[16]:= Apply[f, {{1, 2, 3}, {5, 6, 7}}]
Out[16]= f[{1, 2, 3}, {5, 6, 7}]

In[17]:= Apply[Plus, {{1, 2, 3}, {5, 6, 7}}]
Out[17]= {6, 8, 10}
```

Or it can be applied only to the inner lists within a nested list.

```
In[18]:= Apply[f, {{1, 2, 3}, {5, 6, 7}}, 2]

Out[18]= {f[1, 2, 3], f[5, 6, 7]}


In[19]:= Apply[Plus, {{1, 2, 3}, {5, 6, 7}}, 2]

Out[19]= {6, 18}
```

Exercises

1. A matrix can be rotated by performing a number of successive operations. Rotate the matrix {{1, 2, 3}, {4, 5, 6}} clockwise by 90 degrees, obtaining {{4, 1}, {5, 2}, {6, 3}}, in two steps. Use TableForm to display the results.

2. While matrices can easily be added using Plus, matrix multiplication is more complicated. The Dot function, written as a single period, can be used:

```
In[1]:= {{1, 2}, {3, 4}} . {x, y}

Out[1]= {x + 2 y, 3 x + 4 y}
```

Perform matrix multiplication on {{1, 2}, {3, 4}} and {x, y} without using Dot. (This can be done in three steps.)

Solutions

1.
```
In[1]:= {{1, 2, 3}, {4, 5, 6}} //TableForm

Out[1]//TableForm= 1    2    3
                   4    5    6


In[2]:= Transpose[{{1, 2, 3}, {4, 5, 6}}]

Out[2]= {{1, 4}, {2, 5}, {3, 6}}


In[3]:= Map[Reverse, %] //TableForm

Out[3]//TableForm= 4    1
                   5    2
                   6    3
```

2. This can be done either in three steps, or by using the `Inner` function.

```
In[1]:= Transpose[{{1, 2}, {3, 4}}]
Out[1]= {{1, 3}, {2, 4}}

In[2]:= % * {x, y}
Out[2]= {{x, 3 x}, {2 y, 4 y}}

In[3]:= Apply[Plus, %]
Out[3]= {x + 2 y, 3 x + 4 y}

In[4]:= Inner[Times, {{1, 2}, {3, 4}}, {x, y}, Plus]
Out[4]= {x + 2 y, 3 x + 4 y}
```

3.6 Applying Functions to Lists Repeatedly

A function can be applied to an argument and then applied to the result and then applied to that result and so on, a specified number of times. For example, using an undefined function g and a starting value of a, four applications of g gives,

```
In[1]:= Nest[g, a, 4]
Out[1]= g[g[g[g[a]]]]
```

The `NestList` function retains all of the intermediate values of the `Nest` operation.

```
In[2]:= NestList[g, a, 4]
Out[2]= {a, g[a], g[g[a]], g[g[g[a]]], g[g[g[g[a]]]]}
```

Using a starting value of 0.85, this generates a list of ten iterates of the `Cos` function.

```
In[3]:= NestList[Cos, 0.85, 10]
Out[3]= {0.85, 0.659983, 0.790003, 0.703843, 0.76236,
         0.723208, 0.749687, 0.731902, 0.743904, 0.73583,
         0.741274}
```

The list elements above are the values of 0.85, Cos[0.85], Cos of the result of Cos[0.85], and so on:

```
In[4]:= {0.85, Cos[0.85], Cos[Cos[0.85]], Cos[Cos[Cos[0.85]]]}

Out[4]= {0.85, 0.659983, 0.790003, 0.703843}
```

With the Fold function, a function is applied to a starting value and the first element of a list. The function is then applied to the result and the second element of the list. The function is then applied to that result and the third element of the list, and so on.

```
In[5]:= Fold[f, 0, {a, b, c, d}]

Out[5]= f[f[f[f[0, a], b], c], d]
```

If FoldList is used, then you will see all of the intermediate results of the Fold operation.

```
In[6]:= FoldList[f, 0, {a, b, c, d}]

Out[6]= {0, f[0, a], f[f[0, a], b], f[f[f[0, a], b], c],
           f[f[f[f[0, a], b], c], d]}
```

It is easy to see what is going on with the FoldList function by working with an arithmetic operator.

```
In[7]:= FoldList[Plus, 0, {a, b, c, d}]

Out[7]= {0, a, a + b, a + b + c, a + b + c + d}
```

```
In[8]:= FoldList[Plus, 0, {3, 5, 2, 4}]

Out[8]= {0, 3, 8, 10, 14}
```

```
In[9]:= FoldList[Times, 1, {3, 5, 2, 4}]

Out[9]= {1, 3, 15, 30, 120}
```

You can thread a function onto several lists and then use the result as the argument to another function. Using undefined functions f and g, you can see how this works:

```
In[10]:= Inner[f, {a, b, c}, {d, e, f}, g]
```

```
Out[10]= g[f[a, d], f[b, e], f[c, f]]
```

This function lets you carry out some interesting operations.

```
In[11]:= Inner[Times, {a, b, c}, {d, e, f}, Plus]
```

```
Out[11]= a d + b e + c f
```

```
In[12]:= Inner[List, {a, b, c}, {d, e, f}, Plus]
```

```
Out[12]= {a + b + c, d + e + f}
```

Looking at these two examples, you can see that Inner is really a generalization of the mathematical dot product (and Dot).

Exercises

1. Determine the locations after each step of a ten-step one-dimensional random walk. (Recall that you have already generated the step *directions* in Exercise 3 on page 63 of Section 3.3.)

2. Create a list of the step locations of a ten-step random walk on a square lattice.

Solutions

1.

```
In[1]:= Table[(-1)^Random[Integer], {10}]
```

```
Out[1]= {-1, 1, 1, 1, 1, -1, 1, -1, 1, 1}
```

```
In[2]:= FoldList[Plus, 0, %]
```

```
Out[2]= {0, -1, 0, 1, 2, 3, 2, 3, 2, 3, 4}
```

2. We can use the method of generating a list of step locations that was shown in Exercise 3 on page 71.

```
In[1]:= {{1,0}, {-1,0}, {0,1}, {0,-1}}[[Table[
            Random[Integer, {1,4}], {10}]]]
```

```
Out[1]= {{1, 0}, {1, 0}, {1, 0}, {0, 1}, {1, 0}, {0, 1},
         {-1, 0}, {0, -1}, {0, -1}, {0, 1}}
```

```
In[2]:= FoldList[Plus, {0, 0}, %]

Out[2]= {{0, 0}, {1, 0}, {2, 0}, {3, 0}, {3, 1}, {4, 1},
          {4, 2}, {3, 2}, {3, 1}, {3, 0}, {3, 1}}
```

3.7 Strings and Characters

Characters are the objects that appear on the computer screen like "a", "3", or "!". Upper and lower-case letters, numbers, punctuation marks, and spaces form the basic set of characters. A sequence of characters enclosed in double quotes is called a *string*. When *Mathematica* prints out a string, it appears without the quotes but you can use the InputForm function to see these quotes:

```
In[1]:= "It's the economy, stupid"

Out[1]= It's the economy, stupid
```

```
In[2]:= InputForm["It's the economy, stupid"]

Out[2]//InputForm= "It's the economy, stupid"
```

A string is a value and like other values (such as numbers and lists), there are built-in functions available to manipulate strings, similar to the ones for lists. Their operations are indicated by their names.

```
In[3]:= StringLength["It's the economy, stupid"]

Out[3]= 24
```

```
In[4]:= StringReverse["abcde"]

Out[4]= edcba
```

```
In[5]:= StringTake["abcde", 3]

Out[5]= abc
```

```
In[6]:= StringDrop["abcde", -1]

Out[6]= abcd
```

```
In[7]:= StringPosition["abcde", "bc"]

Out[7]= {{2, 3}}


In[8]:= StringInsert["abcde", "t", 3]

Out[8]= abtcde


In[9]:= StringReplace["abcde", "cd"->"uv"]

Out[9]= abuve
```

In addition to using built-in functions to manipulate a string, you can convert a string to a list of characters with the built-in `Characters` function. You can then use the list-manipulating functions to alter the list. Finally, you can change the resulting list back into a string using the built-in `StringJoin` function. For example,

```
In[10]:= Characters["abcde"]

Out[10]= {a, b, c, d, e}


In[11]:= Take[%, {2, 3}]

Out[11]= {b, c}


In[12]:= StringJoin[%]

Out[12]= bc
```

Another way to manipulate a string is to convert it to a list of character codes. Each character in a computer's character set is assigned a number, called its *character code*. Moreover, by general agreement, almost all computers use the same character codes, called the *ASCII codes*.[1] In this code, the upper-case letters A, B, ..., Z are assigned the numbers 65, 66, ..., 90 while the lower-case letters a, b, ..., z have the numbers 97, 98, ..., 122 (note that the number of an upper-case letter is 32 less than its lower-case version). The numbers 0, 1, ..., 9 are coded as 48, 49, ..., 57 while the punctuation marks period, comma, and exclamation point have the codes 46, 44, and 33, respectively. The space character is represented by the code 32. Table 3.1 shows the characters and their codes.

---

[1] ASCII stands for American Standard Code for Information Interchange.

Characters	ASCII codes
A, B, . . ., Z	65, 66, . . ., 90
a, b, . . ., z	97, 98, . . ., 122
0, 1, . . ., 9	48, 49, . . ., 57
. (period)	46
, (comma)	44
!	33
(space)	32

Table 3.1: ASCII character codes.

To illustrate the use of the character code representation of a character, this changes a word from lower-case to upper-case:

```
In[13]:= ToCharacterCode["darwin"]
Out[13]= {100, 97, 114, 119, 105, 110}

In[14]:= % - 32
Out[14]= {68, 65, 82, 87, 73, 78}

In[15]:= FromCharacterCode[%]
Out[15]= DARWIN
```

Exercises

1. Convert the first character in a string (which you may assume to be a lower-case letter) to upper case.

2. Given a string containing two digits, convert it to its integer value; so the string "73" produces the number 73.

3. Given a string containing two digits, convert it to its value as an integer in base 8; for example, the string "73" will produce the number 59.

4. Given a string of digits of arbitrary length, convert it to its integer value. (Hint: You may find that the Dot function is helpful.)

Solutions

1.

```
In[1]:= "this is a test string"

Out[1]= this is a test string

In[2]:= StringTake[%, 1]

Out[2]= t

In[3]:= ToCharacterCode[%]

Out[3]= {116}

In[4]:= % - 32

Out[4]= {84}

In[5]:= FromCharacterCode[%]

Out[5]= T

In[6]:= StringDrop[%%%%,1]

Out[6]= his is a text string

In[7]:= StringJoin[%%, %]

Out[7]= This is a test string
```

2.

```
In[1]:= "73"

Out[1]= 73

In[2]:= ToCharacterCode[%][[1]]

Out[2]= 55

In[3]:= 10(% - 48)

Out[3]= 70

In[4]:= ToCharacterCode[%%%][[2]]

Out[4]= 51

In[5]:= % - 48

Out[5]= 3

In[6]:= % + %%%

Out[6]= 73
```

3.

```
In[1]:= "73"

Out[1]= 73

In[2]:= ToCharacterCode[%]

Out[2]= {55, 51}

In[3]:= %[[1]]

Out[3]= 55

In[4]:= 8(% - 48)

Out[4]= 56

In[5]:= ToCharacterCode[%%%]

Out[5]= {55, 51}

In[6]:= %[[2]]

Out[6]= 51

In[7]:= % - 48

Out[7]= 3

In[8]:= % + %%%%

Out[8]= 59
```

4.

```
In[1]:= "10495"

Out[1]= 10495

In[2]:= ToCharacterCode[%]

Out[2]= {49, 48, 52, 57, 53}

In[3]:= % - 48

Out[3]= {1, 0, 4, 9, 5}

In[4]:= Reverse[Table[10^j, {j, 0, 4}]]

Out[4]= {10000, 1000, 100, 10, 1}

In[5]:= Dot[%, %%]

Out[5]= 10495
```

4 Functions

Programming in *Mathematica* is essentially a matter of writing user-defined functions that work like mathematical functions; when applied to specific values, they perform computations producing results. In this chapter we will demonstrate, in a step-by-step manner, how to create a user-defined function using a functional programming style. Building upon the nesting of function calls, extensive use is made of higher-order and anonymous functions. The construction of one-liners and compound functions with local names is described.

4.1 Introduction

So far, we have described the simple use of some built-in functions with various data objects. While these are very useful operations, we often want to do more complicated data manipulations. In this chapter we will first demonstrate the use of nested function calls. We will then go on to create our own user-defined functions.

We will be using many of the built-in list manipulating functions discussed earlier as well as introducing new ones. It is well worthwhile to spend time familiarizing yourself with these functions by playing around with them; for example, create various lists and apply built-in functions to them. Having a larger vocabulary of built-in functions will not only make it easier to follow the programs and do the exercises here, but will enhance your own programming skills as well.

4.2 Programs as Functions

A computer program is a set of instructions (*i.e.*, a recipe) for carrying out a computation. When a program is evaluated with appropriate inputs, the computation is performed and the result is returned. In this sense, a program is a mathematical function and the inputs to a program are the arguments of the function. Executing a program is equivalent to applying a function to its arguments or, as it often referred to, making a function call.

4.2.1 | Nesting Function Calls

Earlier, we demonstrated some of the uses of built-in functions. Most of the examples simply applied a built-in function to a data object one time and returned some value. For example, to calculate a remainder, use the Mod function.

```
In[1]:= Mod[5, 3]
Out[1]= 2
```

Sorting the numbers in a list is accomplished with Sort.

```
In[2]:= Sort[{1, 3, 2, 8, 3, 6, 3, 2}]
Out[2]= {1, 2, 2, 3, 3, 3, 6, 8}
```

Some of the examples were a bit more complicated, using higher-order functions that employ repeated function application.

```
In[3]:= Fold[Plus, 0, {5, 2, 9, 12, 4, 23}]
Out[3]= 55
```

The sequential application of several functions is known as a *nested function call*. Nested function calls are not limited to using a single function repeatedly such as with the built-in Nest and Fold functions. We can create our own nested function calls employing different functions, one after the other. For example, here is a nested function call in which the tangent of 4.0 is computed and then the sine of the result is computed and then the cosine of that result is computed.

```
In[4]:= Cos[Sin[Tan[4.0]]]
Out[4]= 0.609053
```

To see this more clearly, we can step through the computation.

```
In[5]:= Tan[4.0]
Out[5]= 1.15782
```

```
In[6]:= Sin[%]
Out[6]= 0.915931
```

```
In[7]:= Cos[%]
```

Out[7]= 0.609053

Or, we can evaluate the entire expression after wrapping it in the `Trace` function:

```
In[8]:= Trace[Cos[Sin[Tan[4.0]]]]
```

Out[8]= {{{Tan[4.], 1.15782}, Sin[1.15782], 0.915931},
        Cos[0.915931], 0.609053}

Stated in general terms, a nested function call is an application of a function to the result of applying another function to some argument value.

A nested function call is read in much the same way that it is created, starting with the innermost functions and working towards the outermost functions. For example, the following expression determines whether all the elements in a list are even numbers:

```
In[9]:= Apply[And, Map[EvenQ, {2, 4, 6, 7, 8}]]
```

Out[9]= False

The operations can be described as follows:

1. apply the predicate `EvenQ` to every element in the list {2, 4, 6, 7, 8}:

    ```
    In[10]:= Map[EvenQ, {2, 4, 6, 7, 8}]
    ```

    Out[10]= {True, True, True, False, True}

2. apply the logical function `And` to the result of the previous step:

    ```
    In[11]:= Apply[And, %]
    ```

    Out[11]= False

Another, more complicated, example returns the elements in a list of positive numbers that are bigger than all of the preceding numbers in the list:

```
In[12]:= Union[Rest[FoldList[Max, 0, {3, 1, 6, 5, 4, 8, 7}]]]
```

Out[12]= {3, 6, 8}

The `Trace` of the function call shows the intermediate steps of the computation:

```
In[13]:= Trace[Union[Rest[FoldList[Max, 0, {3, 1, 6, 5, 4, 8, 7}]]]]

Out[13]= {{{FoldList[Max, 0, {3, 1, 6, 5, 4, 8, 7}], {Max[0, 3], 3},
         {Max[3, 1], Max[1, 3], 3}, {Max[3, 6], 6},
         {Max[6, 5], Max[5, 6], 6}, {Max[6, 4], Max[4, 6], 6},
         {Max[6, 8], 8}, {Max[8, 7], Max[7, 8], 8},
         {0, 3, 3, 6, 6, 6, 8, 8}}, Rest[{0, 3, 3, 6, 6, 6, 8, 8}],
         {3, 3, 6, 6, 6, 8, 8}}, Union[{3, 3, 6, 6, 6, 8, 8}],
         {3, 6, 8}}
```

This computation can be described as follows:

1. The FoldList function is first applied to the function Max, 0, and the list {3, 1, 6, 5, 4, 8, 7}.

2. The Rest function is then applied to the result of the previous step.

3. Finally, the Union function is applied to the result of the previous step.

Notice that in each of the nested function call descriptions given here, the data object that was fed into the first function was explicitly referred to, but the data objects that were manipulated in the succeeding function calls were not identified other than as the results of the previous steps (*i.e.*, as the results of the preceding function applications).

Here is an interesting application of nested function calls—the creation of a deck of cards:

```
In[14]:= Flatten[Outer[List, {c, d, h, s},
          Join[Range[2, 10], {J, Q, K, A}]], 1]

Out[14]= {{c, 2}, {c, 3}, {c, 4}, {c, 5}, {c, 6}, {c, 7}, {c, 8},
          {c, 9}, {c, 10}, {c, J}, {c, Q}, {c, K}, {c, A}, {d, 2},
          {d, 3}, {d, 4}, {d, 5}, {d, 6}, {d, 7}, {d, 8}, {d, 9},
          {d, 10}, {d, J}, {d, Q}, {d, K}, {d, A}, {h, 2}, {h, 3},
          {h, 4}, {h, 5}, {h, 6}, {h, 7}, {h, 8}, {h, 9}, {h, 10},
          {h, J}, {h, Q}, {h, K}, {h, A}, {s, 2}, {s, 3}, {s, 4},
          {s, 5}, {s, 6}, {s, 7}, {s, 8}, {s, 9}, {s, 10}, {s, J},
          {s, Q}, {s, K}, {s, A}}
```

To understand what's going on here, we'll construct this call from scratch. First we form a list of the number and face cards in a suit by combining a list of the numbers two through ten, Range[2, 10], with a four-element list representing the jack, queen, king and ace, {J, Q, K, A}.

```
In[15]:= Join[Range[2, 10], {J, Q, K, A}]

Out[15]= {2, 3, 4, 5, 6, 7, 8, 9, 10, J, Q, K, A}
```

Now we pair each of the 13 elements in this list with each of the 4 elements in the list representing the card suits {c, d, h, s}. This produces a list of 52 ordered pairs representing the cards in a deck, where the king of clubs for example, is represented by {c, K}).

```
In[16]:= Outer[List, {c, d, h, s}, %]

Out[16]= {{{c, 2}, {c, 3}, {c, 4}, {c, 5}, {c, 6}, {c, 7}, {c, 8},
           {c, 9}, {c, 10}, {c, J}, {c, Q}, {c, K}, {c, A}},
          {{d, 2}, {d, 3}, {d, 4}, {d, 5}, {d, 6}, {d, 7}, {d, 8},
           {d, 9}, {d, 10}, {d, J}, {d, Q}, {d, K}, {d, A}},
          {{h, 2}, {h, 3}, {h, 4}, {h, 5}, {h, 6}, {h, 7}, {h, 8},
           {h, 9}, {h, 10}, {h, J}, {h, Q}, {h, K}, {h, A}},
          {{s, 2}, {s, 3}, {s, 4}, {s, 5}, {s, 6}, {s, 7}, {s, 8},
           {s, 9}, {s, 10}, {s, J}, {s, Q}, {s, K}, {s, A}}}
```

While we now have all of the cards in the deck, they are grouped by suit in a nested list. We therefore un-nest the list,

```
In[17]:= Flatten[%, 1]

Out[17]= {{c, 2}, {c, 3}, {c, 4}, {c, 5}, {c, 6}, {c, 7}, {c, 8},
          {c, 9}, {c, 10}, {c, J}, {c, Q}, {c, K}, {c, A}, {d, 2},
          {d, 3}, {d, 4}, {d, 5}, {d, 6}, {d, 7}, {d, 8}, {d, 9},
          {d, 10}, {d, J}, {d, Q}, {d, K}, {d, A}, {h, 2}, {h, 3},
          {h, 4}, {h, 5}, {h, 6}, {h, 7}, {h, 8}, {h, 9}, {h, 10},
          {h, J}, {h, Q}, {h, K}, {h, A}, {s, 2}, {s, 3}, {s, 4},
          {s, 5}, {s, 6}, {s, 7}, {s, 8}, {s, 9}, {s, 10}, {s, J},
          {s, Q}, {s, K}, {s, A}}
```

Voila!

The step-by-step construction that we used here, applying one function at a time, checking each function call separately, is a very efficient way to make a *prototype* program in *Mathematica*. We'll use this technique again in the next example.

We'll perform what is called a *perfect shuffle*, consisting of cutting the deck in half and then interleaving the cards from the two halves. Rather than working with the large list of fifty-two ordered pairs during the prototyping, we'll use a short made-up list. A short list of an even number of ordered integers is a good choice for the task.

```
In[18]:= Range[6]
```

```
Out[18]= {1, 2, 3, 4, 5, 6}
```

We first divide the list into two equal-sized lists.

```
In[19]:= Partition[%, 3]
```

```
Out[19]= {{1, 2, 3}, {4, 5, 6}}
```

We now want to interleave these two lists to form {1, 4, 2, 5, 3, 6}. The first step is to pair the corresponding elements in each of the two lists above. This can be done using the Transpose function:

```
In[20]:= Transpose[%]
```

```
Out[20]= {{1, 4}, {2, 5}, {3, 6}}
```

We now un-nest the interior lists using the Flatten function. We could flatten our simple list using Flatten[...] but since we know that ultimately we'll be dealing with ordered pairs rather than integers, we will use Flatten[..., 1] as we did in creating the card deck.

```
In[21]:= Flatten[%, 1]
```

```
Out[21]= {1, 4, 2, 5, 3, 6}
```

Okay, that does the job. Given this prototype, it's easy to write the actual nested function call to perform a perfect shuffle on a deck of cards.

```
In[22]:= Flatten[Transpose[Partition[Out[14], 26]], 1]
```

```
Out[22]= {{c, 2}, {h, 2}, {c, 3}, {h, 3}, {c, 4}, {h, 4},
          {c, 5}, {h, 5}, {c, 6}, {h, 6}, {c, 7}, {h, 7},
          {c, 8}, {h, 8}, {c, 9}, {h, 9}, {c, 10}, {h, 10},
          {c, J}, {h, J}, {c, Q}, {h, Q}, {c, K}, {h, K},
          {c, A}, {h, A}, {d, 2}, {s, 2}, {d, 3}, {s, 3},
          {d, 4}, {s, 4}, {d, 5}, {s, 5}, {d, 6}, {s, 6},
          {d, 7}, {s, 7}, {d, 8}, {s, 8}, {d, 9}, {s, 9},
          {d, 10}, {s, 10}, {d, J}, {s, J}, {d, Q}, {s, Q},
          {d, K}, {s, K}, {d, A}, {s, A}}
```

4.2.2 | Value Names

It is convenient to be able to refer to a value by a *name* rather than to have to state the value itself. The syntax for naming a value looks like a mathematical assignment; a *left-hand side* and a *right-hand side* separated by an equal sign.

name = *expr*

The left-hand side, *name*, is a symbol. A symbol consists of a letter followed by an uninterrupted succession of letters and numbers. The right-hand side, *expr*, is an expression. When this assignment is entered, the right-hand side is evaluated and its value is returned. For example, we can name the list representing the deck of cards (it is a good idea to use a name that indicates what the value represents):

```
In[23]:= cardDeck =
         Flatten[Outer[List, {c, d, h, s},
                 Join[Range[2, 10], {J, Q, K, A}]], 1]
```

```
Out[23]= {{c, 2}, {c, 3}, {c, 4}, {c, 5}, {c, 6}, {c, 7},
          {c, 8}, {c, 9}, {c, 10}, {c, J}, {c, Q}, {c, K},
          {c, A}, {d, 2}, {d, 3}, {d, 4}, {d, 5}, {d, 6},
          {d, 7}, {d, 8}, {d, 9}, {d, 10}, {d, J}, {d, Q},
          {d, K}, {d, A}, {h, 2}, {h, 3}, {h, 4}, {h, 5},
          {h, 6}, {h, 7}, {h, 8}, {h, 9}, {h, 10}, {h, J},
          {h, Q}, {h, K}, {h, A}, {s, 2}, {s, 3}, {s, 4},
          {s, 5}, {s, 6}, {s, 7}, {s, 8}, {s, 9}, {s, 10},
          {s, J}, {s, Q}, {s, K}, {s, A}}
```

Notice that what is returned is not just the right-hand side that was entered, but the value that results from evaluating the right-hand side.

After a value is named, entering its name returns the value.

```
In[24]:= cardDeck
```

```
Out[24]= {{c, 2}, {c, 3}, {c, 4}, {c, 5}, {c, 6}, {c, 7},
          {c, 8}, {c, 9}, {c, 10}, {c, J}, {c, Q}, {c, K},
          {c, A}, {d, 2}, {d, 3}, {d, 4}, {d, 5}, {d, 6},
          {d, 7}, {d, 8}, {d, 9}, {d, 10}, {d, J}, {d, Q},
          {d, K}, {d, A}, {h, 2}, {h, 3}, {h, 4}, {h, 5},
          {h, 6}, {h, 7}, {h, 8}, {h, 9}, {h, 10}, {h, J},
          {h, Q}, {h, K}, {h, A}, {s, 2}, {s, 3}, {s, 4},
          {s, 5}, {s, 6}, {s, 7}, {s, 8}, {s, 9}, {s, 10},
          {s, J}, {s, Q}, {s, K}, {s, A}}
```

Once a value is named, its name can be used anywhere in a function call.

```
In[25]:= Flatten[Transpose[Partition[cardDeck, 26]], 1]

Out[25]= {{c, 2}, {h, 2}, {c, 3}, {h, 3}, {c, 4}, {h, 4},
          {c, 5}, {h, 5}, {c, 6}, {h, 6}, {c, 7}, {h, 7},
          {c, 8}, {h, 8}, {c, 9}, {h, 9}, {c, 10}, {h, 10},
          {c, J}, {h, J}, {c, Q}, {h, Q}, {c, K}, {h, K},
          {c, A}, {h, A}, {d, 2}, {s, 2}, {d, 3}, {s, 3},
          {d, 4}, {s, 4}, {d, 5}, {s, 5}, {d, 6}, {s, 6},
          {d, 7}, {s, 7}, {d, 8}, {s, 8}, {d, 9}, {s, 9},
          {d, 10}, {s, 10}, {d, J}, {s, J}, {d, Q}, {s, Q},
          {d, K}, {s, K}, {d, A}, {s, A}}
```

Essentially, the name of a value is a *nickname* and whenever it appears, it is replaced by the value itself. The name and the value are functionally equivalent.

A name can only be associated with one value at a time. A value can be removed from a name either by simply entering name =. or Clear[name]. A different value can be assigned to a name by entering name = newvalue.

Exercises

1. Write a nested function call that creates a deck of cards and performs a perfect shuffle on it.

2. Write nested function calls using the ToCharacterCode and FromCharacterCode functions to perform the same operations as the built-in StringJoin and StringReverse functions.

3. Take Exercises 4 and 6 on page 71, Exercises 1 and 2 on page 78, Exercises 1 and 2 on page 81, and Exercises 1 and 2 on page 84, in Chapter 3, and rewrite them as nested function calls.

Solutions

1. After creating the card deck, we cut it in half and interleave the two halves.

```
In[1]:= cardDeck =
        Flatten[Outer[List, {c, d, h, s},
            Join[Range[2, 10], {J, Q, K, A}]], 1];
```

```
In[2]:= Flatten[Transpose[Partition[cardDeck, 26]], 1 ]

Out[2]= {{c, 2}, {h, 2}, {c, 3}, {h, 3}, {c, 4}, {h, 4},
         {c, 5}, {h, 5}, {c, 6}, {h, 6}, {c, 7}, {h, 7},
         {c, 8}, {h, 8}, {c, 9}, {h, 9}, {c, 10}, {h, 10},
         {c, J}, {h, J}, {c, Q}, {h, Q}, {c, K}, {h, K},
         {c, A}, {h, A}, {d, 2}, {s, 2}, {d, 3}, {s, 3},
         {d, 4}, {s, 4}, {d, 5}, {s, 5}, {d, 6}, {s, 6},
         {d, 7}, {s, 7}, {d, 8}, {s, 8}, {d, 9}, {s, 9},
         {d, 10}, {s, 10}, {d, J}, {s, J}, {d, Q}, {s, Q},
         {d, K}, {s, K}, {d, A}, {s, A}}
```

2.

```
In[1]:= FromCharacterCode[Join[ToCharacterCode["To be, "],
            ToCharacterCode["or not to be"]]]

Out[1]= To be, or not to be
```

```
In[2]:= FromCharacterCode[Reverse[ToCharacterCode[%]]]

Out[2]= eb ot ton ro ,eb oT
```

3.

```
In[1]:= Flatten[Partition[Rest[{a, b, c, d, e, f, g}], 1, 2]]

Out[1]= {b, d, f}
```

```
In[2]:= Transpose[Sort[Transpose[{{3,2,4,1},{a,b,c,d}}]]][[2]]

Out[2]= {d, b, a, c}
```

```
In[3]:= FoldList[Plus, 0, Table[(-1)^Random[Integer], {10}]]

Out[3]= {0, 1, 0, 1, 2, 3, 2, 3, 2, 1, 2}
```

```
In[4]:= FoldList[Plus, {0,0}, {{1,0}, {-1,0}, {0,1}, {0,-1}}[[
            Table[Random[Integer, {1, 4}], {10}]]]]

Out[4]= {{0, 0}, {0, 1}, {0, 2}, {-1, 2}, {-1, 3}, {0, 3},
         {1, 3}, {0, 3}, {0, 4}, {0, 5}, {-1, 5}}
```

```
In[5]:= Map[Reverse, Transpose[{{1, 2, 3}, {4, 5, 6}}]]

Out[5]= {{4, 1}, {5, 2}, {6, 3}}
```

```
In[6]:= Apply[Plus, Transpose[{{1, 2}, {3, 4}}] {x, y}]

Out[6]= {x + 2 y, 3 x + 4 y}
```

```
In[7]:= StringJoin[FromCharacterCode[ToCharacterCode[
            StringTake["this is a test string", 1]] - 32],
            StringDrop["this is a test string", 1]]

Out[7]= This is a test string


In[8]:= (ToCharacterCode["73"][[1]] - 48) * 10 +
            (ToCharacterCode["73"][[2]] - 48)

Out[8]= 73
```

4.3 User-Defined Functions

While there are a great many built-in functions that can be called to carry out computations, we invariably find ourselves needing *customized* functions. For example, once we've figured out how to perform a nested function call on a particular value, we might want to perform the same set of operations on different values. We would therefore like to create our own *user-defined* functions that we could then apply in the same way as we call a built-in function—by entering the function name and specific argument values. We'll start with the proper syntax (or grammar) to use when writing a function definition.

The right-hand side function looks very much like a mathematical equation: a left-hand side and a right-hand side separated by a colon-equal sign.

$$\mathbf{name}[arg_1, \ arg_2, \ \ldots, \ arg_n] \ := \ body$$

The left-hand side starts with a symbol. This symbol is referred to as the *function name* (or sometimes just as the function, as in "the sine function"). The function name is followed by a set of square brackets, inside of which are a sequence of symbols ending with *blanks*. These symbols are referred to as the *function argument names*, or just the *function arguments*.

The right-hand side of a user-defined function definition is called the *body* of the function. The body can be either a single expression (a one-liner), or a series of expressions (a compound function), both of which will be discussed in detail shortly. Argument names from the left-hand side appear on the right-hand side without blanks. Basically, the right-hand side is a formula stating what computations are to be done when the function is called with specific values of the arguments.

When a user-defined function is first entered, nothing is returned. Thereafter, calling the function by entering the left-hand side of the function definition with specific values of the arguments causes the body of the function to be computed with the specific argument values substituted where the argument names occur.

A simple example of a user-defined function is `square` which squares a value (it is a good idea to use a function name that indicates the purpose of the function).

```
In[1]:= square[x_] := x^2
```

After entering a function definition, we can call the function in the same way that a built-in function is applied to an argument.

```
In[2]:= square[5]
Out[2]= 25
```

We can quite easily turn the earlier examples of nested function calls into programs. For example,

```
Apply[And, Map[EvenQ, {2, 4, 6, 7, 8}]]
```

becomes

```
areEltsEven[lis_] := Apply[And, Map[EvenQ, lis]]
```

After entering the definition of this user-defined function,

```
In[3]:= areEltsEven[lis_] := Apply[And, Map[EvenQ, lis]]
```

we can apply it to a list.

```
In[4]:= areEltsEven[{2, 8, 14, 6, 16}]
Out[4]= True
```

Another nested function call that we developed earlier,

```
In[5]:= Union[Rest[FoldList[Max, 0, {3, 1, 6, 5, 4, 8, 7}]]]
```

becomes

In[6]:= **maxima[x_] := Union[Rest[FoldList[Max, 0, x]]]**

which, after being entered, can be applied to a list of numbers.

In[7]:= **maxima[{4, 2, 7, 3, 4, 9, 14, 11, 17}]**

Out[7]= {4, 7, 9, 14, 17}

Any definition associated with a name can be removed by simply entering
Clear[*name*]. If this is done, calling the function simply returns the input itself. For
example,

In[8]:= **Clear[maxima]**

In[9]:= **maxima[{6, 3, 9, 2}]**

Out[9]= maxima[{6, 3, 9, 2}]

Once a user-defined function is entered, it can be used in another function,
which can be either a built-in function or another user-defined function. We will
illustrate both of these by writing a function that deals cards from a card deck. We'll
construct this function in stages using the prototyping method we showed earlier,
except that we won't evaluate any code here (the reader might want to try out the
various expressions shown here with specific values to see how they work).

The first thing we will need is the card deck.

In[10]:= **cardDeck =**
          Flatten[Outer[List, {c, d, h, s},
            Join[Range[2, 10], {J, Q, K, A}]], 1];

Next, we need to define a function that removes a single element from a randomly
chosen position in a list.

In[11]:= **removeRand[lis_] :=**
          Delete[lis, Random[Integer, {1, Length[lis]}]]

The user-defined function removeRand first uses the Random function to
randomly choose an integer k between 1 and the length of the list, and then uses the
Delete function to remove the kth element of the list (for example, if a list has ten
elements, an integer between one and ten, say six, is randomly determined and the
element in the sixth position in the list is then removed from the list).

Now we want to make a function call that applies the removeRand function to the cardDeck list, then applies the removeRand function to the resulting list, then applies the removeRand function to the resulting list, and so on, a total of *n* times. The way to carry out this operation is with the Nest function:

```
Nest[removeRand, cardDeck, n]
```

Lastly, we want the cards that are removed from cardDeck rather than those that remain so we write:

```
Complement[cardDeck, Nest[removeRand, cardDeck, n]]
```

Now, we write this up formally into the user-defined deal function.

```
In[12]:= deal[n_] :=
         Complement[cardDeck, Nest[removeRand, cardDeck, n]]
```

Let's try it out:

```
In[13]:= deal[5]

Out[13]= {{d, 6}, {d, 10}, {h, 2}, {h, 5}, {s, 6}}
```

Not a bad hand!

Exercises

1. One of the games in the Illinois State Lottery is based on choosing *n* numbers, each between zero and nine, with duplicates allowed (in practice, a selection is made from containers of numbered ping pong balls). We can model this game using a simple user-defined function which we will call pick (after the official lottery names of *Pick3* and *Pick4*).

```
In[1]:= pick[n_] := Table[Random[Integer, {0,9}], {n}]


In[2]:= pick[4]

Out[2]= {0, 9, 4, 5}
```

Exercises (cont.)

This program can be generalized to perform *random sampling with replacement* on any list. Write a function chooseWithReplacement[*lis_*, *n_*], where *lis* is the list, *n* is the number of elements being chosen and the following is a typical result.

```
In[3]:= chooseWithReplacement[{a, b, c, d, e, f, g, h}, 3]

Out[3]= {c, g, g}
```

2. Write your own user-defined functions using the ToCharacterCode and FromCharacterCode functions to perform the same operations as String-Insert and StringDrop.

3. Create a function distance[{a, b}] that finds the distance between two points a and b in the plane.

4. Rewrite all the nested function calls you created in Exercise 3 in Section 4.2 on page 94 as user-defined functions. Check that your definitions work by running the programs with various input values.

5. Write a user-defined function interleave2 that interleaves the elements of two lists of unequal length. (You've already seen how to interleave lists of equal length on page 92.) Your function should take the lists {1, 2, 3} and {a, b, c, d} as inputs and return {1, a, 2, b, 3, c, d}.

Solutions

1. The obvious way to do this is to take the list and simply pick out elements at random locations in the list (note: the right-most location in the list is given by Length[*lis*]), using the built-in Part and Random functions.

```
In[1]:= chooseWithReplacement[lis_, n_] :=
            lis[[Table[Random[Integer, {1, Length[lis]}], {n}]]]

In[2]:= chooseWithReplacement[{a, b, c, d, e, f, g, h}, 3]

Out[2]= {e, c, c}
```

2.

```
In[1]:= stringInsert[str1_, str2_, pos_] :=
            FromCharacterCode[Join[Take[ToCharacterCode[str1],
                                        pos - 1],
                                   ToCharacterCode[str2],
                              Drop[ToCharacterCode[str1], pos - 1]]]
```

```
In[2]:= stringDrop[str_,pos_]:=
            FromCharacterCode[Drop[ToCharacterCode[str], pos]]
```

3.

```
In[1]:= distance[lis_] :=
           (N[Sqrt[Apply[Plus, {lis[[1, 1]] - lis[[2, 1]],
                          lis[[1, 2]] - lis[[2, 2]]}^2]]])
```

```
In[2]:= distance[{{2, 5}, {6, 8}}]
```

```
Out[2]= 5.
```

4.

```
In[1]:= name1[lis_] := Flatten[Partition[Rest[lis], 1, 2]]
```

```
In[2]:= name2[lis1_, lis2_] :=
            Transpose[Sort[Transpose[{lis1, lis2}]]][[2]]
```

```
In[3]:= name3[n_] :=
           FoldList[Plus, 0, Table[(-1)^Random[Integer], {n}]]
```

```
In[4]:= name3[n_] :=
           FoldList[Plus, {0, 0}, {{1, 0}, {-1, 0}, {0, 1}, {0,
      -1}}[[Table[Random[Integer, {1, 4}], {n}]]]]
```

```
In[5]:= name4[lis_] :=Map[Reverse, Transpose[lis]]
```

```
In[6]:= name5[lis_] :=
           Apply[Plus, Transpose[lis] {x, y}]
```

```
In[7]:= capitalizeFirstLetter[x_] :=
           StringJoin[FromCharacterCode[
            ToCharacterCode[StringTake[x, 1]]- 32],
            StringDrop[x, 1]]
```

```
In[8]:= stringToInteger[x_] :=
          (ToCharacterCode[x][[1]] - 48) * 10 +
          (ToCharacterCode[x][[2]] - 48)
```

5. We assume that lis1 is longer than lis2 and pair off the corresponding elements in the lists and then tack on the leftover elements from lis1.

```
In[1]:= interLeave2[lis1_, lis2_] :=
          Flatten[Join[Transpose[{lis2, Take[lis1, Length[lis2]]}],
             Take[lis1, (Length[lis2] - Length[lis1])]]]

In[2]:= interLeave2[{a, b, c, d}, {1, 2, 3}]
Out[2]= {1, a, 2, b, 3, c, d}
```

4.4 Auxiliary Functions

There is one major drawback to the deal function: In order to use deal, the definition of removeRand and the value of cardDeck must be entered before calling deal. It would be much more convenient if we could incorporate these functions within the deal function definition itself. In the next section, we will show how this can be done.

4.4.1 | Compound Functions

The left-hand side of a *compound function* is the same as that of any user-defined function. The right-hand side consists of consecutive expressions enclosed in parentheses and separated by semicolons (possibly split across several lines).

name[arg_1_, arg_2_, ..., arg_n_] := (*$expr_1$*; *$expr_2$*, ...; *$expr_m$*)

The expressions can be user-defined functions (also known as *auxiliary* functions), value declarations, and function calls. When a compound function is evaluated with particular argument values, these expressions are evaluated in order and the result of the evaluation of the last expression is returned (by adding a semicolon after *$expr_n$*, the display of the final evaluation result can also be suppressed).

We can work with the deal function to illustrate how a compound function is created. We need the following three expressions:

```
cardDeck =
    Flatten[Outer[List, {c, d, h, s},
                   Join[Range[2, 10], {J, Q, K, A}]], 1];

removeRand[lis_] := Delete[lis, Random[Integer, {1, Length[lis]}]]

deal[n_] := Complement[cardDeck, Nest[removeRand, cardDeck, n]]
```

The conversion to a compound function is easily done. We'll first remove the old definitions.

```
In[1]:= Clear[deal, cardDeck, removeRand]
```

Now we can create and enter the new definition,

```
In[2]:= deal[n_] :=
        (cardDeck =
           Flatten[Outer[List, {c, d, h, s},
                         Join[Range[2, 10], {J, Q, K, A}]], 1];
         removeRand[lis_] :=
            Delete[lis, Random[Integer, {1, Length[lis]}]];

         Complement[cardDeck, Nest[removeRand, cardDeck, n]])
```

Let's check that this works.

```
In[3]:= deal[5]
Out[3]= {{c, Q}, {d, 8}, {d, A}, {h, 9}, {s, 10}}
```

A couple of things should be pointed out about the right-hand side of a compound function definition: Since the expressions on the right-hand side are evaluated in order, value declarations and auxiliary function definitions should be given *before* they are used and the argument names used on the left-hand side of auxiliary function definitions *must* differ from the argument names used by the compound function itself.

Finally, when we enter a compound function definition, we are entering not only the function but also the auxiliary functions and the value declarations. If we then remove the function definition using Clear[*name*], the auxiliary function definitions and value declarations remain. This can cause a problem if we subsequently try to use the names of these auxiliary functions and values elsewhere. It is best to isolate the names of auxiliary functions and value declarations inside the main function definition. We'll show how to do this next.

4.4.2 | Localizing Names

When a user-defined function is written, it is generally a good idea to isolate the names of values and functions defined on the right-hand side from the outside world in order to avoid any conflict with the use of a name elsewhere in the session (for example, `cardDeck` might be used elsewhere to represent a pinochle deck). This can be done by wrapping the right-hand side of the function definition in the built-in `Module` function as follows:

```
name[arg₁_, arg₂_, ..., argₙ_] :=
    Module[{name₁, name₂ = value, ...}, expr]
```

The first argument of the `Module` function is a list of the names we want to localize. If we wish, we can assign values to these names, as is shown with *name₂* above (the assigned value is only an initial value and can be changed subsequently). The list is separated from the right-hand side by a comma and so the parentheses enclosing the right-hand side of a compound function are not needed.

We can demonstrate the use of `Module` with the `deal` function:

```
In[4]:= Clear[deal]


In[5]:= deal[n_] := Module[{removeRand, cardDeck},
            cardDeck = Flatten[Outer[List, {c, d, h, s},
                                Join[Range[2, 10], {J,Q,K,A}]],
                        1];
            removeRand[lis_] :=
                Delete[lis, Random[Integer, {1, Length[lis]}]];

            Complement[cardDeck, Nest[removeRand, cardDeck, n]]]
```

It is generally a good idea to wrap the right-hand side of all compound function definitions in the `Module` function. Another way to avoid conflicts in the use of names of auxiliary function definitions is to use a function that can be applied without being given a name. Such functions are called *anonymous functions*, which we discuss in the next section.

Exercises

1. Rewrite the user-defined functions that you created in the exercises for Section 4.3 on page 99 as compound function definitions. Check that your definitions work.

2. Write a compound function definition for the location of steps taken in an *n*-step random walk on a square lattice. Hint: Use the definition for the step increments of the walk as an auxiliary function.

3. The `searchPerfect` function defined in Chapter 2.1 on page 57 is impractical for checking large numbers because it has to check all numbers from 1 through *n*. If you already know the perfect numbers below 500, say, it is inefficient to check all numbers from 1 to 1000 if you are only looking for perfect numbers in the range 500 to 1000. Modify `searchPerfect` so that it accepts two numbers as input and computes all perfect numbers between the inputs. For example, `searchPerfect[a, b]` will produce a list of all perfect numbers in the range from *a* to *b*.

4. Write a function `search3Perfect` that computes all *3-perfect* numbers. (A 3-perfect number is such that the sum of its divisors equals *three* times the number. For example, 120 is 3-perfect since

    ```
    In[1]:= Apply[Plus, Divisors[120]]

    Out[1]= 360
    ```

    which is of course, $3 \cdot 120$. Find the only other 3-perfect number under 1000.

5. Write a function `search4Perfect` and find the three 4-perfect numbers less than 2,200,000.

6. Write a function `searchKPerfect` that accepts as input a number *k*, and two numbers *a* and *b*, and computes all *k*-perfect numbers in the range from *a* to *b*. For example, `searchKPerfect[1, 30, 2]` would compute all 2-perfect numbers in the range from 1 to 30 and hence, would output {6, 28}.

7. If $\sigma(n)$ is defined to be the sum of the divisors of *n*, then a number *n* is called *superperfect* if $\sigma(\sigma(n)) = 2n$. Write a function `searchSuperperfect[a, b]` that finds all superperfect numbers in the range from *a* to *b*.

Solutions

1.

```
In[1]:= name1[lis_] := Flatten[Partition[Rest[lis], 1, 2]]
```

```
In[2]:= name2[lis1_, lis2_] :=
            (pairing[m_, n_] := Transpose[{m, n}];
            Transpose[Sort[pairing[lis1, lis2]]][[2]]])
```

```
In[3]:= name3[n_] :=
            (steps[m_] := Table[(-1)^Random[Integer], {m}];
            FoldList[Plus, 0, steps[n]])
```

```
In[4]:= name3[n_] :=
            FoldList[Plus, {0, 0}, {{1, 0}, {-1, 0}, {0, 1}, {0,
        -1}}[[Table[Random[Integer, {1, 4}], {n}]]]]
```

```
In[5]:= name4[lis_] :=
            Map[Reverse, Transpose[lis]]
```

```
In[6]:= name5[lis_] :=
            Apply[Plus, Transpose[lis] {x, y}]
```

```
In[7]:= capitalizeFirstLetter[str1_] :=
            (ucletter = FromCharacterCode[
                    ToCharacterCode[StringTake[str1, 1]]-32];
            StringJoin[ucletter, StringDrop[str1, 1]])
```

```
In[8]:= stringToInteger[str2_] :=
            (tensplace = (ToCharacterCode[str2][[1]] - 48) * 10;
            onesplace = ToCharacterCode[str2][[2]] - 48;
            tensplace + onesplace)
```

2.

```
In[1]:= latticeWalk1[n_] :=
            (steps[m_] := {{1,0}, {-1,0}, {0,1}, {0,-1}}[[
                    Table[Random[Integer, {1,4}], {m}]]];
            FoldList[Plus, {0,0}, steps[n]])
```

```
In[2]:= latticeWalk1[10]
```

```
Out[2]= {{0, 0}, {0, 1}, {1, 1}, {1, 0}, {0, 0}, {-1, 0},
            {-1, -1}, {-1, 0}, {-1, 1}, {-1, 0}, {-1, -1}}
```

```
In[3]:= latticeWalk2[n_] :=
        (choices = {{1, 0}, {-1, 0}, {0, 1}, {0, -1}};
         steps[m_] := choices[[Table[Random[Integer, {1, 4}],
                      {m}]]];
         FoldList[Plus, {0, 0}, steps[n]])
```

```
In[4]:= latticeWalk2[10]
```

```
Out[4]= {{0, 0}, {0, -1}, {-1, -1}, {-1, 0}, {-2, 0}, {-2, 1},
         {-2, 2}, {-2, 3}, {-2, 4}, {-3, 4}, {-3, 3}}
```

3. The following function creates a local function `perfectQ` using the `Module` construct. It then checks every other number between n and m by using a third argument to the `Range` function.

```
In[1]:= searchPerfect[n_, m_] := Module[{perfectQ},
        perfectQ[j_] := Apply[Plus,Divisors[j]]==2j;
        Select[Range[n,m,2], perfectQ]]
```

```
In[2]:= searchPerfect[2, 1000]
```

```
Out[2]= {6, 28, 496}
```

This function does not guard against the user supplying "bad" inputs. For example, if the user starts with an odd number, then this version of `searchPerfect` will check every other odd number, and since there are no odd numbers below at least 10^{300}, none are reported:

```
In[3]:= searchPerfect[1, 1000]
```

```
Out[3]= {}
```

You can fix this situation by using the (as yet unproved) assumption that there are *no* odd perfect numbers. This next version first checks that the first argument is an even number.

```
In[4]:= Clear[searchPerfect]
```

```
In[5]:= searchPerfect[n_?EvenQ, m_] := Module[{perfectQ},
        perfectQ[j_] := Apply[Plus, Divisors[j]] == 2j;
        Select[Range[n,m,2], perfectQ]]
```

Now, the function only works if the first argument is even.

```
In[6]:= searchPerfect[2, 1000]

Out[6]= {6, 28, 496}


In[7]:= searchPerfect[1,1000]

Out[7]= searchPerfect[1, 1000]
```

4. This only requires a slight change to the code from the searchPerfect function from the previous exercise.

```
In[1]:= search3Perfect[n_, m_] := Module[{perfectQ},
            perfectQ[j_] := Apply[Plus,Divisors[j]] == 3j;
            Select[Range[n,m], perfectQ]]


In[2]:= search3Perfect[1, 1000]

Out[2]= {120, 672}
```

5. Again, this function only requires a slight modification from that for the searchPerfect function above.

```
In[1]:= search4Perfect[n_, m_] := Module[{perfectQ},
            perfectQ[j_] := Apply[Plus,Divisors[j]]==4j;
            Select[Range[n,m], perfectQ]]
```

The following computation can be quite time-consuming and requires a fair amount of memory to run to completion. If your computer's resources are limited, you should split up the search intervals into smaller units.

```
In[2]:= search4Perfect[1, 2200000]

Out[2]= {30240, 32760, 2178540}
```

6. This function requires a third argument.

```
In[1]:= searchKPerfect[n_, m_, k_] := Module[{perfectQ},
            perfectQ[j_] := Apply[Plus,Divisors[j]] == k j;
            Select[Range[n,m], perfectQ]]
```

```
In[2]:= searchKPerfect[1, 100, 2]

Out[2]= {6, 28}
```

7. This function will require two auxiliary functions—the function σ and a predicate to determine whether a number is *super-perfect*.

```
In[1]:= searchSuperperfect[a_, b_] := Module[{sigma, superQ},
            sigma[n_] := Apply[Plus, Divisors[n]];
            superQ[n_]:= Nest[sigma, n, 2] == 2n;
            Select[Range[a,b], superQ]]
```

Here then, are all super-perfect numbers less than 10,000.

```
In[2]:= searchSuperperfect[1, 10000]

Out[2]= {2, 4, 16, 64, 4096}
```

4.5 Anonymous Functions

An *anonymous function* is a function that does not have a name and that can be used "on the spot;" *i.e.*, at the moment it is created. This is often convenient, especially if the function is only going to be used once or as an argument to a higher-order function, such as Map, Fold, or Nest. The built-in function Function is used to create an anonymous function.

The basic form of an anonymous function is Function[x, *body*] for an anonymous function with a single variable x (note that any symbol can be used for the variable), and Function[{x, y, ...}, *body*] for an anonymous function with more than one variable. The body looks like the right-hand side of a user-defined function definition, with the variables x, y, ..., where argument names would be.

As an example, the first user-defined function we created, square, can be written as an anonymous function:

```
In[1]:= Function[z, z^2]

Out[1]= Function[z, z ]
                      2
```

There is also a standard input form that can be used in writing an anonymous function which is easier to write and read than the `Function` notation. In this form, the right-hand side of the function definition is placed inside parentheses which are followed by the ampersand symbol (&), and the argument name is replaced by the pound symbol (#). If there is more than one variable, #1, #2, and so on are used. Using this notation, the `square` function,

```
square[x_] := x^ 2
```

becomes

```
(#^ 2)&
```

An anonymous function is applied in the usual way, by following the function with the argument values enclosed in square brackets. For example,

```
In[2]:= (#^2)&[6]
Out[2]= 36
```

We can, if we wish, give an anonymous function a name and then use that name to call the function later. This has the same effect as defining the function in the more traditional manner.

```
In[3]:= squared = (#^2)&;
```

```
In[4]:= squared[6]
Out[4]= 36
```

The best way to become comfortable with anonymous functions is to see them in action, so we will convert some of the functions we defined earlier into anonymous functions (we'll show both the (... # ...) & and the `Function` forms so that you can decide which you prefer to use):

- The function that tests whether all the elements of a list are even:

```
areEltsEven[lis_] := Apply[And, Map[EvenQ, lis]]
```

becomes

```
      (Apply[And, Map[EvenQ, #]])&
```

or

```
      Function[x, Apply[And, Map[EvenQ, x]]]
```

- The function that returns each element in the list greater than all previous elements,

```
      maxima[x_] := Union[Rest[FoldList[Max, 0, x]]]
```

becomes

```
      (Union[Rest[FoldList[Max, 0, #]]])&
```

or

```
      Function[y, Union[Rest[FoldList[Max, 0, y]]]]
```

- The function that removes a randomly-chosen element from a list,

```
      removeRand[lis_] := Delete[lis, Random[Integer, {1, Length[lis]}]]
```

becomes

```
      (Delete[#, Random[Integer, {1, Length[#]}]])&
```

or

```
      Function[x, Delete[x, Random[Integer, {1, Length[x]}]]]
```

We can also create nested anonymous functions. For example, this maps the anonymous squaring function over the three-element list {3, 2, 7}:

```
In[5]:= (Map[(#^2)&, #])&[{3, 2, 7}]
Out[5]= {9, 4, 49}
```

When dealing with nested anonymous functions, the shorthand notation can be used for each of the anonymous functions but care needs to be taken to avoid confusion as to which # variable belongs to which anonymous function. This can be avoided by using Function, in which case different variable names can be used.

```
In[6]:= Function[y, Map[Function[x, x^2], y]][{3, 2, 7}]

Out[6]= {9, 4, 49}
```

We can use the anonymous function version of the removeRand function in the deal function definition.

```
In[7]:= Clear[deal]
```

```
In[8]:= deal[n_] := Module[{cardDeck},
          cardDeck = Flatten[Outer[List, {c, d, h, s},
                                Join[Range[2,10], {J,Q,K,A}]], 1];
          Complement[cardDeck,
                  Nest[(Delete[#, Random[Integer, {1, Length[#]}]])&,
                       cardDeck, n]]
              ]
```

We see that this works fine:

```
In[9]:= deal[5]

Out[9]= {{c, J}, {h, 8}, {h, 10}, {h, K}, {s, 7}}
```

Finally, we can generalize the deal function to work with any list.

```
chooseWithoutReplacement[lis_, n_] :=
   Complement[lis,
           Nest[(Delete[#, Random[Integer, {1,Length[#]}]])&,
                lis, n]]
```

Notice that it is not necessary to use the Module function in the above definition because the only quantities on the right-hand side of the function definition are anonymous functions, built-in functions, and the names of the arguments of the function. Functions that have this form are called *one-liners*.

Exercises

1. Write a function to sum the squares of the elements of a numeric list.

2. Write a function to sum the digits of any integer. You will need the `In-tegerDigits` function (use `?IntegerDigits`, or look in the Function Browser or the manual [Wolfram 1991] to find out about this function).

3. Write a function `setOfDistances[1]` that, given a list `1` of points in the plane, finds the set of all distances between the points. So for example, given a list of 5 points in the plane

```
In[1]:= points = Table[{Random[], Random[]}, {5}]

Out[1]= {{0.0560708, 0.6303}, {0.359894, 0.871377},
           {0.858645, 0.584579}, {0.742906, 0.391461},
           {0.357224, 0.380514}}
```

`setOfDistances` will return the $\binom{5}{2}$ = 10 distances between the 5 points (omitting of course the 0-distance from any point to itself).

```
In[2]:= setOfDistances[points]

Out[2]= {0.379543, 0.605334, 0.545621, 0.52146, 0.319283,
           0.358581, 0.322219, 0.646454, 0.134257, 0.677717}
```

For an interesting discussion of the many facts concerning the set of distances determined by n points in the plane, see Chapter 12 of [Honsberger 1976].

4. Write an anonymous function that moves a random walker from one location on a square lattice to one of the four adjoining locations with equal probability. For example, starting at `{0, 0}`, the function should return either `{0, 1}`, `{0, -1}`, `{1, 0}` or `{-1, 0}` with equal likelihood. Now, use this anonymous function with `NestList` to generate the list of step locations for an n-step random walk starting at `{0, 0}`.

Solutions

1.

```
In[1]:= elementsSquared[x_] := Apply[Plus, x^2]
```

```
In[2]:= elementsSquared[{3, 29, 2, 17}]
Out[2]= 1143
```

2.

```
In[1]:= name7[x_Integer] := Apply[Plus, IntegerDigits[x]]


In[2]:= name7[629]

Out[2]= 17
```

3. Consider adapting the code from Exercise 3 on page 100 by making it an anonymous function inside your setOfDistances function.

```
In[1]:= setOfDistances[x_] :=
          Module[{pairOff, distance},
            pairOff[lis_] :=
              Flatten[Table[{lis[[i]], lis[[j]]},
                {i, 1, Length[lis] - 1}, {j, i + 1, Length[lis]}], 1];
            distance = Sqrt[Apply[Plus,(#[[1]] - #[[2]])^2]]&;
          Map[distance, pairOff[x]]
              ]


In[2]:= points = {{0.78078, 0.807045}, {0.541564, 0.512377},
                  {0.226454, 0.56383}, {0.797958, 0.261694},
                  {0.27265, 0.689889}};


In[3]:= setOfDistances[points]

Out[3]= {0.379544, 0.605335, 0.545621, 0.521461, 0.319283,
         0.35858, 0.322219, 0.646454, 0.134257, 0.677716}
```

4. Using the list of the step increments in the the north, south, east, and west directions, this ten-step walk starts at the origin.

```
In[1]:= NestList[(# + {{1,0}, {-1,0}, {0,1}, {0,-1}}[[
                  Random[Integer, {1,4}]]])&,
            {0,0}, 10]

Out[1]= {{0, 0}, {0, 1}, {1, 1}, {1, 2}, {0, 2}, {0, 1},
         {0, 0}, {0, 1}, {1, 1}, {0, 1}, {1, 1}}
```

4.6 One-Liners

In the simplest version of a user-defined function, there are no value declarations or auxiliary function definitions; the right-hand side is a single nested function call whose arguments are the names of the arguments on the left-hand side, without the blanks. What we want to show now is how to construct one-liners from scratch.

4.6.1 | The Josephus Problem

Flavius Josephus was a Jewish historian during the Roman-Jewish war of the first century A.D. Through his writings comes the following story:

> The Romans had chased a group of ten Jews into a cave and were about to attack. Rather than die at the hands of their enemy, the group chose to commit suicide one by one. Legend has it though, that they decided to go around their circle of ten individuals and eliminate every other person until no-one was left. Who was the last to survive?

Although a bit macabre, this problem has a definite mathematical component and lends itself well to a functional style of programming. We'll start by changing the problem a bit (the importance of rewording a problem can hardly be overstated; the key to most problem solving resides in turning something we can't work with into something we can work with). We'll restate the problem as follows: n people are lined up. The first person is moved to the end of the line, the second person is removed from the line, the third person is moved to the end of the line, the fourth person is ... and so on until only one person remains in the line.

The statement of the problem indicates that there is a repetitive action, performed over and over again. It involves the use of the `RotateLeft` function (move the person at the front of the line to the back of the line) followed by the use of the `Rest` function (remove the next person from the line). We can write an anonymous function to make this nested function call:

```
(Rest[RotateLeft[#]])&
```

At this point it is already pretty clear where this computation is headed. We want to take a list and using the `Nest` function, perform the anonymous function call `(Rest[RotateLeft[#]])&` on the list until only one element remains. A list of n elements will need $n-1$ calls. So we can now write the function, to which we give the apt name `survivor`:

```
In[1]:= survivor[lis_] :=
            Nest[(Rest[RotateLeft[#]])&, lis, Length[lis] - 1]
```

Trying out the survivor function on a list of ten,

```
In[2]:= survivor[Range[10]]
```

```
Out[2]= {5}
```

we see that the fifth position will be the position of the survivor.

Tracing the applications of RotateLeft in this example gives a very clear picture of what is going on. The following form of TracePrint shows only the results of the applications of RotateLeft that occur during evaluation of the expression survivor[Range[10]].

```
In[3]:= TracePrint[survivor[Range[10]], RotateLeft]
            RotateLeft
            {2, 3, 4, 5, 6, 7, 8, 9, 10, 1}
            RotateLeft
            {4, 5, 6, 7, 8, 9, 10, 1, 3}
            RotateLeft
            {6, 7, 8, 9, 10, 1, 3, 5}
            RotateLeft
            {8, 9, 10, 1, 3, 5, 7}
            RotateLeft
            {10, 1, 3, 5, 7, 9}
            RotateLeft
            {3, 5, 7, 9, 1}
            RotateLeft
            {7, 9, 1, 5}
            RotateLeft
            {1, 5, 9}
            RotateLeft
            {9, 5}
```

```
Out[3]= {5}
```

4.6.2 | Pocket Change

As another example, we'll write a program to perform an operation most of us do every day. We'll calculate how much change we have in our pocket. Suppose we have the following collection of coins,

```
coins = {p, p, q, n, d, d, p, q, q, p}
```

where p, n, d, and q represent pennies, nickels, dimes, and quarters, respectively. Let's start by using the Count function to determine the number of pennies we have,

```
In[1]:= Count[coins, p]
Out[1]= 4
```

This works. So let's do the same thing for all of the coin types.

```
In[2]:= {Count[coins, p], Count[coins, n],
        Count[coins, d], Count[coins, q]}
Out[2]= {4, 1, 2, 3}
```

Looking at this list, it's apparent that there ought to be a more compact way of writing the list. If we Map an anonymous function involving Count and coins onto the list {p, n, d, q}, it should do the job.

```
In[3]:= Map[(Count[coins, #])&, {p, n, d, q}]
Out[3]= {4, 1, 2, 3}
```

Now that we know how many coins of each type we have, we want to calculate how much change we have. We first do the calculation *manually* to see what we get for an answer (so we'll know when our program works).

```
In[4]:= (4 1) + (1 5) + (2 10) + (3 25)
Out[4]= 104
```

From the above computation we see that the lists {4, 1, 2, 3} and {1, 5, 10, 25} are first multiplied together element-wise and then the elements of the result are added. This suggests a few possibilities:

```
In[5]:= Apply[Plus, {4, 1, 2, 3} * {1, 5, 10, 25}]
Out[5]= 104
```

```
In[6]:= {4, 1, 2, 3} . {1, 5, 10, 25}
Out[6]= 104
```

These two operations are both suitable for the job (to coin a phrase, "there's not a penny, nickel, quarter, or dime's worth of difference"). We'll write the one-liner using the first method.

```
In[7]:= pocketChange[x_] :=
          Apply[Plus, Map[(Count[x, #])&,
                    {p, n, d, q}] {1, 5, 10, 25}]


In[8]:= pocketChange[coins]

Out[8]= 104
```

Exercises

One of the best ways to learn how to write programs is to practice reading code. We list below a number of one-liner function definitions along with a very brief explanation of what these user-defined functions *do* and a typical input and output. The reader should deconstruct these programs to see what they do and then reconstruct them as compound functions without any anonymous functions.

1. Determine the frequencies with which the distinct elements in a list appear in the list.

```
In[1]:= frequencies[lis_] :=
          Map[({#, Count[lis, #]})&, Union[lis]]


In[2]:= frequencies[{a, a, b, b, b, a, c, c}]

Out[2]= {{a, 3}, {b, 3}, {c, 2}}
```

2. Divvy up a list into parts:

```
In[1]:= split1[lis_, parts_] :=
          Inner[Take[lis, {#1, #2}]&,
                    Drop[#1, -1] + 1,
              Rest[#1],
              List]&[FoldList[Plus, 0, parts]]


In[2]:= split1[Range[10], {2, 5, 0, 3}]

Out[2]= {{1, 2}, {3, 4, 5, 6, 7}, {}, {8, 9, 10}}
```

Exercises (cont.)

3. This is the same as the previous program, done in a different way.

```
split2[lis_, parts_] :=
   Map[(Take[lis, # + {1, 0}])&,
      Partition[FoldList[Plus, 0, parts], 2, 1]]
```

4. Another game in the Illinois State Lottery is based on choosing *n* numbers, each between 0 and *s* with no duplicates allowed. Write a user-defined function called *lotto* (after the official lottery names of *Little Lotto* and *Big Lotto*) to perform *sampling without replacement* on an arbitrary list. (Note: The difference between this function and the function `chooseWithoutReplacement` is that the order of selection is needed here.)

```
In[1]:= lotto1[lis_, n_] :=
           (Flatten[Rest[MapThread[Complement,
              {RotateRight[#], #}, 1]]])&[
           NestList[Delete[#,
                       Random[Integer,{1,Length[#]}]]&,
              lis, n]]
```

```
In[2]:= lotto1[Range[10], 5]
```

```
Out[2]= {8, 7, 9, 10, 2}
```

5. This is the same as the previous program, done in a different way.

```
In[1]:= lotto2[lis_, n_] :=
           Take[Transpose[Sort[
              Transpose[{Table[Random[], {Length[lis]}],
                 lis}]]][[2]], n]
```

As the `split` and `lotto` programs above illustrate, user-defined functions can be written in several ways. In the final analysis, the choice as to which version of a program to use has to be based on efficiency. A program whose development time (how long it took to create) was shorter and which runs faster (gives a result sooner) is *better* than a program which took more time to develop and which runs more slowly. As a rough guide, the most concise version of a *Mathematica* program, using as many built-in functions as possible, runs fastest. However, when execution speed is a primary concern (*e.g.*, when dealing with very large lists) it is a good idea to take various programming approaches and perform `Timing` tests to determine the fastest program.

Exercises (cont.)

6. Use the `Timing` function to determine when (in terms of the relative sizes of the list and the number of elements being chosen) it is preferable to use the different versions of the `lotto` function.

7. Make change with quarters, dimes, nickels and pennies, using the fewest number of coins.

```
In[1]:= makeChange[x_] :=
            Quotient[Drop[FoldList[Mod, x, {25, 10, 5, 1}], -1],
                {25, 10, 5, 1}]


In[2]:= makeChange[119]

Out[2]= {4, 1, 1, 4}
```

8. Write a one-liner to create a list of the step locations of a two-dimensional random walk that is not restricted to a lattice. Hint: Each step length must be the same, so the sum of the squares of the x- and y-components of each step should be equal to 1.

Solutions

1.
```
In[1]:= frequencies[lis_] :=
            Module[{pair},
                pair[x_] := {x, Count[lis, x]};
                Map[pair, Union[lis]]]


In[2]:= frequencies[{a, a, b, b, b, a, c, c}]

Out[2]= {{a, 3}, {b, 3}, {c, 2}}
```

2.
```
In[1]:= split1[lis_, parts_] :=
            Module[{lis1, lis2},
                lis1[y_, z_] := Take[lis, {y, z}];
                lis2[x_] := Inner[lis1, Drop[x, -1]+1, Rest[x], List];
                lis2[FoldList[Plus, 0, parts]] ]


In[2]:= split1[Range[10], {2,5,0,3}]

Out[2]= {{1, 2}, {3, 4, 5, 6, 7}, {}, {8, 9, 10}}
```

3.

```
In[1]:= split2[lis_, parts_] :=
          Module[{lis1},
            lis1[x_] := Take[lis, x + {1, 0}];
            Map[lis1, Partition[FoldList[Plus, 0, parts], 2, 1]] ]
```

```
In[2]:= split2[Range[10], {2,5,0,3}]
```

```
Out[2]= {{1, 2}, {3, 4, 5, 6, 7}, {}, {8, 9, 10}}
```

4.

```
In[1]:= lotto1[lis_, n_] :=
          Module[{lis1, lis2, lis3},
            lis1[x_] :=
              Flatten[Rest[
                MapThread[Complement, {RotateRight[x], x}, 1]]];
            lis2[y_] := Delete[y, Random[Integer, {1, Length[y]}]];
            lis3[z_] := NestList[lis2, z, n];
            lis1[lis3[lis]]]
```

```
In[2]:= lotto1[Range[10], 5]
```

```
Out[2]= {5, 3, 9, 10, 1}
```

5.

```
In[1]:= lotto2[lis_, n_] :=
          Take[Transpose[Sort[Transpose[{Table[Random[],
                      {Length[lis]}], lis}]]][[2]], n]
```

```
In[2]:= lotto2[Range[10], 5]
```

```
Out[2]= {1, 9, 3, 10, 8}
```

6.

```
In[1]:= {Timing[lotto1[Range[500], 3];],
          Timing[lotto2[Range[500], 3];]}
```

```
Out[1]= {{0.116667 Second, Null}, {0.166667 Second, Null}}
```

```
In[2]:= {Timing[lotto1[Range[500], 60];],
          Timing[lotto2[Range[500], 60];]}
```

```
Out[2]= {{1.53333 Second, Null}, {0.183333 Second, Null}}
```

7.

```
In[1]:= makeChange[x_] :=
          Module[{denominations = {25, 10, 5, 1}},
            Quotient[Drop[FoldList[Mod, x, denominations], -1],
              denominations] ]
```

```
In[2]:= makeChange[119]

Out[2]= {4, 1, 1, 4}
```

8.

```
In[1]:= offLattice[n_] :=
           Map[({Sin[#], Cos[#]})&, Table[Random[Real, {0, N[2Pi]}],
       {n}]]


In[2]:= offLattice[n_] :=
           Module[{step},
           step[x_] := {Sin[x], Cos[x]};
           Map[step, Table[Random[Real, {0, N[2Pi]}], {n}]]]
```

5 | Evaluation of Expressions

"A program is a set of rewrite rules. A rewrite rule consists of two parts: the pattern on the left side and the replacement text on the right side. Computation proceeds by evaluation of expressions. An expression is evaluated by finding those rewrite rules whose pattern matches part of the expression. That part is then replaced by the replacement text of that rule. Evaluation then proceeds by searching for further matching rules until no more are found."

5.1 Introduction

In previous chapters, we performed computations with both built-in and user-defined functions. We now turn to a description of the general mechanism underlying all computation in *Mathematica*. A succinct statement of the evaluation process used by *Mathematica* has been given by Roman Maeder, one of the original authors of *Mathematica*, in the quote above [Maeder 1992].

This evaluation mechanism is actually quite familiar to anyone who has ever used a handbook of mathematical formulas to solve an equation. For example, if you want to carry out an integration, you might consult a handbook to locate an integration formula whose left-hand side has the same form as your integral except that instead of having specific values, it has unspecified variables. You then replace your integral with the right-hand side of the formula in the handbook, substituting the appropriate specific values from your original integral for the symbols in the right-hand side of the formula. Once you have the result, you then look around in the handbook to see if there are any other formulas, perhaps a trigonometric identity, that can be used to further simplify the result by the same sort of replacement procedure.

In this chapter we will elaborate upon this description. We'll explain and illustrate *rewrite rules*, *expressions*, and *patterns* and show how these quantities are used together to perform evaluation by *term rewriting*.

5.2 Creating Rewrite Rules

There are two kinds of rewrite rules: *built-in functions* and *user-defined rewrite rules*. There are nearly a thousand built-in functions in *Mathematica*, a few of which we have already described.

A user-defined rewrite rule is created using either the Set or SetDelayed function. The SetDelayed function is written SetDelayed[*left-hand side, right-hand side*] or in its standard input form:

```
left-hand side := right-hand side
```

The Set function is written Set[*left-hand side, right-hand side*] or in its special notation:

```
left-hand side = right-hand side
```

The left-hand side of a Set or SetDelayed function starts with a name. The name may be followed by a set of square brackets containing a sequence of patterns. The most common pattern appearing within the square brackets is called a *pattern variable* or *labeled pattern* which consists of a symbol ending with a blank. The right-hand side of a Set or SetDelayed function is an expression which may contain the names of arguments on the left-hand side, without blanks.

We developed a number of relatively sophisticated Set or SetDelayed functions in the previous chapter when we created function definitions (*e.g.*, deal) and value names (*e.g.*, cardDeck). Here, we will use some very simple examples to illustrate the use of the SetDelayed and Set functions to create rewrite rules. One difference between these two functions becomes apparent when they are entered. For example, consider the function definition, rand1, and the value name, rand2:

```
In[1]:= rand1[x_] := Random[Real, x]
```

```
In[2]:= rand2 = Random[]
Out[2]= 0.0560708
```

When a SetDelayed function is entered, nothing is returned. When a Set function is entered, the value resulting from evaluating the right-hand side (which we call the *evaluated right-hand side*) is returned. This difference in output is indicative of a more fundamental difference in what happens when the two kinds of functions are entered and rewrite rules are thereby created. To see this, we need to look at the *global rule base*, wherein reside rewrite rules.

5.2.1 | The Global Rule Base

The global rule base is composed of the two kinds of rewrite rules: the built-in functions, which are part of every *Mathematica* session, and the user-defined rewrite rules which are entered during a particular session.

We can get information about both kinds of rules in the global rule base by entering ?name. In the case of a built-in function, the syntax for entering the function and a statement of what is computed when the function is called are printed. For example,

```
In[1]:= ?Apply

Apply[f, expr] or f @@ expr replaces the head of expr
    by f. Apply[f, expr, levelspec] replaces heads in
    parts of expr specified by levelspec.
```

In the case of a user-defined rewrite rule, the rule itself is printed. For the simple examples above, the crucial difference between rewrite rules created with the SetDelayed and Set functions becomes apparent by querying the rule base for the rewrite rules associated with the symbols rand1 and rand2.

```
In[2]:= ?rand1

Global`rand1
rand1[x_] := Random[Real, x]
```

Comparing this with the original SetDelayed function, we see that a rewrite rule created using the SetDelayed function looks exactly like the function that created it. This is because both the left-hand side and right-hand side of a SetDelayed function are placed in the rule base without being evaluated.

```
In[3]:= ?rand2

Global`rand2
rand2 = 0.05607079483486153
```

Comparing this with the original Set function, we see that a rewrite rule created using the Set function has the same left-hand side as the function that created it but the right-hand side of the rule may differ from the right-hand side of the function. This is because the right-hand side of the rule is the evaluated right-hand side of the function.

In view of this difference between the SetDelayed and Set functions, the question is when should we use one or the other function to create a rewrite rule?

When we define a function, we don't want either the left-hand side or the right-hand side to be evaluated; we just want to make it available for use when the appropriate function call is made. This is precisely what occurs when a SetDelayed function is entered, so the SetDelayed function is commonly used in writing function definitions.

When we make a value declaration, we don't want the left-hand side to be evaluated; we just want to make it a nickname to serve as a shorthand for a value. This is what happens when a Set function is entered and so the Set function is commonly used to make value declarations.

It is necessary to be aware of how the global rule base treats compound functions. When a compound function definition is entered, a rewrite rule corresponding to the entire definition is created. Each time the compound function is subsequently called, rewrite rules are created from the auxiliary function definitions and value declarations within the compound function. To prevent these additional rewrite rules from being placed in the global rule base, we can localize their names by using the Module construct in the compound function definition.

A new rewrite rule overwrites (replaces) an older rule with the same left-hand side. However, keep in mind that if two left-hand sides are the same except for the names of their pattern variables, they are considered different by *Mathematica*. Clear[*name*] can be used to remove a rewrite rule from the global rule base.

Exercises

1. What rewrite rules do each of the following functions create? Check your predictions by entering them and then querying the rule base.

   (a) randLis1[n_] := Table[Random[ ], {n}]

   (b) randLis2[n_] := (x = Random[ ]; Table[x, {n}])

   (c) randLis3[n_] := (x := Random[ ]; Table[x, {n}])

   (d) randLis4[n_] = Table[Random[ ], {n}]

Solutions

1.a

```
In[1]:= randLis1[n_] := Table[Random[], {n}]


In[2]:= ?randLis1

Global`randLis1
randLis1[n_] := Table[Random[], {n}]
```

1.b   *In[1]:=* **randLis2[n_] := (x = Random[]; Table[x, {n}])**

In[2]:= **?randLis2**

Global`randLis2

randLis2[n_] := (x = Random[]; Table[x, {n}])

1.c   *In[1]:=* **randLis3[n_] := (x := Random[]; Table[x, {n}])**

In[2]:= **?randLis3**

Global`randLis3

randLis3[n_] := (x := Random[]; Table[x, {n}])

1.d   *In[1]:=* **randLis4[n_] = Table[Random[], {n}]**

Out[1]= Table[Random[], {n}]

In[2]:= **?randLis4**

Global`randLis4

randLis4[n_] = Table[Random[], {n}]

5.3 Expressions

The evaluation of an expression involves making a *pattern match* with the left-hand side of a rewrite rule. We therefore turn to a discussion of patterns in *Mathematica*. Since a pattern is defined syntactically (that is to say, the patterns that an expression matches are determined by the internal representation of the expression), we start our discussion of patterns by explaining the expression structure used to represent quantities in *Mathematica*.

Every quantity that is entered into *Mathematica*, regardless of how it may appear, is represented internally in the same way—as an expression. An expression has the form,

head[*arg*$_1$, *arg*$_2$, ..., *arg*$_n$]

where the head and the arguments, *arg$_i$*, of the expression can be other expressions, including *atoms* (atoms are a special kind of expression that we will describe in a moment).

We have already seen that built-in functions (*e.g.*, Plus, List) can be cast in this form. Other quantities can also be shown to have this form using the built-in FullForm function.

```
In[1]:= FullForm[x + 2 y + z^2]

Out[1]//FullForm= Plus[x, Times[2, y], Power[z, 2]]
```

The parts of an expression are indexed with positive values counting from the first argument and negative values counting from the last argument. The parts can be extracted from an expression (except for an atom) using Part[*expression, i*]. For example,

```
In[2]:= Part[{a, b, c, d, e}, 2]
Out[2]= b

In[3]:= Part[{a, b, c, d, e}, -2]
Out[3]= d

In[4]:= Part[x + 2 y + z^2, 3]
          2
Out[4]= z

In[5]:= Part[x + 2 y + z^2, -3]
Out[5]= x
```

The head of any expression (including an atom) has index 0. The head can be extracted either using Part or by applying the Head function.

```
In[6]:= Part[a^3, 0]
Out[6]= Power

In[7]:= Head[{a, b}]
Out[7]= List
```

The head of any expression (except an atom) is the name of the outermost function.

Because everything in *Mathematica* has the common structure of an expression, most of the built-in functions that are used for list manipulation (such as the Part function just shown) can also be used to manipulate the arguments of any other kind of expression (except atoms). For example,

```
In[8]:= Append[w + x y,   z]

Out[8]= w + x y + z
```

This result can be understood by looking at the FullForm of the following two expressions.

```
In[9]:= FullForm[w + x y]

Out[9]//FullForm= Plus[w, Times[x, y]]
```

```
In[10]:= FullForm[w + x y + z]

Out[10]= Plus[w, Times[x, y], z]
```

We see that appending z to w + x y is equivalent to adding z as an argument to the Plus function.

We previously described the built-in Apply function as applying a function to the arguments of another function, specifically to the elements of a list. Thinking now in terms of the parts of an expression, we can reinterpret the operation of the Apply function. Here is an example of how Apply works with an expression.

```
In[11]:= FullForm[Apply[Power, Plus[a, b]]]

Out[11]//FullForm= Power[a, b]
```

We can say that the Apply function replaces the head of an expression (this agrees with what we were told when we queried the rule base about Apply earlier on page 125). Thus, the expression Apply[*func, expr*] is equivalent to the expression ReplacePart[*expr, func, 0*].

Having said above that atoms are like expressions but somewhat different, we need to look in detail at this special kind of expression.

5.3.1 | Atoms

The three basic building blocks of *Mathematica*—the *atoms*—from which all other quantities are ultimately constructed are: symbol, number, and string.

We defined a symbol before; it consists of a letter followed without interruption by letters and numbers.

The four kinds of numbers—integers, real numbers, complex numbers and rational numbers—are shown in the list below:

```
{4, 5.201, 3 + 4I, 5/7}
```

A string is composed of characters, as discussed in Section 3.7.

As we said previously, an atom is an expression and therefore has a head and argument(s). However, there are several important differences between atomic expressions and nonatomic expressions.

While the heads of all expressions are extracted in the same way—using the Head function—the head of an atom provides different information than the head of other expressions. For example, the head of a symbol or string is the kind of atom that it is.

```
In[12]:= Map[Head, {a, List, "give me a break"} ]

Out[12]= {Symbol, Symbol, String}
```

The head of a number is the specific kind of number that it is.

```
In[13]:= Map[Head, {2, 5.201, 3 + 4I, 5/7}]

Out[13]= {Integer, Real, Complex, Rational}
```

The FullForm of an atom (except a complex or rational number) is the atom itself.

```
In[14]:= FullForm[{darwin, "read my lips", 1, 3 + 4I, 5/7}]

Out[14]//FullForm= List[darwin, "read my lips", 1, Complex[3, 4],
                   Rational[5, 7]]
```

Atoms have no parts (which is why they are called atoms).

```
In[15]:= "read my lips"[[1]]

        Part::partd:
          Part specification read my lips[[1]]
            is longer than depth of object.
        Part::partd:
          Part specification read my lips[[1]]
            is longer than depth of object.

Out[15]= read my lips[[1]]
```

This error message indicates that "read my lips" has no first part, which is true since an atomic expression has no parts.

```
In[16]:= Part[3 + 4I, 1]

        Part::partd:
          Part specification (3 + 4 I)[[1]]
            is longer than depth of object.

Out[16]= (3 + 4 I)[[1]]
```

Finally, we want to emphasize that regardless of how an atomic or nonatomic expression may appear on the computer screen, its structure is uniquely determined by its head and parts as seen using FullForm. This is important for understanding the *Mathematica* evaluation mechanism which depends on the matching of patterns based on their FullForm representation. In the next section, we will look at the various kinds of patterns that are recognized in *Mathematica*.

Exercises

1. What do you expect to be the result of the following operations? Use the FullForm of the expressions to understand what is going on.

   (a) ((x^2 + y) z / w) [[2, 1, 2]]

   (b) (a/b) [[2, 2]]

   (c) What is the part specification of the b in a x^2 + b x^2 + c x^2?

Solutions

1.a

```
In[1]:= ((x^2 + y) z / w)[[2, 1, 2]]

Out[1]= 2

In[2]:= FullForm[((x^2 + y) z / w)]

Out[2]//FullForm= Times[Power[w, -1], Plus[Power[x, 2], y], z]
```

1.b

```
In[1]:= (a/b)[[2, 2]]

Out[1]= -1

In[2]:= FullForm[(a/b)]

Out[2]//FullForm= Times[a, Power[b, -1]]
```

1.c

```
In[1]:= (x^2 + b x^2 + c x^2)[[2, 1]]

Out[1]= b

In[2]:= FullForm[(x^2 + b x^2 + c x^2)]

Out[2]//FullForm= Plus[Power[x, 2], Times[b, Power[x, 2]],
                       Times[c, Power[x, 2]]]
```

5.4 Patterns

A pattern is a *form*, and *pattern matching* in *Mathematica* consists of finding a fit between a form and an expression. To make this abstract statement more concrete, we can, as an example, fit the expression x^2 to the following forms, which will be explained more fully in the next section.

- x^2 matches "x raised to the power of two"

- x^2 matches "x raised to the power of a number"

- x^2 matches "x raised to the power of something"

- x^2 matches "a symbol raised to the power of two"

- x^2 matches "a symbol raised to the power of a number"

- x^2 matches "a symbol raised to the power of something"

- x^2 matches "something raised to the power of two"

- x^2 matches "something raised to the power of a number"

- x^2 matches "something raised to the power of something"

- x^2 matches "something"

We now want to learn how patterns are defined in *Mathematica*.

5.4.1 | Blanks

While any specific expression can be pattern matched (because any object must pattern match itself), we want to be able to pattern match larger classes of expressions (*i.e.*, a sequence of expressions or expressions having Integer as the head). For this purpose, *patterns* are defined as expressions that may contain *blanks*. That is to say, a pattern may contain one of the following: a single (_) blank, a double (__) blank, or a triple (___) blank.

We often want to identify the pattern that an expression is matched to (*i.e.*, on the left-hand side of a function definition) so that it can be referred to by name elsewhere (*i.e.*, on the right-hand side of the function definition). A pattern can be labeled by *name_*, or *name__*, or *name___* (which can be read as "a pattern called *name*") and the labeled pattern will be matched by the same expression that matches its unlabeled counterpart. (The matching expression is given the name used in the labeled pattern.)

The best way to understand the uses of blanks is to actually see them used. We can do this, using the built-in MatchQ function which returns True if a pattern matches an expression and False if it does not.

5.4.2 | Expression Pattern-Matching

A single blank (sometimes referred to as a *wild card* pattern) represents an individual expression. It matches any data object. Blank[h] or _h stands for an expression with the head h.

To see how the single blank pattern match works, we'll work with the expression x^2 mentioned earlier. After each application of MatchQ, we describe the pattern match in words.

```
In[1]:= MatchQ[x^2, x^2]

Out[1]= True
```

x^2 pattern matches "x^2" (of course).

> *In[2]:=* **MatchQ[x^2, x^_Integer]**
>
> *Out[2]=* True

x^2 pattern matches "x raised to the power of an integer" (to put it more formally, "x raised to the power of an expression whose head is Integer").

> *In[3]:=* **MatchQ[x^2, x^_Real]**
>
> *Out[3]=* False

x^2 does not pattern match "x raised to the power of a real number".

> *In[4]:=* **MatchQ[x^2, x^_]**
>
> *Out[4]=* True

x^2 pattern matches "x raised to the power of an expression".

> *In[5]:=* **MatchQ[x^2, x^y_]**
>
> *Out[5]=* True

x^2 pattern matches "x raised to the power of an expression" (the label y is irrelevant to the match).

> *In[6]:=* **MatchQ[x^2, _Symbol^2]**
>
> *Out[6]=* True

x^2 pattern matches "an expression whose head is Symbol, raised to the power two".

> *In[7]:=* **MatchQ[x^2, _List^2]**
>
> *Out[7]=* False

x^2 does not pattern match "an expression whose head is List, raised to the power two".

> *In[8]:=* **MatchQ[x^2, _Symbol^_Integer]**
>
> *Out[8]=* True

x^2 pattern matches "an expression whose head is Symbol, raised to the power of an expression whose head is Integer" (or stated less formally, "a symbol raised to the power of an integer").

```
In[9]:= MatchQ[x^2, _Symbol^_]
Out[9]= True
```

x^2 pattern matches "a symbol raised to the power of an expression".

```
In[10]:= MatchQ[x^2, _^2]
Out[10]= True
```

x^2 pattern matches "an expression raised to the power two".

```
In[11]:= MatchQ[x^2, _^_Integer]
Out[11]= True
```

x^2 pattern matches "an expression raised to the power of an integer".

```
In[12]:= MatchQ[x^2, _^_]
Out[12]= True
```

x^2 pattern matches "an expression raised to the power of an expression".

```
In[13]:= MatchQ[x^2, _Power]
Out[13]= True
```

x^2 pattern matches "an expression whose head is Power".

```
In[14]:= MatchQ[x^2, _]
Out[14]= True
```

x^2 pattern matches "an expression".

5.4.3 | Sequence Pattern-Matching

A *sequence* consists of a number of expressions separated by commas. For example, the arguments of expressions are written as sequences.

A double blank represents a sequence of one or more expressions and __h represents a sequence of one or more expressions, each of which has head h. An expression that matches a blank will also match a double blank.

A triple blank represents a sequence of zero or more expressions and ___h represents a sequence of zero or more expressions, each of which has a head h. An expression that matches a blank will also match a triple blank and a sequence that matches a double blank pattern will also match a triple blank pattern. For example,

```
In[15]:= MatchQ[x^2, __]

Out[15]= True


In[16]:= MatchQ[x^2, ___]

Out[16]= True
```

Since x^2 is an expression, it matches both the pattern "one or more expressions" and the pattern "zero or more expressions".

A list {a, b, c} is matched by pattern _ (an expression), as well as by List[__] (a list of one or more expressions) and List[___] (a list of zero or more expressions). However, the list {a, b, c} is not matched by the pattern List[_] (a list of one expression) because for the purposes of pattern matching, a sequence is not an expression. Here are some other examples of successful pattern matches:

```
In[17]:= MatchQ[{a, b, c}, __]

Out[17]= True


In[18]:= MatchQ[{a, b, c}, {___}]

Out[18]= True


In[19]:= MatchQ[{a, b, c}, x__]

Out[19]= True


In[20]:= MatchQ[{a, b, c}, {x___}]

Out[20]= True
```

In the last two cases above, the labels on the blanks do not affect the success or failure of the pattern match. The labels simply serve to identify different parts of the expression. In MatchQ[{a, b, c}, x_], x names the list {a, b, c}, but in `MatchQ[{a, b, c}, {x___}]`, x names the *sequence* a, b, c, which is quite different. This is illustrated on page 138.

Finally, note that the discussion about lists here applies equally to any function. For example, `MatchQ[Plus[a, b, c], Plus[x___]]` returns `True`, with x naming the sequence a, b, c.

5.4.4 | Conditional Pattern-Matching

In addition to specifying the head of an expression, other requirements can be placed on an expression for it to match a specific pattern.

Attaching a predicate

If the blanks of a pattern are followed with ? *test*, where *test* is a predicate, then a match is only possible if *test* returns `True` when applied to the entire expression. For example, we can use a built-in predicate,

```
In[21]:= MatchQ[{a, b, c}, _?ListQ]

Out[21]= True
```

```
In[22]:= MatchQ[{a, b, c}, _?NumberQ]

Out[22]= False
```

or we can use an anonymous predicate.

```
In[23]:= MatchQ[{a, b, c}, _List?(Length[#] >2&)]
Out[23]= True
```

```
In[24]:= MatchQ[{a, b, c}, _List?(Length[#] > 4&)]
Out[24]= False
```

Note that when using an anonymous function in ? *test*, it is necessary (because of the precedence *Mathematica* gives to evaluating various quantities) to enclose the entire function, including the &, in parentheses. We use *test* to place a constraint on an entire expression.

Attaching a condition

If part of a labeled pattern is followed with an expression such as / ; *condition*, where *condition* contains labels appearing in the pattern, then a match is possible only if *condition* returns `True`. We use *condition* to place a constraint on the labeled parts of an expression. The use of labels in *condition* is useful for narrowing down the requirements for a pattern match. For example,

```
In[25]:= MatchQ[x^2, _^y_ /; EvenQ[y]]

Out[25]= True
```

```
In[26]:= MatchQ[x^2, _^y_ /; OddQ[y]]

Out[26]= False
```

We mentioned above that matching a list like {a, b, c} with the pattern x_ is different from matching it with {x___} because of the various expressions that are associated with x.

```
In[27]:= MatchQ[{4, 6, 8}, x_ /; Length[x] > 4]

Out[27]= False
```

```
In[28]:= MatchQ[{4, 6, 8}, {x___} /; Length[x] > 4]

        Length::argx:
           Length called with 3
              arguments; 1 argument is expected.

Out[28]= False
```

```
In[29]:= MatchQ[{4, 6, 8}, {x___} /; Plus[x] > 10]

Out[29]= True
```

In the first example, x was associated with the entire list {4, 6, 8}; since Length[{4, 6, 8}] is not greater than 4, the match failed. In the second example, x became the *sequence* 4, 6, 8, so that the condition was Length[4, 6, 8] > 4; but Length can only have one argument, hence the error. In the last example, x was again associated with 4, 6, 8, but now the condition was Plus[4, 6, 8] > 10, which is perfectly legal, and true.

5.4.5 | Alternatives

A final type of pattern uses *alternatives*. This is a pattern consisting of several independent patterns, which matches an expression whenever any one of those independent patterns matches it. In such a pattern, the alternatives are separated by a single vertical bar, |. Here are some examples:

```
In[30]:= MatchQ[x^2, x^_Real | x^_Integer]

Out[30]= True
```

x^2 matches "an expression which is either the symbol x raised to a real number or the symbol x raised to an integer".

```
In[31]:= MatchQ[x^2, x^(_Real | _Integer)]

Out[31]= True
```

x^2 matches "x raised to an expression which is either a real number or an integer".

Exercises

1. Find as many patterns as possible that match the expression $x^3 + yz$. (In general, if you take an expression and replace any part of it with _ (or name_), the resulting pattern will match the expression; furthermore, if the head of that part is h, the patterns _h and name_h will also match.

2. Find as many pattern matches as possible for the following expression:

   ```
   {5, erina, {}, "give me a break"}
   ```

3. Using both forms (predicate and condition), write down five conditional patterns that match the expression {4, {a, b}, "g"}.

Solutions

1. Using the FullForm of the expression, we can find many pattern matches. Here are just a few:

```
In[1]:= FullForm[x^3 + y z]

Out[1]//FullForm= Plus[Power[x, 3], Times[y, z]]


In[2]:= MatchQ[(x^3 + y z), _Plus]

Out[2]= True


In[3]:= MatchQ[(x^3 + y z), _Power + _Times]

Out[3]= True
```

2.
```
In[1]:= FullForm[{5, erina, "give me a break"}]

Out[1]//FullForm= List[5, erina, "give me a break"]

In[2]:= MatchQ[{5, erina, "give me a break"}, _List]
Out[2]= True

In[3]:= MatchQ[{5, erina, "give me a break"},
            {_Integer, _Symbol, _String}]

Out[3]= True
```

3.
```
In[1]:= FullForm[{4, {a, b}, "g"}]
Out[1]//FullForm= List[4, List[a, b], "g"]

In[2]:= MatchQ[{4, {a, b}, "g"}, x_List /; Length[x] == 3]
Out[2]= True

In[3]:= MatchQ[{4, {a, b}, "g"}, _List?(Length[#] == 3&)]
Out[3]= True

In[4]:= MatchQ[{4, {a, b}, "g"}, {_, y_, _} /; y[[0]] == List]
Out[4]= True

In[5]:= MatchQ[{4, {a, b}, "g"}, {x_, y_, z_} /; AtomQ[z]]
Out[5]= True

In[6]:= MatchQ[{4, {a, b}, "g"}, {x_, _, _} /; EvenQ[x]]
Out[6]= True
```

5.5 Term Rewriting

Evaluation takes place whenever an expression is entered. Here is the general procedure followed by *Mathematica* when evaluating an expression (with a few exceptions):

1. If the expression is a number or a string, it isn't changed.

2. If the expression is a symbol, it is rewritten if there is an applicable rewrite rule in the global rule base; otherwise, it is unchanged.

3. If the expression is not a number, string or symbol, its parts are evaluated in a specific order.

    (a) The head of the expression is evaluated.

    (b) The arguments of the expression are evaluated in order (except when the head is a symbol with a `Hold` attribute. In this case, some of its arguments are left in their unevaluated forms; this is discussed in the first exercise at the end of this section).

4. After the head and arguments of an expression are each completely evaluated, the expression consisting of the evaluated head and arguments is rewritten (after making any necessary changes to the arguments based on the `Attributes` of the head) if there is an applicable rewrite rule in the global rule base.

5. After carrying out the previous steps, the resulting expression is evaluated in the same way and then the result of that evaluation is evaluated, and so on until there are no more applicable rewrite rules.

The term-rewriting process done in steps 2 and 4 above can be described as follows:

- pattern match parts of an expression and the left-hand side of a rewrite rule.

- substitute the values which match labeled blanks in the pattern into the right-hand side of the rewrite rule and evaluate it.

- replace the matched part of the expression with the evaluated result.

Both built-in and user-defined rewrite rules are available for use in evaluation. When more than one rewrite rule is found to match an expression, the rule used for term rewriting is selected based on the following priority:

- A user-defined rule is used before a built-in rule.

- A more specific rule is used before a more general rule. One rule is more specific than another if its left-hand side matches fewer expressions; for example, the rule f[0] := ... is more specific than f[_] := .... This is discussed further in Chapter 6.

The evaluation process can be illustrated with a simple case. We first enter the square rewrite rule into the global rule base.

```
In[1]:= square[x_?OddQ] := x^2
```

If we now enter the expression,

```
In[2]:= square[3]
Out[2]= 9
```

the number 9 is returned as the result. We can step through the details of the evaluation process that took place above.

1. The Head, square, was evaluated first. The global rule base was searched for a rewrite rule whose left-hand side was the symbol square. No matching rewrite rule was found and so the symbol was left unchanged.

2. The argument 3 was evaluated. Since 3 is a number, it was left unchanged.

3. The expression square[3] was evaluated. The global rule base was searched for a rewrite rule whose left-hand side pattern matched square[3]. The pattern square[3] was found to match square[x_?OddQ] and so the value of 3 was substituted for x in the right-hand side of the rewrite rule, Power[x, 2], to give Power[3, 2].

4. Power[3, 2] was then evaluated (by the same general procedure) to give 9.

5. The value 9 was evaluated. Since 9 is a number, it was left unchanged.

6. Since there were no more rules to use, the final value 9 was returned.

These steps can be seen in detail by using Trace with the TraceOriginal option set to True.

```
In[3]:= Trace[square[3], TraceOriginal->True]
Out[3]= {square[3], {square}, {3}, square[3],
          {OddQ[3], {OddQ}, {3}, OddQ[3], True}, 3^2, {Power},
          {3}, {2}, 3^2, 9}
```

As a final note, term rewriting can be used to create rewrite rules during evaluation. In a process known as *dynamic programming*, a SetDelayed function whose right-hand side is a Set function of the same name is defined.

```
f[x_] := f[x] = right-hand side
```

When an expression is pattern matched to this rewrite rule, term rewriting creates a Set function with the specific argument value which, upon evaluation of the right-hand side, becomes a rewrite rule. Since the global rule base is always consulted during evaluation, storing results as rewrite rules can cut down on computation time, especially in recursive computations, as we will see in Section 7.7.

Exercises

1. There is function called Thread which on first glance, seems to perform the same operation as MapThread.

    ```
    In[1]:= Thread[List[{1, 3, 5}, {2, 4, 6}]]
    Out[1]= {{1, 2}, {3, 4}, {5, 6}}

    In[2]:= MapThread[List, {{1, 3, 5}, {2, 4, 6}}]
    Out[2]= {{1, 2}, {3, 4}, {5, 6}}

    In[3]:= Thread[Plus[{1, 3, 5}, {2, 4, 6}]]
    Out[3]= {3, 7, 11}

    In[4]:= MapThread[Plus, {{1, 3, 5}, {2, 4, 6}}]
    Out[4]= {3, 7, 11}
    ```

    There is, however, a major difference between the operation of these two functions as you can see by using Trace on their two evaluations. What the trace reveals is that the first argument of Thread is evaluated before it is applied while the first argument of MapThread is applied in its unevaluated form.
    In many cases, the difference between these two functions is not important and either function can be used, but there are cases where the difference becomes crucial.

Exercises (cont.)

One example is the determination of the Hamming distance which is the number of nonmatching elements in lists of 0s and 1s. We can do this by applying `MatchQ` to each pair of corresponding elements in the lists. Using `MapThread`, we get

```
In[5]:= MapThread[MatchQ, {{1, 0, 0, 1, 1}, {0, 1, 0, 1, 0}}]

Out[5]= {False, False, True, True, False}
```

Trying to do this with the `Thread` function doesn't work, however.

```
In[6]:= Thread[MatchQ[{1, 0, 0, 1, 1}, {0, 1, 0, 1, 0}]]

        Thread::normal:
            Normal expression expected at position 1 in
            Thread[False].

Out[6]= Thread[False]
```

What has happened here is that `MatchQ[{1, 0, 0, 1, 1}, {0, 1, 0, 1, 0}]` was first evaluated to `False` which was then passed to the `Thread` function. When the expression `Thread[False]` was evaluated it produced an error message.

There is a way to make `Thread` work for the Hamming distance calculation. Essentially, we need to prevent the evaluation of the expression `MatchQ[{1, 0, 0, 1, 1}, {0, 1, 0, 1, 0}]`. This can be done by using the `Hold` function which leaves an expression unevaluated:

```
Thread[Hold[MatchQ][{1, 0, 0, 1, 1}, {0, 1, 0, 1, 0}]]
```

Use `Trace` to see what happens when this is entered.

Once we have threaded the unevaluated `Hold[MatchQ]` onto the two lists, we can finish the evaluation by applying the `ReleaseHold` function to remove `Hold` from the expression, thereby allowing the evaluation to take place.

```
In[7]:= ReleaseHold[Thread[Hold[MatchQ]
                    [{1,0,0,1,1}, {0,1,0,1,0}]]]

Out[7]= {False, False, True, True, False}
```

Explain what is occurring in the evaluation of the following expressions. Use `Trace` to confirm your analysis.

Exercises (cont.)

(a)
```
In[1]:= ReleaseHold[Thread[Hold[MatchQ[{1,0,0,1,1},
                                       {0,1,0,1,0}]]]]

Out[1]= False
```

(b)
```
In[1]:= Apply[Times, Thread[Plus[{1, 2}, {3, 4}]], 2]
Out[1]= {4, 6}
```

(c)
```
In[1]:= ReleaseHold[Apply[Times,
             Thread[Hold[Plus][{1, 2}, {3, 4}]], 2]]
Out[1]= {3, 8}
```

Solutions

1.a When Thread is applied to Hold[*expr*], it yields Hold[*expr*].

```
In[1]:= Trace[ReleaseHold[Thread[Hold[MatchQ[{1,0,0,1,1},
                                             {0,1,0,1,0}]
        ]]]]

Out[1]= {{Thread[Hold[MatchQ[{1, 0, 0, 1, 1},
            {0, 1, 0, 1, 0}]]],
         Hold[MatchQ[{1, 0, 0, 1, 1}, {0, 1, 0, 1, 0}]]},
         ReleaseHold[Hold[MatchQ[{1, 0, 0, 1, 1},
            {0, 1, 0, 1, 0}]]],
         MatchQ[{1, 0, 0, 1, 1}, {0, 1, 0, 1, 0}], False}
```

1.b
```
In[1]:= Trace[Apply[Times, Thread[Plus[{1, 2}, {3, 4}]], 2]]
Out[1]= {{{{1, 2} + {3, 4}, {1 + 3, 2 + 4}, {1 + 3, 4},
            {2 + 4, 6}, {4, 6}}, Thread[{4, 6}], {4, 6}},
         Apply[Times, {4, 6}, 2], {4, 6}}
```

1.c
```
In[1]:= Trace[ReleaseHold[Apply[Times,
             Thread[Hold[Plus][{1, 2}, {3, 4}]], 2]]]
```

```
Out[1]= {{{Thread[Hold[Plus][{1, 2}, {3, 4}]],
         {Hold[Plus][1, 3], Hold[Plus][2, 4]}},
       Apply[Times, {Hold[Plus][1, 3], Hold[Plus][2, 4]},
       2], {1 3, 2 4}, {1 3, 3}, {2 4, 8}, {3, 8}},
     ReleaseHold[{3, 8}], {3, 8}}
```

5.6 Transformation Rules

A rewrite rule stored in the global rule base is automatically used by *Mathematica* whenever there is an appropriate pattern match during evaluation. If we wish to restrict the use of a rule to a specific expression, we can use the `ReplaceAll` function with the expression as the first argument and a user-defined `Rule` or `RuleDelayed` function as the second argument. In standard input form, the transformation rule (or local rewrite rule) appears immediately after the expression.

expr **/. Rule** [*left-hand side*, *right-hand side*}]

expr **/. RuleDelayed** [*left-hand side*, *right-hand side*}]

A transformation rule is applied to the expression to which it is attached only *after* the expression is evaluated. While the left-hand side of a transformation rule is *always* evaluated before it is used (in contrast to a rewrite rule), we can specify whether the right-hand side of the rule is evaluated before it is used or only after it is used, by using `Rule` or `RuleDelayed`, respectively.

The `Rule` function is written in standard input form as

left-hand side -> right-hand side

When the `Rule` function is used with an expression, the expression itself is first evaluated. Then *both* the left-hand side and right-hand side of the rule are evaluated, except for those parts of the right-hand side that are held unevaluated by the `Hold` attribute. Finally, everywhere that the evaluated left-hand side of the rule appears in the evaluated expression, it is replaced by the evaluated right-hand side of the rule.

```
In[1]:= Table[x, {5}] /. x -> Random[]
Out[1]= {0.6303, 0.6303, 0.6303, 0.6303, 0.6303}
```

Using `Trace`, we can see the way the transformation rule works.

```
In[2]:= Trace[Table[x, {5}] /. x -> Random[]]

Out[2]= {{Table[x, {5}], {x, x, x, x, x}},
            {{Random[], 0.871377}, x -> 0.871377,
             x -> 0.871377}, {x, x, x, x, x} /. x -> 0.871377,
            {0.871377, 0.871377, 0.871377, 0.871377, 0.871377}}
```

The `RuleDelayed` function is written in shorthand notation as

```
left-hand side :> right-hand side
```

When the `RuleDelayed` function is used with an expression, the expression itself is first evaluated. Then the left-hand side (but *not* the right-hand side) of the rule is evaluated. Next, everywhere that the evaluated left-hand side of the rule appears in the evaluated expression, it is replaced by the unevaluated right-hand side of the rule, which is then evaluated.

```
In[3]:= Table[x, {5}] /. x :> Random[]

Out[3]= {0.858645, 0.584579, 0.742906, 0.391461, 0.357224}
```

Using `Trace`, we can see the way this transformation rule works.

```
In[4]:= Trace[Table[x, {5}] /. x :> Random[]]

Out[4]= {{Table[x, {5}], {x, x, x, x, x}},
            {x, x, x, x, x} /. x :> Random[],
            {Random[], Random[], Random[], Random[], Random[]},
            {Random[], 0.277896}, {Random[], 0.649997},
            {Random[], 0.200244}, {Random[], 0.17906},
            {Random[], 0.853949},
            {0.277896, 0.649997, 0.200244, 0.17906, 0.853949}}
```

The observant reader might notice that the value of x in the first `Table` does not appear in the second `Table`. This is because the transformation rule `x :> Random[]` was not placed in the global rule base.

Transformation rules can be written using symbols. For example,

```
In[5]:= {a, b, c} /. List -> Plus

Out[5]= a + b + c
```

or using labeled patterns.

```
In[6]:= {{3, 4}, {7, 2}, {1, 5}} /. {x_, y_} -> {y, x}
Out[6]= {{4, 3}, {2, 7}, {5, 1}}
```

We can use multiple rules with an expression by enclosing them in a list.

```
In[7]:= {a, b, c} /. {c -> b, b -> a}
Out[7]= {a, a, b}
```

A transformation rule is applied only once to each part of an expression (in contrast to a rewrite rule) and multiple transformation rules are used in parallel. Hence, in the above example, the symbol c is transformed into b but it is not further changed into a. In order to apply one or more transformation rules repeatedly to an expression until the expression no longer changes, the ReplaceRepeated function is used.

```
In[8]:= {a, b, c} //. {c -> b, b -> a}
Out[8]= {a, a, a}
```

We want to conclude this discussion of the evaluation process in *Mathematica* with a demonstration of a sophisticated rewrite rule which employs most of the things discussed here: the repeated use of a transformation rule with delayed evaluation, sequence patterns, and conditional pattern matching.

Recall the maxima function that we defined in the previous chapter which returns the elements in a list of positive numbers that are bigger than all of the preceding numbers in the list.

```
In[9]:= maxima1[x_List] := Union[Rest[FoldList[Max, 0, x]]]
```

```
In[10]:= maxima1[{3, 5, 2, 6, 1, 8, 4, 9, 7}]
Out[10]= {3, 5, 6, 8, 9}
```

We can also write this function using a pattern matching transformation rule.

```
In[11]:= maxima2[x_List] := x //. {a___, b_, c___, d_, e___} /;
            d <= b :> {a, b, c, e}
```

Basically, the transformation rule repeatedly looks through the list for two elements, separated by a sequence of zero or more elements, such that the second selected element is no greater than the first selected element. It then eliminates the second element. The process stops when there are no two elements such that the second is less than or equal to the first.

```
In[12]:= maxima2[{3, 5, 2, 6, 1, 8, 4, 9, 7}]

Out[12]= {3, 5, 6, 8, 9}
```

Exercises

1. Using `Trace` on `maxima2` and `maxima1`, explain why the functional version is much faster than the pattern matching version of the `maxima` function.

2. The following compound expression input returns a value of 14:

   ```
   y = 11; a = 9; (y + 3 /. y->a)
   ```

   Predict the evaluation sequence that was followed. Use the `Trace` function to check your answer.

3. Use the `Hold` function in the compound expression in the previous exercise to obtain a value of 12.

4. The function definition `f[x_Plus] := Apply[Times, x]` works as follows:

   ```
   In[1]:= f[a + b + c]

   Out[1]= a b c
   ```

   The rewrite rule `g[x_] := x /. Plus[z___] -> Times[z]` does not work. Use `Trace` to see why and then modify this rule so that it performs the same operation as the function `f` above.

5. Create a rewrite rule that uses a repeated replacement to "unnest" the nested lists within a list:

   ```
   In[1]:= unNest[{{a, a, a}, {a}, {{b, b, b}, {b, b}}, {a, a}}]

   Out[1]= {{a, a, a}, {a}, {b, b, b}, {b, b}, {a, a}}
   ```

Exercises (cont.)

6. Define a function using a pattern matching repeated replacement to sum the elements of a list.

Solutions

1. The pattern matched function is slower because it repeatedly applies transformation rules.

```
In[1]:= maxima1[x_] := Union[Rest[FoldList[Max, -Infinity, x]]]
```

```
In[2]:= maxima2[x_List] :=
            x//.{a___, b_, c___, d_, e___} /;
                d <= b :> {a, b, c, e}
```

```
In[3]:= Trace[maxima1[{3, 5, 2, 6, 1, 8, 4, 9, 7}]]
```

```
Out[3]= {maxima1[{3, 5, 2, 6, 1, 8, 4, 9, 7}],
         Union[Rest[FoldList[Max, -Infinity,
            {3, 5, 2, 6, 1, 8, 4, 9, 7}]]],
         {{{{Infinity, Infinity}, -(Infinity), -Infinity},
         FoldList[Max, -Infinity,
            {3, 5, 2, 6, 1, 8, 4, 9, 7}],
         {Max[-Infinity, 3], Max[3, -Infinity], 3},
         {Max[3, 5], 5}, {Max[5, 2], Max[2, 5], 5},
         {Max[5, 6], 6}, {Max[6, 1], Max[1, 6], 6},
         {Max[6, 8], 8}, {Max[8, 4], Max[4, 8], 8},
         {Max[8, 9], 9}, {Max[9, 7], Max[7, 9], 9},
         {-Infinity, 3, 5, 5, 6, 6, 8, 8, 9, 9}},
         Rest[{-Infinity, 3, 5, 5, 6, 6, 8, 8, 9, 9}],
         {3, 5, 5, 6, 6, 8, 8, 9, 9}},
         Union[{3, 5, 5, 6, 6, 8, 8, 9, 9}], {3, 5, 6, 8, 9}}
```

```
In[4]:= Trace[maxima2[{3, 5, 2, 6, 1, 8, 4, 9, 7}]]
```

Out[4]= {maxima2[{3, 5, 2, 6, 1, 8, 4, 9, 7}],

        {3, 5, 2, 6, 1, 8, 4, 9, 7} //.

        {a___, b_, c___, d_, e___} /; d <= b :> {a, b, c, e}

        , {5 <= 3, False}, {2 <= 3, True}, {5 <= 3, False},

        {6 <= 3, False}, {6 <= 5, False}, {1 <= 3, True},

        {5 <= 3, False}, {6 <= 3, False}, {6 <= 5, False},

        {8 <= 3, False}, {8 <= 5, False}, {8 <= 6, False},

        {4 <= 3, False}, {4 <= 5, True}, {5 <= 3, False},

        {6 <= 3, False}, {6 <= 5, False}, {8 <= 3, False},

        {8 <= 5, False}, {8 <= 6, False}, {9 <= 3, False},

        {9 <= 5, False}, {9 <= 6, False}, {9 <= 8, False},

        {7 <= 3, False}, {7 <= 5, False}, {7 <= 6, False},

        {7 <= 8, True}, {5 <= 3, False}, {6 <= 3, False},

        {6 <= 5, False}, {8 <= 3, False}, {8 <= 5, False},

        {8 <= 6, False}, {9 <= 3, False}, {9 <= 5, False},

        {9 <= 6, False}, {9 <= 8, False}, {3, 5, 6, 8, 9}}}

2.

In[1]:= **Trace[y = 11; a = 9; (y + 3 /. y -> a)]**

Out[1]= {y = 11; a = 9; y + 3 /. y -> a, {y = 11, 11},

     {a = 9, 9}, {{{y, 11}, 11 + 3, 3 + 11, 14},

     {{y, 11}, {a, 9}, 11 -> 9, 11 -> 9}, 14 /. 11 -> 9,

     14}, 14}

3.

In[1]:= **Hold[y = 11]; a = 9; (y + 3 /. y -> a)**

Out[1]= 12

4. Hint: You need to maintain the left-hand side of the transformation rule un-evaluated for purposes of pattern matching and the right-hand side of the rule unevaluated until the rule is used.

In[1]:= **Trace[g[x_] = x /. Plus[z___] -> Times[z]]**

Out[1]= {{{{Plus[z___], z___}, {Times[z], z}, z___ -> z,

      z___ -> z}, x /. z___ -> z, x}, g[x_] = x, x}

In[2]:= **Clear[a, g]**

In[3]:= **g[x_] := x /. Literal[Plus[z___]] :> Times[z]**

In[4]:= **g[a + b + c]**

Out[4]= a b c

5. The transformation rule unnests lists within a list.

In[1]:= **unNest[lis_] := Map[(# //.{x__List}->x)&, lis]**

In[2]:= **unNest[{{a, a, a}, {a}, {{b, b, b}, {b, b}}, {a, a}}]**

Out[2]= {{a, a, a}, {a}, {b, b, b}, {b, b}, {a, a}}

6. *In[1]:=* **sum[lis_] := First[lis //. {x_, y___} -> x + {y}]**

In[2]:= **sum[{1, 5, 8, 3, 9, 3}]**

Out[2]= 29

6 | Conditional Function Definitions

An essential aspect of computer programs is their ability to make decisions—to proceed differently depending upon properties of the data they are operating upon. In this chapter, we explore two different ways of building decision-making into *Mathematica* programs: *multiclause definitions*, in which a function is defined by more than one rule; and *conditional functions*, which return one of several values depending upon a condition. These features will greatly expand what we can do with *Mathematica*; they are also a crucial step toward learning the basic techniques of recursion and iteration.

6.1 Introduction

Consider the following problem: Given a list of characters L, if the first character is a "whitespace" character—space or newline—return `Rest[L]`; otherwise, return L.

```
In[1]:= removeWhiteSpace[{"a", "b", "c"}]

Out[1]= {a, b, c}


In[2]:= removeWhiteSpace[{" ", "a", "b", "c"}]

Out[2]= {a, b, c}


In[3]:= removeWhiteSpace[{" ", " ", "a", "b", "c"}]

Out[3]= { , a, b, c}
```

None of the built-in functions, like `Map` or `Fold`, help us do this, but in fact the answer was given, indirectly, in Chapter 5. Here's how: We know that function definitions are really just rewrite rules, that a rewrite rule consists of a *pattern* (on the left side) and a *body* (on the right side), and that evaluation of an expression consists of applying such rewrite rules to parts of the expression that match the pattern.

This implies that, to define a function f, we can give *several* rules for applying f in specific cases. Thus, we could write:

```
f[0] := ...
f[n_Integer] := ...
f[x_Real] := ...
f[{0, 0}] := ...
f[{x_}] := ...
f[z_] := ...
```

These rules give the value of f when its argument is zero, or any integer, or a real number, or the list of two zeros, or an arbitrary one-element list, or an arbitrary argument. As stated in Chapter 5, whenever more than one rule can be applied to an expression, the most specific rule (*i.e.*, the one that matches the fewest expressions) is used. So, the expression f[0] is rewritten using the first rule, even though the second and the last rules also match; f[2] is rewritten using the second rule, even though the last rule also matches.

Such a definition is called a *multiclause definition*. We can solve the problem with removeWhiteSpace easily using one such definition[1]:

```
removeWhiteSpace[{" ", r___}] := {r}
removeWhiteSpace[{" \n", r___}] := {r}
removeWhiteSpace[L_] := L
```

To summarize: the definition of removeWhiteSpace is given in three clauses. The first applies to lists beginning with a space, the second to lists beginning with a newline character, and the third to any other argument; since *Mathematica* gives priority to rules with more specific patterns, the third clause will be used for lists that don't begin with a space or newline.

There are a number of ways to do this using the different types of patterns introduced in Chapter 5. The third clause never changes, but the first two can be changed and possibly replaced by a single one, in several ways. Here are some alternative definitions:

```
removeWhiteSpace[{x_, r___} /; x == " "] := {r}
removeWhiteSpace[{x_, r___} /; x == " \n"] := {r}
removeWhiteSpace[L_] := L
```

```
removeWhiteSpace[{x_, r___} /; x == " " || x == " \n"] := {r}
removeWhiteSpace[L_] := L
```

---

[1] Important note: When using the sequence patterns ___ or __, the item that is matched is a *sequence* rather than a list (the difference is explained on page 136). Since the result of the function is supposed to be a list, we need to apply the List function to that sequence. That is why r on the right appears inside list braces. Omitting them is a very common mistake.

```
removeWhiteSpace[{" " | " \n", r___}] := {r}
removeWhiteSpace[L_] := L


removeWhiteSpace[{x_, r___} /; MemberQ[{" ", " \n"}, x]] := {r}
removeWhiteSpace[L_] := L

stringMemberQ[str_, ch_] := StringPosition[str, ch] != {}
whiteSpaceQ[ch_] := stringMemberQ[" \n", ch]


removeWhiteSpace[{x_?whiteSpaceQ, r___}] := {r}
removeWhiteSpace[L_] := L


removeWhiteSpace[{x_, r___} /; whiteSpaceQ[x]] := {r}
removeWhiteSpace[L_] := L


removeWhiteSpace[{x_, r___}] := {r} /; whiteSpaceQ[x]
removeWhiteSpace[L_] := L
```

The last of these is essentially an alternative syntax for the one before it. This alternative syntax—putting the pattern-match condition on the right-hand side of the rewrite rule—is heavily used in *Mathematica* because it existed in earlier versions that lacked the other form. The newer form should be used when the clause occupies multiple lines; that way, someone looking at the clause can see immediately the conditions under which it is applicable. The older syntax is fine when the clause is only one or two lines long.

As another example, the function vowelQ[*ch*] tests whether the character (a one-element string) *ch* is a lower-case vowel.

```
vowelQ["a"] := True
vowelQ["e"] := True
vowelQ["i"] := True
vowelQ["o"] := True
vowelQ["u"] := True
vowelQ[c_] := False
```

As a final example, canFollow[c_1, c_2] is true if c_2 can follow c_1 when the latter appears as the first character of a word in English.

```
canFollow["a", ch_] := True
canFollow["b", ch_] := True /; stringMemberQ["aeiloruy", ch]
canFollow["c", ch_] := True /; stringMemberQ["aehiloruy", ch]
canFollow["d", ch_] := True /; stringMemberQ["aeioruy", ch]
                               .
                               .
                               .
canFollow["z", ch_] := True /; stringMemberQ["aeiouy", ch]
canFollow[ch1_, ch2_] := False
```

6.2 Conditional Functions

Though multiclause function definitions and pattern-matching are convenient, it is not necessary to use complex patterns. An alternative is to use *conditional functions*, functions which return the values of different expressions depending upon a condition. Of these, the simplest and most important is If.

$$\text{If}[cond,\ expr_1,\ expr_2]$$

returns the value of *expr_1* if *cond* is true, and returns the value of *expr_2* otherwise. For example, If[n > 0, 5, 9] returns 5 if n has a value greater than 0, and returns 9 otherwise.

```
If[lis == {}, lis, Rest[lis]]
```

returns the Rest of lis, after making sure it is non-empty.

All our examples in the previous section can be expressed in *single-clause* definitions using If, for example.

```
removeWhiteSpace[lis_] :=
    If[stringMemberQ["\n", First[lis]], Rest[lis], lis]
```

We can nest If expressions just like any other function calls. This can give it some advantages over multiclause function definitions. For example, suppose we change our removeWhiteSpace example to removeTwoSpaces: removeTwoSpaces[*lis*] returns *lis* if it does not start with a blank or newline, Rest[*lis*] if it starts with exactly one, and Rest[Rest[*lis*]] if it starts with two. Here is the multiclause definition.

```
removeTwoSpaces[{x_, y_, r___}] := {r} /;
        whiteSpaceQ[x] && whiteSpaceQ[y]
removeTwoSpaces[{x_, r___}] := {r} /; whiteSpaceQ[x]
removeTwoSpaces[lis_] := lis
```

What is inelegant and inefficient about this code is that it may test the first element of lis *twice*. For example, in evaluating removeTwoSpaces[{"a", "b"}], *Mathematica* uses the first (most specific) rule, and tests whether the expression whiteSpaceQ["a"] is true; since it isn't, it tries the second (less specific) rule, which requires it to again evaluate whiteSpaceQ["a"]; since it is (still) false, *Mathematica* goes on to the third (most general) rule, which, of course, applies. By using an If expression, this inefficiency can be avoided.

```
removeTwoSpaces[lis_] :=
    If[whiteSpaceQ[First[lis]],
        If[whiteSpaceQ[First[Rest[lis]]],
            Rest[Rest[lis]],
            Rest[lis]],
        lis]
```

Here, the first element is tested once, at the beginning. If the second element turns out not to be zero, we return Rest[lis] without having to retest the first element.

As another example of If, the function applyChar has, as its one argument, a list. This list must contain, first, a character, which must be one of "+", "-", "*", or "/"; that character must be followed by all numbers. ApplyChar applies the function named by the character to the elements of the rest of the list:

```
In[1]:= applyChar[lis_] :=
        Module[{op = First[lis], nums = Rest[lis]},
            If[op == "+", Apply[Plus, nums],
                If[op == "-", Apply[Subtract, nums],
                    If[op == "*", Apply[Times, nums],
                        If[op == "/", Apply[Divide, nums],
                            Print["Bad argument to applyChar"]]]]]]
```

```
In[2]:= applyChar[{"+", 1, 2, 3, 4}]

Out[2]= 10
```

(Recall the Module function, which permits us to introduce local variables. In this case, it saves us from having to write First[lis] and Rest[lis] several times each.)

Even though the argument list *must* contain one of the four operators as its first element, it is still best to check for it explicitly; otherwise, if the condition is ever violated, the results may be very mysterious. We've used the Print function, which prints all of its arguments (of which it can have an arbitrary number) and then skips to a new line.

Notice that what we have in this code is several nested If's, each occurring in the false part of the previous one. Thus, the structure of the computation is a sequence of tests of predicates until one is found to be true, at which point a result can be computed. Such a sequence of *cascaded* If statements can get quite long, and the indentation can become unmanageable, so it is conventional to violate the usual rule for indenting If expressions and indent this type of structure as follows:

```
If[cond₁,  result₁,
  If[cond₂,  result₂,
      ⋮
    If[condₙ,  resultₙ,
              resultₙ₊₁]  ...  ]]
```

But, in fact, cascaded Ifs are so common that *Mathematica* provides a more direct way of writing them, using the function Which. Its general form is,

```
Which[cond₁,  result₁,
      cond₂,  result₂,
        ⋮
      condₙ,  resultₙ,
      True,  resultₙ₊₁]
```

which has exactly the same effect as the cascaded If expression above: It tests each condition in turn, and when it finds an *i* such that *cond*$_i$ is true, it returns *result*$_i$ as the result of the Which expression itself. If none of the conditions turns out to be true, then it will test the final "condition", namely the expression True, which always evaluates to true, and it will then return *result*$_{n+1}$.

ApplyChar can now be written more neatly.

```
In[3]:= applyChar[lis_] :=
      Module[{op = First[lis], nums = Rest[lis]},
          Which[op == "+", Apply[Plus, nums],
              op == "-", Apply[Subtract, nums],
              op == "*", Apply[Times, nums],
              op == "/", Apply[Divide, nums],
              True, Print["Bad argument to applyChar"]]]
```

One last form deserves mention. Our use of the Which command is still quite special, in that it consists of a simple sequence of comparisons between a variable and a constant. Since this is also a common form, *Mathematica* again provides a special function for it, called Switch. Its general form is,

```
Switch[e, pattern₁,  result₁,
          pattern₂,  result₂,
              ⋮
          patternₙ,  resultₙ,
          _,  resultₙ₊₁]
```

which evaluates *e* and then checks each pattern, in order, to see whether *e* matches; as soon as *e* matches one, say *patternᵢ*, it returns the value of *resultᵢ*. Of course, if none of the patterns *pattern₁*, ..., *patternₙ* matches, the _ certainly will.

If all the patterns happen to be constants, the Switch expression is equivalent to the following Which expression.

```
Which[e == pattern₁,  result₁,
      e == pattern₂,  result₂,
              ⋮
      e == patternₙ,  resultₙ,
      True,  resultₙ₊₁]
```

Here then, is our final version of applyChar.

```
In[4]:= applyChar[lis_] :=
        Module[{op = First[lis], nums = Rest[lis]},
            Switch[op, "+", Apply[Plus, nums],
                       "-", Apply[Subtract, nums],
                       "*", Apply[Times, nums],
                       "/", Apply[Divide, nums],
                       _,   Print["Bad argument to ApplyChar"]]]
```

Notice that Switch uses the blank character, _, for the final, or *default* case, just as Which uses the always-true expression True.

If can also be used in conjunction with the higher-order functions of Chapter 4 to achieve greater flexibility. For example, given a list, the following function adds one to all the numbers occurring in it.

```
incrementNumbers[L_] := Map[If[NumberQ[#], #+1, #]&, L]
```

Here is a function that divides 100 by every number in a numerical list, except zeroes.

```
divide100By[L_] := Map[If[#==0, #, 100/#]&, L]
```

And here is a function to remove consecutive occurrences of the same value, *e.g.*, removeRepetitions[{0, 1, 1, 2, 2, 2, 1, 1}] will return {0, 1, 2, 1}.

```
removeRepetitions[L_] :=
    Fold[If[#2==Last[#1], #1, Append[#1, #2]]&,
        {First[L]}, Rest[L]]
```

Summary of Conditionals

When writing a function whose result must be computed differently depending upon the values of its arguments, you have a choice:

- Use a multiclause definition:

$$f[\textit{pattern}_1\_] \; /; \; \textit{cond}_1 \; := \; \textit{rhs}_1$$
$$\vdots$$
$$f[\textit{pattern}_n\_] \; /; \; \textit{cond}_n \; := \; \textit{rhs}_n$$

  where the conditions are optional, and may appear after the right-hand sides.

- Use a single-clause definition with a conditional expression:

$$f[x\_] \; := \; If[\textit{cond}_1, \; \textit{rhs}_1,$$
$$\vdots$$
$$If[\textit{cond}_n, \; \textit{rhs}_n,$$
$$\textit{rhs}_{n+1}] \; \ldots]$$

In the latter case, if n is greater than two, use the equivalent Which expression; and if all conditions have the form x == $\textit{const}_i$, for a given variable x and some constants $\textit{const}_i$, use the Switch function.

The next section contains a final example, which we will solve using various combinations of the features we have learned in this chapter.

6.3 Example—Classifying Points

Quadrants on the Euclidean plane are conventionally numbered counterclockwise from quadrant 1 (x and y positive) to quadrant 4 (x positive, y negative). The function pointLoc[{x, y}] has the job of computing the classification of point (x,y), according to Table 6.1.

Point	Classification
$(0, 0)$	0
$y = 0$ (i.e. on the x-axis)	-1
$x = 0$ (i.e. on the y-axis)	-2
Quadrant 1	1
Quadrant 2	2
Quadrant 3	3
Quadrant 4	4

Table 6.1: Quadrant classification

We will use this problem to illustrate the features covered in this chapter, by giving a number of different solutions, using multiclause function definitions with predicates, single-clause definitions with If and its relatives, and combinations of the two.

Perhaps the first solution that suggests itself is one that uses a clause for each of the cases above:

```
pointLoc[{0, 0}] := 0
pointLoc[{x_, 0}]   := -1
pointLoc[{0, y_}]   := -2
pointLoc[{x_, y_}]    := 1 /; x>0 && y>0
pointLoc[{x_, y_}]    := 2 /; x<0 && y>0
pointLoc[{x_, y_}]    := 3 /; x<0 && y<0
pointLoc[{x_, y_}]    := 4   (* /; x>0 && y<0 *)
```

(It is a good idea to include the last condition as a comment, rather than as a condition in the code, because *Mathematica* would not realize that the condition has to be true at that point and would check it anyway.)

Translated directly to a one-clause definition using If, this would be:

```
pointLoc[{x_, y_}] :=
    If[x == 0 && y == 0, 0,
    If[y == 0, -1,
    If[x == 0, -2,
    If[x > 0 && y > 0, 1,
    If[x < 0 && y > 0, 2,
    If[x < 0 && y < 0, 3,
      (* x > 0 && y < 0 *) 4]]]]]]
```

Actually, a more likely solution here uses Which.

```
pointLoc[{x_, y_}] :=
  Which[
      x == 0 && y == 0, 0,
      y == 0, -1,
      x == 0, -2,
      x > 0 && y > 0, 1,
      x < 0 && y > 0, 2,
      x < 0 && y < 0, 3,
      True (* x > 0 && y < 0 *) , 4]
```

All of our solutions so far suffer from a certain degree of inefficiency, because of repeated comparisons of a single value with zero. Take the last solution as an example, and suppose the argument is $(-5, -9)$. It will require five comparisons of -5 with zero and three comparisons of -9 with zero to obtain this result. Specifically:

1. evaluate x == 0; since it is false, the associated y == 0 will not be evaluated, and we next

2. evaluate y == 0 on the following line; since it is false

3. evaluate x == 0 on the third line; since it is false

4. evaluate x > 0 on next line; since it is false, the associated y > 0 will not be evaluated, and we next

5. evaluate x < 0 on the next line; since it is true, we do

6. the y > 0 comparison, which is false, so we next

7. evaluate x < 0 on the next line; since it is true, we (8) evaluate y < 0, which is also true, so we return the answer 3.

How can we improve this? By nesting conditional expressions inside other conditional expressions. In particular, as soon as we discover that x is less than, greater than, or equal to zero, we should make maximum use of that fact without rechecking it. That is what the following pointLoc function does.

```
pointLoc[{x_, y_}] :=
Which[
    x == 0, If[y == 0, 0, -2],
    x > 0, Which[y > 0, 1,
                 y < 0, 4,
                 True (* y == 0 *), -1],
    True (* x<0 *),
    Which[y < 0, 3,
          y > 0, 2,
          True (* y == 0 *), -1]
    ]
```

Let's count up the comparisons for (−5, −9) this time: (1) evaluate x == 0; since it is false, we next, (2) evaluate x > 0; since it is false, we go to the third branch of the Which, evaluate True, which is, of course, true; then, (3) evaluate y < 0, which is true, and we return 3. Thus, we made only three comparisons—a substantial improvement.

When pattern-matching is used, as in our first, multiclause solution, efficiency calculations are more difficult. It would be inaccurate to say that *Mathematica* has to compare x and y to zero to tell whether the first clause applies; what actually happens is more complex. What is true, however, is that it will do the comparisons indicated in the last four clauses. So even if we discount the first three clauses, with argument (−5, −9), some extra comparisons are done. Specifically: (1) the comparison x > 0 is done; then, (2) x < 0 and (3) y > 0; then, (4) x < 0 and (5) y < 0. This can be avoided by using conditional expressions *within* clauses.

```
pointLoc[{0, 0}] := 0
pointLoc[{x_, 0}] := -1
pointLoc[{0, y_}] := -2
pointLoc[{x_, y_}] := If[x < 0, 2, 1] /; y > 0
pointLoc[{x_, y_}] := If[x < 0, 3, 4] (* /; y < 0 *)
```

Now, no redundant comparisons are done. For (−5, −9), since y > 0 fails, the fourth clause is not used, so the x > 0 comparison in it is not done. Only the single x < 0 comparison in the final clause is done, for a total of two comparisons.

Having done all these versions of pointLoc, we would be remiss if we did not remind the reader of a basic fact of life in programming: your time is more valuable than your computer's time. You shouldn't be worrying about how slow a function is until there is a demonstrated need to worry. Far more important is the clarity and simplicity of the code, since this will determine how much time you (or another programmer) will have to spend when it comes time to modify it. In the case of pointLoc, we would argue that we got lucky and found a version (the final one) that wins on both counts (if only programming were always like that!).

Finally, a technical, but potentially important, point: Not all of the versions of pointLoc work exactly the same. The integer 0, as a pattern, does not match the real number 0.0, since they have different Heads. Thus, using the last version as an example, pointLoc[{0.0, 0.0}] returns 4. On the other hand, the single-clause versions using If and Which return 0, because 0.0 == 0 is true. How can we fix this? There are a number of possibilities. A simple way is to change the rules involving zeroes to

```
pointLoc[{0 | 0.0, 0 | 0.0}] := 0
pointLoc[{x_, 0 | 0.0}] := -1
pointLoc[{0 | 0.0, y_}] := -2
```

Exercises

1. Write the function `signum[x]` which, applied to an integer x, returns -1, 0, or 1, according as x is less than, equal to, or greater than, zero. Write it in three ways: using three clauses; using a single clause with `If`; and using a single clause with `Which`.

2. Extend `signum` from Exercise 1 to apply to both integers and reals; again, write it in three ways (though you may use more than three clauses for the multiclause version).

3. `swapTwo[L_List]` returns L with its first two elements interchanged; *e.g.*, `swapTwo[{a, b, c, d, e}]` is `{b, a, c, d, e}`. If L has fewer than two elements, `swapTwo` just returns it. Write `swapTwo` using three clauses: one for the empty list, one for one-element lists, and one for all other lists. Then write it using two clauses: one for lists of length zero or one, and another for all longer lists.

4. Write `vowelQ` and `canFollow` using single-clause definitions.

5. Our one-clause definition of `removeTwoSpaces` has a flaw: it doesn't handle the empty list or one-element lists. Correct these omissions.

6. Convert this definition to one that has no conditional parts (`/;`), but instead uses pattern-matching in the argument list:

   ```
   f[x_, y_] := x - y /; IntegerQ[x]

   f[x_, y_] := x[[1]] + y /;
       Head[x] == List && IntegerQ[First[x]] && y == 1
   ```

7. Write `applyChar` in multiclause form, using pattern-matching on the first element of its argument.

8. Extend `pointLoc` to three dimensions, following this rule: For point (x, y, z), if $z \geq 0$, then give the same classification as (x, y), with the exception that zero is treated as a positive number (so the only classifications are 1, 2, 3, and 4); if $z < 0$, add 4 to the classification of (x, y) (with the same exception). For example, $(1, 0, 1)$ is in octant 1, and $(0, -3, -3)$ is in octant 8. `pointLoc` should work for points in two *or* three dimensions.

9. Use `If` in conjunction with `Map` or `Fold` to define the following functions:

   (a) In a list of numbers, double all the positive ones, but leave the negative ones alone.

Exercises (cont.)

(b) `remove3Repetitions` is like `removeRepetitions` except that it only alters three or more consecutive occurrences, changing them to two occurrences; if there are only two occurrences to begin with, they are left alone. For example, `remove3Repetitions[{0, 1, 1, 2, 2, 2, 1}]` will return `{0, 1, 1, 2, 2, 1}`.

(c) Add the elements of a list in consecutive order, but never let the sum go below zero. For example:

```
In[1]:= positiveSum[{5, 3, -13, 7, -3, 2}]

Out[1]= 6
```

Since the -13 caused the sum to go below zero, it was instead put back to zero and the summation continued from there.

Solutions

1.

```
In[1]:= signum1[x_ /; x<0] := -1
        signum1[x_ /; x>0] := 1
        signum1[0]         := 0
```

```
In[2]:= Map[signum1, {-2, 0, 1}]

Out[2]= {-1, 0, 1}
```

```
In[3]:= signum2[x_] :=
          If[x<0, -1,
            If[x==0, 0, 1]]
```

```
In[4]:= Map[signum2, {-2, 0, 1}]

Out[4]= {-1, 0, 1}
```

```
In[5]:= signum3[x_] :=
          Which[x<0, -1,
                x==0, 0,
                True, 1]
```

```
In[6]:= Map[signum3, {-2, 0, 1}]

Out[6]= {-1, 0, 1}
```

2.
```
In[1]:= signum1[x_ /; x<0] := -1
        signum1[x_ /; x>0] := 1
        signum1[0]         := 0
        signum1[0.0]       := 0


In[2]:= Map[signum1, {-2, 0, 2}]

Out[2]= {-1, 0, 1}


In[3]:= signum2[x_] :=
          If[x<0, -1,
            If[x>0, 1, 0]]


In[4]:= Map[signum2, {-2, 0, 2}]

Out[4]= {-1, 0, 1}


In[5]:= signum3[x_] :=
          Which[x<0, -1,
                x>0, 1,
                True, 0]


In[6]:= Map[signum3, {-2, 0, 2}]

Out[6]= {-1, 0, 1}
```

3.
```
In[1]:= swapTwo1[{}] := {}
        swapTwo1[{x_}] := {x}
        swapTwo1[{x_, y_, r___}] := {y, x, r}


In[2]:= swapTwo2[{x_, y_, r___}] := {y, x, r}
        swapTwo2[x_] := x


In[3]:= Map[swapTwo1, {{}, {a}, {a,b,c,d}}]

Out[3]= {{}, {a}, {b, a, c, d}}
```

4.

```
In[1]:= vowelQ[c_] :=
          Switch[c, "a", True,
                    "e", True,
                    "i", True,
                    "o", True,
                    "u", True,
                    _,  False]
```

```
In[2]:= canFollow[ch1_, ch_] :=
          Switch[ch1, "a", True,
                     "b", StringMemberQ["aeiloruy", ch],
                     "c", StringMemberQ["aehiloruy", ch],
                     "d", StringMemberQ["aeioruy", ch],
                     "z", StringMemberQ["aeiouy", ch],
                     _,  False]
```

5.

```
In[1]:= removeTwoSpaces[lis_] :=
          If[lis == {}, {},
            If[whiteSpaceQ[First[lis]],
              If[Length[lis]==1, {},
                If[whiteSpaceQ[First[Rest[lis]]],
                  Rest[Rest[lis]],
                  Rest[lis]]],
              lis]]
```

6.

```
In[1]:= f[{x1_Integer, ___}, 1] := x1 + 1
          f[x_Integer, y_] := x - y
```

7.

```
In[1]:= applyChar[{"+", nums__}] := Apply[Plus, {nums}]
          applyChar[{"-", nums__}] := Apply[Minus, {nums}]
          applyChar[{"*", nums__}] := Apply[Times, {nums}]
          applyChar[{"/", nums__}] := Apply[Divide, {nums}]
          applyChar[_] := Print["Bad argument to applyChar"];
```

8.

```
In[1]:= pointLoc[{0, 0}] := 0
          pointLoc[{x_, 0}] := -1
          pointLoc[{0, y_}] := -2
          pointLoc[{x_, y_}] := If[x<0, 2, 1] /; y>0
          pointLoc[{x_, y_}] := If[x<0, 3, 4] (* /; y<0 *)
          pointLoc[{x_, y_, z_}] := If[x<0, 2, 1] /; y>=0 && z>=0
          pointLoc[{x_, y_, z_}] := If[x<0, 3, 4] /; y<0 && z>=0
          pointLoc[{x_, y_, z_}] := If[x<0, 6, 5] /; y>=0 && z<0
          pointLoc[{x_, y_, z_}] := If[x<0, 7, 8] /; y<0 && z<0
```

9.a

```
In[1]:= doublePos[lis_] := Map[If[#>0, 2 * #, #]&, lis]
```

9.b

```
In[1]:= remove3Repetitions[lis_] :=
           Fold[If[Length[#1]>2 && #2 == #1[[-1]] == #1[[-2]],
                 #1, Join[#1, {#2}]]&,
              {},
              lis]
```

9.c

```
In[1]:= positiveSum[L_] :=
           Fold[If[#1+#2 < 0, 0, #1+#2]&, 0, L]
```

7 Recursion

A function is *defined using recursion* if in its definition, it makes calls to itself. Though this sounds like a circular definition, the use of recursion in *Mathematica* is perfectly legal and extremely useful. In fact, many of the built-in operations of *Mathematica* could be written in *Mathematica* itself using recursion. In this chapter, we will present many examples of recursion and explain how recursive functions are written.

7.1 Fibonacci Numbers

Recursive definitions of mathematical quantities were used by mathematicians for centuries before computers even existed. One famous example is the definition of a special sequence of numbers first studied by the thirteenth century Italian mathematician Leonardo Fibonacci. The *Fibonacci numbers* have since been studied extensively, finding application in many areas; see [Knuth 1973] for a detailed discussion.

The Fibonacci numbers are obtained as follows: write down 0 and 1, then continue writing numbers computed by adding the last two numbers you've written down:

$$
\begin{array}{ccccccccc}
0 & 1 & 1 & 2 & 3 & 5 & 8 & 13 & 21 & \ldots \\
F_0 & F_1 & F_2 & F_3 & F_4 & F_5 & F_6 & F_7 & F_8 & \ldots
\end{array}
$$

The neatest way to define these numbers is with recursion:

$$
\begin{aligned}
F_0 &= 0 \\
F_1 &= 1 \\
F_n &= F_{n-2} + F_{n-1}, \text{ for } n > 1
\end{aligned}
$$

If we think of this sequence as a function, we would just change this to

$$
\begin{aligned}
F(0) &= 0 \\
F(1) &= 1 \\
F(n) &= F(n-2) + F(n-1), \text{ for } n > 1
\end{aligned}
$$

In this form, we can translate the definition directly into *Mathematica*.

```
In[1]:= F[0] := 0

In[2]:= F[1] := 1

In[3]:= F[n_] := F[n-2] + F[n-1] /; n > 1

In[4]:= F[6]
Out[4]= 8

In[5]:= Table[F[i], {i, 0, 8}]
Out[5]= {0, 1, 1, 2, 3, 5, 8, 13, 21}
```

It is somewhat amazing that this works, but note that whenever we want to compute F[*n*] for some *n* > 1, we only apply F to numbers *smaller than n*. A look at the Trace of F[4] makes the point well. For example, the first two lines indicate that F[4] is rewritten to F[4 - 2] + F[4 - 1], and the lines that are indented one space show the calls of F[2] and F[3]. The lines showing calls to F[0] and F[1] don't have any indented lines under them, since those values are computed directly by a single rewrite rule, without making any recursive calls[1]:

```
In[6]:= TracePrint[F[4], F[_Integer] | F[_] + F[_]]
        F[4]
        F[4 - 2] + F[4 - 1]
         F[2]
         F[2 - 2] + F[2 - 1]
          F[0]
          F[1]
         F[3]
         F[3 - 2] + F[3 - 1]
          F[1]
          F[2]
          F[2 - 2] + F[2 - 1]
           F[0]
           F[1]

Out[6]= 3
```

The key thing to understand about recursion is this: *you can always apply a function within its own definition, so long as you apply it only to* smaller *values.*

---

[1] For a fuller explanation of this use of TracePrint, see Section 7.9.

We will see this principle used repeatedly in this chapter.

There is one other key point as well: we can apply the function to smaller and smaller values, but we must eventually reach a value that can be computed *without* recursion. In the case of the Fibonacci numbers, the numbers that can be computed without recursion—the *base cases*—are 0 and 1.

We will return to the Fibonacci numbers later in this chapter, in Section 7.7, where we will see what can be done about the terrible inefficiency of our definition of F (see Exercise 2).

Exercises

Before doing the exercises in this chapter, you may want to take a look at Section 7.9, which discusses some common programming errors, and how to debug recursive functions.

1. For each of the following sequences of numbers, see if you can deduce the pattern and write a *Mathematica* function to compute the ith value:

   (a)
2,	3,	6,	18,	108,	1944,	209952,	...
A_0,	A_1,	A_2,	A_3,	A_4,	A_5,	A_6,	...

   (b)
0,	1,	-1,	2,	-3,	5,	-8,	13,	-21,	...
B_0,	B_1,	B_2,	B_3,	B_4,	B_5,	B_6,	B_7,	B_8,	...

   (c)
0,	1,	2,	3,	6,	11,	20,	37,	68,	...
C_0,	C_1,	C_2,	C_3,	C_4,	C_5,	C_6,	C_7,	C_8,	...

2. The numbers FA_n represent the number of additions that are done in the course of evaluating F[n]:

0,	0,	1,	2,	4,	7,	12,	20,	33,...
FA_0,	FA_1,	FA_2,	FA_3,	FA_4,	FA_5,	FA_6,	FA_7,	FA_8,	...

   Write a function FA such that FA[n] = FA_n.

3. The Collatz sequence for n is defined as follows: The first number, c_0^n, is n, and after that each number c_i^n is derived from c_{i-1}^n by the following rule: if c_{i-1}^n is odd, then $c_i^n = 3c_{i-1}^n + 1$; if it is even, $c_i^n = c_{i-1}^n/2$. Define c[i_, n_] to be

Exercises (cont.)

c_i^n. (This sequence presents a seemingly elementary but in fact very difficult mathematical problem; see Exercise 2, page 217.)

Solutions

1.a
```
a[0] := 2
a[1] := 3
a[i_] := a[i-1] * a[i-2]
```

1.b
```
b[0] := 0
b[1] := 1
b[i_] := b[i-2] - b[i-1]
```

1.c
```
c[0] := 0
c[1] := 1
c[2] := 2
c[i_] := c[i-3] + c[i-2] + c[i-1]
```

2.
```
FA[0] := 0
FA[1] := 0
FA[i_] := FA[i-2] + FA[i-1] + 1
```

3.
```
c[0_, n_] := n
c[i_, n_] := c[i-1, n]/2 /; EvenQ[c[i-1, n]]
c[i_, n_] := 3 c[i-1, n] + 1 /; OddQ[c[i-1, n]]
```

7.2 List Functions

There are few list-oriented functions that cannot be defined using the built-in operations, as was done in Chapter 4; but since they can be used to exhibit simple examples of recursion, we'll look at them now.

We noted in our discussion of Fibonacci numbers that recursion works if the arguments of recursive calls are smaller than the original argument. The same principle applies to functions on lists. One common case is when the argument in the recursive call is the "tail" (*i.e.*, Rest) of the original argument. An example is length, our recursively-defined version of the built-in function Length. (We call it length instead of Length because Length is Protected.) The idea is that the length of a list is always one greater than the length of its tail.

```
In[1]:= length[lis_] := length[Rest[lis]] + 1
```

Applying `length` to a list, however, leads to trouble.

```
In[2]:= length[{a, b, c}]

    Rest::norest:
        Cannot take Rest of expression {} with length zero.

    Rest::argx:
        Rest called with 0 arguments; 1 argument is
            expected.

    Rest::norest:
        Cannot take Rest of expression Rest[]
            with length zero.

    $RecursionLimit::reclim:
        Recursion depth of 256 exceeded.

Out[2]= 250 + Hold[length[Rest[Hold[Rest[Hold[]]]]]]
```

Well, perhaps it's already obvious, but what we are experiencing is one of the most common errors in defining functions recursively: we forgot the base cases. For `length`, there is just one base case, the empty list.

```
In[3]:= length[{}] := 0
```

Now `length` works as we had intended it to.

```
In[4]:= length[{a, b, c}]
Out[4]= 3
```

Here is another simple example (for which we again have better solutions using built-in operations): adding the elements of a list. We know several ways to do this, for example,

```
sumElts[lis_] := Apply[Plus, lis]
```

```
sumElts[lis_] := Fold[Plus, 0, lis]
```

But for now we're just trying to get some practice with recursion. Here's the most obvious recursive solution.

```
sumElts[{}] := 0
sumElts[{x_, r___}] := x + sumElts[{r}]
```

We can use recursion for functions with multiple arguments as well. addPairs[L, M] is given two lists of numbers of equal length and returns a list containing the pairwise sums.

```
In[5]:= addPairs[{1, 2, 3}, {4, 5, 6}]

Out[5]= {5, 7, 9}
```

Here, the idea is to apply addPairs recursively to the tails of both lists:

```
addPairs[{}, {}] := {}
addPairs[{x1_, r1___}, {x2_, r2___}] :=
    Join[{x1 + x2}, addPairs[{r1}, {r2}]]
```

The recursive calls don't always have to be on the tail of the original argument. Any smaller list will do. The function multPairwise multiplies together every pair of elements in a list:

```
In[6]:= multPairwise[{3, 9, 17, 2, 6, 60}]

Out[6]= {27, 34, 360}
```

The trick is to make the recursive call on the tail of the tail.

```
multPairwise[{}] := {}
multPairwise[{x_, y_, r___}] := Join[{x y}, multPairwise[{r}]]
```

As a last simple example, consider the function deal defined on page 99. deal[n] produces a list of n playing cards randomly chosen from a 52-card deck (stored as the value of cardDeck, a 52-element list). Here's how we might write this function recursively.

First, dealing zero cards is easy:

```
deal[0] := {}
```

Now, suppose we have dealt $n - 1$ cards; how do we deal n? Just randomly deal a card from the remaining $52 - (n - 1) = 53 - n$ cards. To do this, randomly choose an integer r between 1 and $53 - n$, remove the rth card, and add it to the list of cards already dealt.

```
deal[n_] := Module[{dealt = deal[n-1]},
   Append[dealt,
        Complement[cardDeck,dealt][[Random[Integer,{1,53-n}]]]]]]
```

Exercises

1. In Chapter 6, we gave the examples of functions removeWhiteSpace and removeTwoSpaces. The obvious generalization of these two functions is removeLeadingSpaces, which removes *all* leading space or newline characters. Program removeLeadingSpaces using recursion.

2. Write a function sumOddElts[L] which adds up only the elements of L that are odd integers. L may contain even integers and non-integers. (Use IntegerQ to determine if a given element is an integer.)

3. Write a function sumEveryOtherElt[L] which adds up L[[1]], L[[3]], L[[5]], etc. Each of these elements is a number. L may have any number of elements.

4. Write a function addTriples[L1, L2, L3] which is like addPairs in that it adds up the corresponding elements of the three equal-length lists of numbers.

5. Write a function multAllPairs[L] which multiplies every consecutive pair of integers in the numerical list L:

```
In[1]:= multAllPairs[{3, 9, 17, 2, 6, 60}]

Out[1]= {27, 153, 34, 12, 360}
```

6. Write the function maxPairs[L, M] which, for L and M numerical lists of equal length, returns a list of the greater value in each corresponding pair.

7. The function interleave[L, M], which merges two lists of equal length, can be defined as follows:

```
interleave[L_, M_] := Flatten[Transpose[{L, M}]]
```

Rewrite interleave using recursion.

1.
```
removeLeadingSpaces[{}] := {}
removeLeadingSpaces[{" " | "\n", r___}] :=
    removeLeadingSpaces[{r}]
removeLeadingSpaces[L_] := L
```

2.
```
sumOddElts[{}] := 0
sumOddElts[{x_, y___}] := x + sumOddElts[{y}] /;
    IntegerQ[x] && OddQ[x]
sumOddElts[{x_, y___}] := sumOddElts[{y}]
```

3.
```
sumEveryOtherElt[{}] := 0
sumEveryOtherElt[{x_}] := x
sumEveryOtherElt[{x_, y_, r___}] := x + sumEveryOtherElt[{r}]
```

4.
```
addTriples[{}, {}, {}] := {}
addTriples[{x1_, y1___}, {x2_, y2___}, {x3_, y3___}] :=
    Join[{x1+x2+x3}, addTriples[{y1}, {y2}, {y3}]]
```

5.
```
multAllPairs[{}] := {}
multAllPairs[{_}] := {}
multAllPairs[{x_, y_, r___}] :=
        Join[{x y}, multAllPairs[{y, r}]]
```

6.
```
maxPairs[{}, {}] := {}
maxPairs[{x_, r___}, {y_, s___}] :=
        Join[{Max[x, y]}, maxPairs[{r}, {s}]]
```

7.
```
interleave[{}, {}] := {}
interleave[{x_, r___}, {y_, s___}] :=
        Join[{x, y}, interleave[{r}, {s}]]
```

7.3 Thinking Recursively

The rewriting evaluation process of *Mathematica* completely explains how recursion works, and it can be seen using Trace or TracePrint, as we did above. But that knowledge is of only limited usefulness in writing recursive functions.

Indeed, the real trick is to *forget* the evaluation process and simply *assume* that the function you're defining will return the correct answer when applied to smaller values. Suspend disbelief—you'll begin to see how simple recursion really is.

doubleUptoZero

Consider this simple function: doubleUptoZero[L] doubles all the numbers in the numeric list L until it finds a zero in L; all the numbers after that are left alone.

```
In[1]:= doubleUptoZero[{2, 3, 0, 4, 5}]

Out[1]= {4, 6, 0, 4, 5}


In[2]:= doubleUptoZero[{2, 3, 4, 5}]

Out[2]= {4, 6, 8, 10}
```

Let's apply the basic principle we've just learned. Given L, *assume* that double-UptoZero[Rest[L]] will give the correct result, and call that result R. How can we then compute doubleUptoZero[L]? Consider the possibilities:

1. If First[L] is not zero, then double it and prepend it to R. So use this clause:

```
doubleUptoZero[{x_, y___}] :=
    Join[{2x}, doubleUptoZero[{y}]] /; x != 0
```

2. If First[L] *is* zero, then we should ignore the value of R entirely, and just return L itself:

```
doubleUptoZero[{x_, y___}] := {x, y} /; x == 0
```

3. For the base case, we need to consider when L is empty:

```
doubleUptoZero[{}] := {}
```

And that's all there is to it!

We can clean this code up a little, by using a specific pattern to handle the x == 0 case, leaving the following definition:

```
doubleUptoZero[{}] := {}
doubleUptoZero[{0, y___}] := {0, y}
doubleUptoZero[{x_, y___}] := Join[{2x}, doubleUptoZero[{y}]]
```

where we've also reordered the rules to put more specific ones first, as is more conventional (though not required), and have omitted the test x != 0 on the last clause, since it is not needed.

runEncode

We now turn to another, somewhat more involved, example—the function runEncode. runEncode implements a method commonly used to compress large amounts of data, in those cases where the data is likely to contain long sequences of the same value ("runs"). A good example is the representation of video images in a computer as collections of color values for the individual dots, or "pixels," in the image. Since video pictures often contain large areas of a single color, this representation may lead to lists of hundreds, or even thousands, of occurrences of the identical color value, one after another. Such a sequence can be represented very compactly, using just two numbers: the color value and the length of the run.

runEncode compresses a list by dividing it into runs of occurrences of a single element, and returns a list of the runs, each represented as a pair containing the element and the length of its run:

```
In[1]:= runEncode[{9, 9, 9, 9, 9, 4, 3, 3, 3, 3, 5, 5, 5, 5, 5, 5}]

Out[1]= {{9, 5}, {4, 1}, {3, 4}, {5, 6}}
```

Given list L, we just *assume* that runEncode[Rest[L]] gives the compressed form of the tail of L (call it R), and ask ourselves: Given the list L and the list R, how can we compute runEncode[L]? Let x be $L[[1]]$, and consider the cases:

1. R might be { }, if L has one element. In this case, $L = \{x\}$ and runEncode[L] = {{x, 1}}.

2. If the length of L is greater than 1, R has the form {{y, k}, ...}, and there are two cases:

   • $y = x$: Then runEncode[L] = {{y, $k + 1$}, ...}.

   • $y \neq x$: runEncode[L] = {{x, 1}, {y, k}, ...}.

If we allow L itself to be empty, this leads to these clauses:

```
runEncode[{}] := {}
runEncode[{x_}] := {{x, 1}}
runEncode[{x_, r___}] :=
   Module[{R = runEncode[{r}]},            (* R = {{y, k},...} *)
      Module[{p = First[R], rst = Rest[R]},   (* p = {y, k}        *)
         If[x == First[p],                     (* First[p] = y      *)
            Join[{{x, p[[2]]+1}}, rst],        (* p[[2]] = k        *)
            Join[{{x, 1}}, R]]]]
```

This can be made a lot clearer by using a transformation rule; replace the last clause by:

```
runEncode[{x_, r__}] :=
    runEncode[{r}] /. {{y_, k_}, s___} ->
                      If[x==y,
                         {{x, k+1}, s},
                         {{x, 1}, {y, k}, s}]
```

Incidentally, a program for this problem, due to Frank Zizza of Willamette College, won an honorable mention in the programming contest at the 1990 *Mathematica* Conference. It uses no recursion, just repeated substitution.

```
runEncode[L_] := Map[({#, 1})&, L] //.
    {x___, {y_, i_}, {y_, j_}, z___} -> {x, {y, i+j}, z}
```

Impressively clever, but our recursive version is much more efficient on most examples. Wolfram gives a different recursive program for this problem [Wolfram 1991, p. 13].

maxima

Consider the function maxima (page 97), which, given a list of numbers, produces a list of those numbers greater than all those that precede them.

```
In[1]:= maxima[{9, 2, 10, 3, 14, 9}]

Out[1]= {9, 10, 14}
```

We again start by assuming that we can easily compute maxima[Rest[*L*]] for any list *L*, and then ask ourselves: How can we compute maxima[*L*] starting from maxima[Rest[*L*]]? The answer is: remove any values not greater than First[*L*], then put First[*L*] at the beginning of the result.

```
In[2]:= maxima[Rest[{9, 2, 10, 3, 14, 9}]]

Out[2]= {2, 10, 14}


In[3]:= Select[%, (#>9)&]

Out[3]= {10, 14}


In[4]:= Join[{9}, %]

Out[4]= {9, 10, 14}
```

Again, the base case ({ }) needs to be accounted for, and we end up with:

```
maxima[{}] := {}
maxima[{x_, r___}] := Join[{x}, Select[maxima[{r}], (#>x)&]]
```

The lesson of this section—and it is an important one—is not to worry about how the recursive cases are computed: *assume* that they work, and just think about how to compute the value you want from the result of the recursive call.

interleave

One kind of problem to which *Mathematica*'s built-in operations cannot be easily applied is when we need to deal with lists of differing length. For example, consider the function `interleave2`, introduced on page 100, which interleaves two lists, appending the remainder of whichever list is longer to the end of the result:

```
In[1]:= interleave2[{a, b, c}, {1, 2, 3, 4, 5}]

Out[1]= {a, 1, b, 2, c, 3, 4, 5}


In[2]:= interleave2[{a, b, c, d, e, f}, {1, 2, 3}]

Out[2]= {a, 1, b, 2, c, 3, d, e, f}
```

We can apply the same principle of recursive thinking to functions of two arguments: *assume* that the function works whenever we apply it to arguments that, *taken together*, are smaller than the original arguments, and then see how to compute the final answer from the result of the recursive call.

For example, in computing `interleave2[`L`, `M`]`, we can safely *assume* the ability to compute either of the three functions `interleave2[Rest[`L`], `M`]`, `interleave2[`L`, Rest[`M`]]`, or `interleave2[Rest[`L`], Rest[`M`]]`. Not all will turn out to be useful—which is why these problems require some thought—but we can use any of them if we wish.

So, let's assume for the moment that both L and M are non-empty lists. Then, the most useful recursive call to make is `interleave2[Rest[`L`], Rest[`M`]]`, for this reason: given that value, we can find `interleave2[`L`, `M`]` by prepending `First[`L`]` and `First[`M`]` to it, in that order. So, one clause of the definition is,

```
interleave2[{x_, r___}, {y_, s___}] :=
   Join[{x, y}, interleave2[{r}, {s}]]
```

We are almost done. We only need to account for the base cases: when L or M is the empty list. These are immediate:

```
interleave2[{}, M_] := M
interleave2[L_, {}] := L
```

and we're done. Note that the case where both arguments are empty lists is handled correctly.

Exercises

1. Write the function `prefixMatch[L, M]` that finds the starting segments of L and M that match.

   ```
   In[1]:= prefixMatch[{1, 2, 3, 4}, {1, 2, 5}]

   Out[1]= {1, 2}
   ```

2. Modify `runEncode` so that it leaves single elements as they are.

   ```
   In[1]:= runEncode2[{9, 9, 9, 4, 3, 3, 5}]

   Out[1]= {{9, 3}, 4, {3, 2}, 5}
   ```

   For this version, you need to assume that the argument is a list of atoms; otherwise, the output would be ambiguous.

3. A slightly more efficient version of `runEncode` uses a three-argument auxiliary function:

   ```
   runEncode[{}] := {}
   runEncode[{x_, r___}] := runEncode[x, 1, {r}]
   ```

   `runEncode[x, k, {r}]` computes the compressed version of $\{\underbrace{x, x, \ldots, x}_{k \text{ times}}, r\}$. Define this three-argument function. (Note that it is legal to have a function be defined for different numbers of arguments; rules in which `runEncode` appears on the left-hand side with two arguments will only be applied when `runEncode` is called with two arguments, and likewise for the three-argument version.) Using the `Timing` function, compare the efficiency of this version with our earlier version; be sure to try a variety of examples, including lists that

Exercises (cont.)

have many short runs and ones that have fewer, longer runs. You'll need to use `Table` to generate lists long enough to see any difference in speed.

4. `maxima` can also be computed more efficiently with an auxiliary function:

```
maxima[{}] := {}
maxima[{x_, r___}] := maxima[x, {r}]
```

The two-argument version has this meaning: `maxima[x, L]` gives the maxima of the list `Join[{x}, L]`. Define it. (Hint: the key point about this is that `maxima[x, L]` = `maxima[x, Rest[L]]` if $x \geq$ `First[L]`.) Compare its efficiency to the version in the text.

5. Write a function `interleave3`, which interleaves three lists, possibly of different lengths. When it gets to the end of any of the three lists, it should append the interleaving of the remainders of the other two lists; when it gets to the end of either of those two, it should append the remainder of the last list:

```
In[1]:= interleave3[{a, b}, {c, d, e, f, g}, {h, i, j}]

Out[1]= {a, c, h, b, d, i, e, j, f, g}
```

6. Write the function `runDecode`, which takes an encoded list produced by `runEncode` and returns its unencoded form:

```
In[1]:= runDecode[{{9, 5}, {4, 1}, {3, 4}, {5, 6}}]

Out[1]= {9, 9, 9, 9, 9, 4, 3, 3, 3, 3, 5, 5, 5, 5, 5, 5}
```

7. The function `kSublists[k, L]` produces a list of the sublists of L of length k. For example:

```
In[1]:= L = {a, b, c, d}
        kSublists[2, L]

Out[1]= {{a, b}, {a, c}, {a, d}, {b, c}, {b, d}, {c, d}}
```

A possible approach to defining `kSublists` is to take `kSublists[k, Rest[L]]` together with the result of prepending `First[L]` to all the elements in `kSublists[k1, Rest[L]]`:

```
In[1]:= L1 = kSublists[2, Rest[L]]}

Out[1]= {{b, c}, {b, d}, {c, d}}}
```

Exercises (cont.)

```
In[2]:= kSublists[1, Rest[L]]}

Out[2]= {{b}, {c}, {d}}}

In[3]:= Map[(Join[{First[L]}, #])&, %]}

Out[3]= {{a, b}, {a, c}, {a, d}}}

In[4]:= Join[%, L1]}

Out[4]= {{a, b}, {a, c}, {a, d}, {b, c}, {b, d}, {c, d}}}
```

Program kSublists.

Solutions

1.
```
prefixMatch[L_, {}] := {}
prefixMatch[{}, M_] := {}
prefixMatch[{x_, r___}, {x_, s___}] :=
    Join[{x}, prefixMatch[{r}, {s}]]
prefixMatch[{x_, r___}, {y_, s___}] := {}
```

2.
```
runEncode2[{}] := {}
runEncode2[{x_}] := {x}
runEncode2[{x_, r__}] :=
    runEncode2[{r}] /. {{{y_, k_}, s___} ->
                        If[x==y, {{x, k+1}, s}, {x, {y, k}, s}],
                        {y_, s___} ->
                        If[x==y, {{x, 2}, s}, {x, y, s}]]
```

3.
```
runEncode[{}] := {}
runEncode[{x_, r___}] := runEncode[x, 1, {r}]

runEncode[x_, k_, {}] := {{x, k}}
runEncode[x_, k_, {x_, r___}] := runEncode[x, k+1, {r}]
runEncode[x_, k_, {y_, r___}] :=
    Join[{{x, k}}, runEncode[y, 1, {r}]]
```

4.
```
maxima[{}] := {}
maxima[{x_, r___}] := maxima[x, {r}]

maxima[x_, {}] := {x}
maxima[x_, {y_, r___}] := maxima[x, {r}] /; x >= y
maxima[x_, {y_, r___}] := Join[{x}, maxima[y, {r}]]
```

5. *Note:* The two-argument form of `interleave` is defined above.

```
interleave2[{}, M_] := M
interleave2[L_, {}] := L
interleave2[{x_, r___}, {y_, s___}] :=
    Join[{x, y}, interleave2[{r}, {s}]]

interleave3[{}, M_, N_] := interleave2[M, N]
interleave3[L_, {}, N_] := interleave2[L, N]
interleave3[L_, M_, {}] := interleave2[L, M]
interleave3[{x1_, r1___}, {x2_, r2___}, {x3_, r3___}] :=
        Join[{x1, x2, x3}, interleave3[{r1}, {r2}, {r3}]]
```

6.
```
runDecode[{}] := {}
runDecode[{{x_, k_}, r___}] :=
    Join[Table[x, {k}], runDecode[{r}]]
```

7.
```
kSublists[0, L_] := {{}}
kSublists[k_, {}] := {}
kSublists[k_, L_] :=
    Module[{ksubs = kSublists[k-1, Rest[L]]},
        Join[Map[(Join[{First[L]}, #])&, ksubs],
            kSublists[k, Rest[L]]]]
```

7.4 Recursion and Symbolic Computations

Chapter 5 emphasized the idea that expressions and data are really the same things in *Mathematica*. All that distinguishes an expression like 2 + 3 from one like x + y is that *Mathematica* has rules for rewriting 2 + 3 but not for x + y.

Symbolic computations are those that transform expressions into other expressions. Programming symbolic computations is no different from any other type of computation: you write rewrite rules, use local transformations, built-in operations, and recursion.

We will illustrate symbolic computation with what may be the most famous recursive definition of them all: the differential calculus. Every elementary calculus book includes rules for finding derivatives of functions. Generally, they assume that there are expressions *u* containing the variable *x*, and they show how to find the derivative of u with respect to x, $\frac{du}{dx}$, by giving rules like these:

$$\frac{d(c)}{dx} \quad = \quad 0, \text{ for } c \text{ a constant}$$

$$\frac{d(x^n)}{dx} \quad = \quad nx^{n-1}$$

$$\frac{d(u+v)}{dx} \quad = \quad \frac{du}{dx} + \frac{dv}{dx}$$

If we think of $\frac{du}{dx}$ as a function $\frac{d}{dx}$ being applied to an expression u, then these rules would be written:

$$\frac{d}{dx}(c) \quad = \quad 0, \text{ for } c \text{ a constant}$$

$$\frac{d}{dx}(x^n) \quad = \quad nx^{n-1}$$

$$\frac{d}{dx}(u + v) \quad = \quad \frac{d}{dx}(u) + \frac{d}{dx}(v)$$

In this form, it is clear that $\frac{d}{dx}$ is just a recursively-defined function from expression to expression, and we can render this function in *Mathematica* directly.

```
In[1]:= ddx[c_]  := 0
        ddx[x^n_]  := n x^(n-1)
        ddx[u_ + v_]  := ddx[u] + ddx[v]


In[2]:= ddx[x^2 + x^3]

Out[2]= 2 x + 3 x
                  2
```

So far, so good, but there are two problems with this—one big and the other bigger. The bigger one is that this function gives completely wrong answers for many expressions.

```
In[3]:= ddx[5 x^3]

Out[3]= 0
```

We haven't been careful enough about our base cases. Specifically, the first rule handles *all* expressions not specifically treated elsewhere, instead of just those for which it was intended: constants. This is easily remedied, by replacing that rule with one that makes sure its argument is a number.

```
In[4]:= ddx[c_]  =.     (* remove the rule ddx[c_]  := 0 *)


In[5]:= ddx[c_?NumberQ]  := 0
```

Now, ddx always gives an answer that is correct, but it still misses a lot of cases.

```
In[6]:= ddx[5 x^3]
```

$$Out[6]= ddx[5\ x^3]$$

At this point, we need to take a close look at the cases we want to cover; that is, the precise set of expressions we want `ddx` to differentiate. We can define this set using recursion.

An expression (that `ddx` can differentiate) is one of the following:

- A number.

- The variable x.

- A sum $u + v$, where u and v are expressions.

- A difference $u - v$ of two expressions.

- A product $u\ v$ of two expressions.

- A quotient $u\ /\ v$ of two expressions.

- A power $u \hat{}\ n$ of an expression and a number.

Now, let's start from scratch, dealing systematically with all the cases.

```
In[7]:= Clear[ddx]
        ddx[c_?NumberQ] := 0
        ddx[x] := 1
        ddx[u_ + v_] := ddx[u] + ddx[v]
        ddx[u_ - v_] := ddx[u] - ddx[v]
        ddx[u_ v_] := u ddx[v] + v ddx[u]
        ddx[u_ / v_] := (v ddx[u] - u ddx[v]) / v^2
        ddx[u_^c_?NumberQ] := c u^(c-1) ddx[u]
```

```
In[8]:= ddx[5 x^3]
```

$$Out[8]= 15\ x^2$$

One interesting point to note here is that one of the cases from our first definition—x^n—does not appear here in that form. Still, this case is handled correctly, as we've just seen. The trace makes it clear why:

```
In[9]:= Trace[ddx[x^3], ddx]
```

$$Out[9]= \{ddx[x^3],\ 3\ x^{3\ -\ 1}\ ddx[x],\ \{ddx[x],\ 1\}\}$$

In other words, it is handled as part of a more general case, namely u^n for arbitrary u. Our new rule works in additional cases:

```
In[10]:= ddx[(x + 2x^2)^4]
```

$$Out[10]= 4 \ (1 + 4 \ x) \ (x + 2 \ x^2)^3$$

It is very common to make the mistake of covering cases in more ways than one; for example, many calculus books include both the case cx^n and, separately, the cases for c, x, u^n, and uv, which together can handle expressions of the form cx^n. It is harmless, but a more systematic treatment of the cases avoids giving extra rules, while also ensuring that all cases are covered.

Finally, we might want to make use of simple algebraic identities to simplify this code. For example, the rule for quotients is already covered by the rules for products and powers, since $u/v = uv^{-1}$. Similarly, $u - v = u + (-1)v$. So change those two rules:

```
In[11]:= ddx[u_ - v_] := ddx[u + -1 v]
         ddx[u_ / v_] := ddx[u (v^-1)]
```

```
In[12]:= ddx[x^2/(x-1)]

         $IterationLimit::itlim:
             Iteration limit of 4096 exceeded.
                     2
                    x
Out[12]= Hold[ddx[———]]
                  -1 + x
```

In other words, this computation was going on forever. Alas, here *Mathematica*'s own simplification rules defeated us, as we can see by looking at the rules for ddx.

```
In[13]:= ?ddx

Global`ddx

ddx[x] := 1

ddx[(c_)?NumberQ] := 0

ddx[(u_) - (v_)] := ddx[u - v]

ddx[(u_) + (v_)] := ddx[u] + ddx[v]

ddx[(u_)/(v_)] := ddx[u/v]

ddx[(u_)*(v_)] := u*ddx[v] + v*ddx[u]

ddx[(u_)^(c_)?NumberQ] := c*u^(c - 1)*ddx[u]
```

When we entered the new rules, *Mathematica rewrote the right-hand sides*, so that the rules just say, in effect, "rewrite ddx[*uv*] to ddx[*uv*]" and "rewrite ddx[*u/v*] to ddx[*u/v*]." This fails to satisfy our rule that recursive calls can only be made to *smaller* values.

On the other hand, let's try just deleting those rules entirely,

```
In[14]:= ddx[u_ - v_] =.
         ddx[u_ / v_] =.
```

and see what happens.

```
In[15]:= Simplify[ddx[x/(x-1)]]

Out[15]= -(-1 + x)^-2
```

Again, we need to take into account what *Mathematica* is doing with the expressions we enter. It turns out that it actually reads expressions of the form u/v as $u(v^{-1})$ and expressions of the form $u - v$ as $u + (-1v)$. So when we entered ddx[x/(x-1)], *Mathematica* read it as ddx[x ((x + -1) ^ -1)], as we can see here:

```
In[16]:= FullForm[Hold[ddx[x/(x-1)]]]

Out[16]//FullForm= Hold[ddx[Times[x, Power[Plus[x, -1], -1]]]]
```

In this form, the existing rules apply.

Exercises

1. Add rules to ddx for the trigonometric functions sine, cosine, and tangent.

2. When variables other than *x* are present in an expression, the rules for differentiation with respect to *x* actually don't change. That is, expressions that have no occurrences of *x* are treated like constants. So there should be a rule that says ddx[*u*] = 0, if *x* doesn't occur anywhere in *u*. Define the function nox[*e*] to return True if x does not occur within *e*, then add the new rule for those expressions. You will need to use the comparison function =!=, called UnsameQ, which tests whether two symbols are unequal; the usual Unequal comparison (!=) cannot be used to compare symbols.

3. Define a two-argument version of ddx whose second argument is the variable with respect to which the derivative of the expression is to be computed. Thus, ddx[*u*, x] will be the same as our current ddx[*u*]. You'll need to determine when an expression has no occurrences of a variable; you can use the built-in function FreeQ.

Solutions

1.

```
ddx[c_?NumberQ] := 0
ddx[x] := 1
ddx[u_ + v_] := ddx[u] + ddx[v]
ddx[u_ - v_] := ddx[u] - ddx[v]
ddx[u_ v_] := u ddx[v] + v ddx[u]
ddx[u_ / v_] := (v ddx[u] - u ddx[v]) / v^2
ddx[u_ ^ c_?NumberQ] := c u^(c-1) ddx[u]

ddx[sin[u_]] := cos[u] * ddx[u]
ddx[cos[u_]] := -sin[u] * ddx[u]
ddx[tan[u_]] := ddx[sin[u]/cos[u]]
```

2.

```
ddx[c_?NumberQ] := 0
ddx[x] := 1
ddx[u_ + v_] := ddx[u] + ddx[v]
ddx[u_ - v_] := ddx[u] - ddx[v]
ddx[u_ v_] := u ddx[v] + v ddx[u]
ddx[u_ / v_] := (v ddx[u] - u ddx[v]) / v^2
ddx[u_ ^ c_?NumberQ] := c u^(c-1) ddx[u]

ddx[u_] := 0 /; nox[u]
```

```
nox[c_?NumberQ] := True
nox[x] := False
nox[y_] := True /; Head[y] == Symbol && y =!= x
nox[u_ + v_] := nox[u] && nox[v]
nox[u_ - v_] := nox[u] && nox[v]
nox[u_ v_] := u nox[v] && v nox[u]
nox[u_ / v_] := nox[u] && nox[v]
nox[u_ ^ c_?NumberQ] := nox[u]
```

3.
```
ddx[c_?NumberQ, y_] := 0
ddx[x_, x_] := 1
ddx[y_, x_] := 0 /; FreeQ[y, x]
ddx[u_ + v_, x_] := ddx[u, x] + ddx[v, x]
ddx[u_ - v_, x_] := ddx[u, x] - ddx[v, x]
ddx[u_ v_, x_] := u ddx[v, x] + v ddx[u, x]
ddx[u_ / v_, x_] := (v ddx[u, x] - u ddx[v, x]) / v^2
ddx[u_ ^ c_?NumberQ, x_] := c u^(c-1) ddx[u, x]
```

7.5 Gaussian Elimination

An extremely common problem in mathematical computation is to solve a linear system S of the form:

$$
\begin{aligned}
E_1 : \quad & a_{11}x_1 \;+\; \cdots \;+\; a_{1n}x_n \;=\; b_1 \\
E_2 : \quad & a_{21}x_1 \;+\; \cdots \;+\; a_{2n}x_n \;=\; b_2 \\
& \;\;\vdots \qquad\qquad\qquad\;\; \vdots \qquad\;\; \vdots \\
E_n : \quad & a_{n1}x_1 \;+\; \cdots \;+\; a_{nn}x_n \;=\; b_n
\end{aligned}
$$

for the values of $x_1, \ldots, x_n$ (called the 'unknowns'), where the a_{ij} and b_i are constants.

Mathematica has a built-in function LinearSolve that will usually give the correct answer. For example, the system:

$$
\begin{aligned}
x_1 \;+\; 2x_2 \;&=\; 3 \\
4x_1 \;+\; 5x_2 \;&=\; 6
\end{aligned}
$$

has the solution $x_1 = -1$, $x_2 = 2$. Here is how to invoke LinearSolve (remember that, as discussed on page 64, matrices in *Mathematica* are represented as lists of lists):

```
In[1]:= m = {{1, 2}, {4, 5}}

Out[1]= {{1, 2}, {4, 5}}

In[2]:= b = {3, 6}

Out[2]= {3, 6}
```

```
In[3]:= LinearSolve[m, b]
```

```
Out[3]= {-1, 2}}
```

So why learn to program it yourself? Because LinearSolve—like any algorithm for this problem—doesn't always work, and when confronted with a system for which it fails, your only recourse will be to write your own program.

The Hilbert matrices, containing entries $h_{ij} = \frac{1}{i+j-1}$, give LinearSolve fits:

```
In[4]:= m = N[Table[1/(i+j-1), {i,15}, {j,15}]];
```

```
In[5]:= b = Table[Random[], {15}];
```

```
In[6]:= xs = LinearSolve[m, b]
```

```
        LinearSolve::luc:
           Warning: Result for LinearSolve
              of badly conditioned matrix {<<15>>}
              may contain significant numerical errors.
```

$$Out[6]= \{8.25197 \times 10^{7},\ -8.57592 \times 10^{9},\ 1.63091 \times 10^{11},$$

$$4.74192 \times 10^{11},\ -4.0761 \times 10^{13},\ 4.95164 \times 10^{14},$$

$$-3.14358 \times 10^{15},\ 1.24735 \times 10^{16},\ -3.31147 \times 10^{16},$$

$$6.05298 \times 10^{16},\ -7.65659 \times 10^{16},\ 6.59206 \times 10^{16},$$

$$-3.68956 \times 10^{16},\ 1.21104 \times 10^{16},\ -1.76954 \times 10^{15}\}$$

```
In[7]:= m . xs - b
```

```
Out[7]= {0.0000357725, 0.0460308, -0.0369039, -0.168,
           0.480565, 0.0967798, -0.089265, -0.640904,
           0.0760878, -0.225912, 0.321558, -0.154709,
           -0.242561, 0.125225, 0.279043}
```

The last result should, of course, contain all zeros.

In this section, we will show a simple and classic method, called *Gaussian elimination*, to solve such a system. Our method, unfortunately, will also fail on the

Hilbert matrix, but we'll revisit the problem in Chapter 9 and show how a variant of this method can solve it. Before that, we will also cover it again in Chapter 8 as an example of programming with iteration.

For now, consider that we have the system shown above. By the principle of recursion, we can assume the ability to solve any 'smaller' system—in particular, any system of $n - 1$ equations in $n - 1$ unknowns—and ask our usual question: how can the ability to solve smaller systems be used to solve this system?

The idea behind Gaussian elimination is to *eliminate* all occurrences of x_1 from the equations $E_2, \ldots, E_n$. For example, here's how to eliminate it from E_2:

1. Take the product of E_1 by $\frac{a_{21}}{a_{11}}$.

$$\left(\frac{a_{21}}{a_{11}}\right)(a_{11}x_1 + a_{12}x_2 + \cdots + a_{1n}x_n) = \left(\frac{a_{21}}{a_{11}}\right)b_1$$

which simplifies to:

$$a_{21}x_1 + \left(\frac{a_{21}}{a_{11}}\right)a_{12}x_2 + \cdots + \left(\frac{a_{21}}{a_{11}}\right)a_{1n}x_n = \left(\frac{a_{21}}{a_{11}}\right)b_1$$

2. and subtract it from equation E_2:

$$
\begin{array}{ccccccccc}
 & a_{21}x_1 & + & a_{22}x_2 & + & \cdots & + & a_{2n}x_n & = & b_2 \\
- (& a_{21}x_1 & + & \left(\frac{a_{21}}{a_{11}}\right)a_{12}x_2 & + & \cdots & + & \left(\frac{a_{21}}{a_{11}}\right)a_{1n}x_n & = & \left(\frac{a_{21}}{a_{11}}\right)b_1 &)
\end{array}
$$

$$\left(a_{22} - \left(\tfrac{a_{21}}{a_{11}}\right)a_{12}\right)x_2 + \cdots + \left(a_{2n} - \left(\tfrac{a_{21}}{a_{11}}\right)a_{1n}\right)x_n = b_2 - \left(\tfrac{a_{21}}{a_{11}}\right)b_1$$

We have obtained an equation having only $n - 1$ variables. Now do this for every equation: Transform E_i, for all i from 2 to n, to $E_i' = E_i - \left(\frac{a_{i1}}{a_{11}}\right)E_1$. Call this new system of equations S'.

We're almost there: We can (recursively) find the solution to the system S', obtaining the values of $x_2, \ldots, x_n$. Now compute x_1 by:

$$x_1 = \frac{b_1 - (a_{12}x_2 + \cdots + a_{1n}x_n)}{a_{11}}$$

In programming this procedure, the system will be represented by the $n \times n$ matrix of coefficients, together with the vector of the b_i. In fact, it is somewhat more convenient to represent the entire system as one $(n + 1) \times n$ matrix (called the **augmented matrix**), with the b_i included as the last column. We will define solve[S], where S is such an $(n + 1) \times n$ matrix, to return a list of the values of the n unknowns $x_1, \ldots, x_n$. Once we understand the algorithm, the programming is just a lot of list manipulating.

```
solve[S_] :=
  Module[{E1 = First[S], x2toxn = solve[elimx1[S]]},
    Module[{b1 = Last[E1], a11 = First[E1],
           a12toa1n = Drop[Rest[E1], -1]},
      Join[{(b1 - a12toa1n . x2toxn) / a11}, x2toxn]]]
```

We need to define `elimx1[S]`, which produces the smaller system. But first, let's not forget the base case, $n = 1$ (i.e., $a_{11}x_1 = b_1$), which is trivial to solve.

```
solve[{{a11_, b1_}}] := {b1/a11}
```

Again, the elimination phase takes each row of the matrix.

$$a_{i1}, \quad a_{i2}, \quad \ldots, \quad a_{in}, \quad b_i$$

and transforms it to:

$$a_{i2} - \left(\frac{a_{i1}}{a_{11}}\right)a_{12}, \quad \cdots, \quad a_{in} - \left(\frac{a_{i1}}{a_{11}}\right)a_{1n}, \quad b_i - \left(\frac{a_{i1}}{a_{11}}\right)b_1$$

Here is the code:

```
elimx1[S_] := Map[subtractE1[S[[1]], #]&, Rest[S]]
subtractE1[E1_, Ei_] :=
  Module[{z = Ei[[1]]/E1[[1]]},
    Module[{newE1 = z * Rest[E1]},
      Rest[Ei] - newE1]]
```

Finally, we will overload this version of `solve` so that it works like the built-in `LinearSolve`—that is, it accepts a matrix of coefficients and a column vector as arguments. This will avoid having to compute the transposition manually.

```
In[8]:= solve[A_, B_] := solve[Transpose[Join[Transpose[A], {B}]]]
```

Now, `solve` works as advertised:

```
In[9]:= solve[{{1, 2}, {4, 5}}, {3,6}]
Out[9]= {-1, 2}
```

Exercises

1. The Gaussian elimination procedure can fail for a variety of reasons. We've already mentioned that it will not give good results for the Hilbert matrix, but the reason for this is quite subtle and we'll postpone our explanation to Section 9.4.2. Another reason it can fail is that there may be no unique solution at all; consider, for example, the system:

$$
\begin{aligned}
x_1 + x_2 &= 0 \\
2x_1 + 2x_2 &= 0
\end{aligned}
$$

Here, the two equations are essentially the same, so we don't have enough information to determine x_1 and x_2 uniquely. This problem is inherent in this system and cannot be solved, no matter how fancy an algorithm we devise.

There is, however, one kind of problem we should be able to overcome, illustrated by this system:

$$
\begin{aligned}
x_1 + x_2 + x_3 &= 1 \\
x_1 + x_2 + 2x_3 &= 2 \\
x_1 + 2x_2 + 2x_3 &= 1
\end{aligned}
$$

Our elimination procedure will produce the smaller system:

$$
\begin{aligned}
x_3 &= 1 \\
x_2 + x_3 &= 0
\end{aligned}
$$

which corresponds to the call `solve[{{0, 1, 1}, {1, 1, 0}}]`. This system obviously does have a solution, but `solve` will fail because, in attempting to eliminate x_2, it will compute the new coefficient of x_3 as $1 - (\frac{1}{0})1$, which involves a division by zero.

The solution to this problem is easily found by observing that in any system of equations, changing the order of the equations does not change the solution. Thus, the above system is equivalent to:

$$
\begin{aligned}
x_2 + x_3 &= 0 \\
x_3 &= 1
\end{aligned}
$$

with which `solve` has no difficulty at all.

Modify `solve` such that it reorders the rows of its argument to ensure that a_{11} is non-zero. (If *every* row has zero as its first element, the system cannot be solved.) This process of reordering the equations is called *pivoting*.

Exercises (cont.)

2. In Exercise 1, suppose A is known to be *upper-triangular*, meaning it has zeroes below the diagonal (formally: $a_{ij} = 0$ for all $i > j$). Define `solveUpper`, having the same arguments as `solve2`, but under the assumption that A is upper-triangular. (This is much simpler than `solve`, since it requires no elimination.) Then define `solveLower`, with the same arguments, but for the case where A is *lower-triangular* (has zeroes *above* the diagonal). `solveLower` should work by manipulating A so as to make it upper triangular, and then calling `solveUpper`.

3. Suppose we could find lower-triangular and upper-triangular matrices, L and U, such that $A = LU$. Then for any vector B, we could easily compute `solve2[A, B]` by computing `solveUpper[U, solveLower[L, B]]`. (Note that for a vector X to be a solution to the original system just means $AX = B$. But this implies that $LUX = B$, which implies that there is a vector Y such that $LY = B$ and $UX = Y$; `solveLower[L, B]` is Y, and `solveUpper[U, Y]` is X.) So, given a square matrix A, if we can find such a decomposition of A, then we can efficiently solve $AX = B$ for any given B. In fact, finding this so-called **LU-decomposition** of A is very similar to doing Gaussian elimination. Specifically, suppose that A' is the smaller matrix produced by the elimination process (that is, the coefficients in the system S'), and suppose further that $A' = L'U'$, where L' is lower-triangular and U' is upper-triangular (so L' and U' can be computed recursively). Then consider the following two matrices U and L:

- U is U' with the first row of coefficients of A added as the top row, and zeroes added as the left column:

$$U = \begin{array}{|cccc|} \hline a_{11} & a_{12} & \cdots & a_{1n} \\ 0 & & & \\ \vdots & & U' & \\ 0 & & & \\ \hline \end{array}$$

U is, of course, upper-triangular.

- L is L' with the following changes: add the row $(1, 0, 0, \ldots, 0)$ as the top row. For the left column, add the *multipliers* computed in the elimination process, that is, the quotients a_{i1}/a_{11}:

$$L = \begin{array}{|cccc|} \hline 1 & 0 & \cdots & 0 \\ a_{21}/a_{11} & & & \\ \vdots & & L' & \\ a_{n1}/a_{11} & & & \\ \hline \end{array}$$

Exercises (cont.)

It can be shown that, when this construction works (which it does in the same situations in which `solve` works), $LU = A$.

Program two versions of LU-decomposition:

(a) `LUdecomp1 [A]` returns two matrices L and U, as just described. That is, it returns a list containing these two matrices.

(b) `LUdecomp2 [A]` returns one matrix which contains both L and U, specifically, the matrix $(L - I) + U$, where I is the identity matrix. In other words, forget the diagonal elements of L (which are all 1s) and then just place the elements of L below the diagonal and the elements of U at or above the diagonal in a single matrix.

Solutions

1. The solution to this problem also appears in Section 9.4.2, page 288.

```
solve[S_] :=
   Module[{PS = pivot[S]},
      Module[{E1 = First[PS],
               x2toxn = solve[elimx1[PS]]},
            Module[{b1 = Last[E1],
                    a11 = First[E1],
                    a12toa1n = Drop[Rest[E1], -1]},
               Join[{(b1 - a12toa1n . x2toxn) / a11}, x2toxn]]]]

solve[{{a11_, b1_}}] := {b1/a11}

elimx1[S_] := Map[subtractE1[S[[1]], #]&, Rest[S]]

subtractE1[E1_, Ei_] :=
   Module[{z = Ei[[1]]/E1[[1]]},
            Module[{newE1 = z * Rest[E1]},
                      Rest[Ei] - newE1]]

pivot[Q_] :=
   Module[{p, ST1, pivotrow},
      ST1 = Transpose[Q][[1]];  (* first elmts of rows of Q *)
      p = Position[ST1, x_/; x!= 0];
      If[p == {},
         Print["Matrix is singular"]; Q,
         pivotrow = p[[1]][[1]];
         Join[{Q[[pivotrow]]}, Delete[Q, pivotrow]]
      ]]
```

3. To compute `solveUpper[A, B]`, first recursively compute `solveUpper[A'`, `B']`, where `A'` is the lower-right square submatrix of A, and `B'` is the Rest

of B. This solution gives the values of $x_2, \ldots, x_n$. B[[1]] is equal to the dot product of the top row of A (*i.e.*, A[[1]]) and the vector $x_1, \ldots, x_n$, that is, B[[1]] is equal to A[[1]]*x1 + ... + A[[n]]*xn; it is easy to compute x1 from this formula.

```
solveUpper[{{ann_}}, {bn_}] := {bn/ann}
solveUpper[{A1_, rA__}, {b1_, rB__}] :=
   Module[{subsoln = solveUpper[Map[Rest, {rA}], {rB}]},
      Join[{(b1 - Dot[Rest[A1], subsoln]) / First[A1]}, subsoln]]
```

It is easy to show that if you rotate a matrix by 90 degrees, and turn the vector B upside down, the solution to the resulting system is the same as the solution to the original system, but turned upside-down.

```
rotateMatrix[A_] := Reverse[Map[Reverse, A]]

solveLower[A_, B_] :=
    Reverse[solveUpper[rotateMatrix[A], Reverse[B]]]
```

4.
```
LUdecomp1[S_] :=
   Module[{mults = multipliers[S[[1,1]], Rest[S]]},
      Module[{Sprime = elimx1[mults, Map[Rest, S]]},
         Module[{LU = LUdecomp1[Sprime]},
            {expandL[mults, LU[[1]]], expandU[First[S], LU[[2]]]}]]]

LUdecomp1[{{a11_}}] := { {{1}}, {{a11}} }

expandU[S1_, U_] := Join[{S1}, Map[Join[{0},#]&, U]]

expandL[mults_, L_] :=
      Transpose[expandU[Join[{1}, mults], Transpose[L]]]

elimx1[mults_, subS_] :=
      Table[subS[[i+1]] - mults[[i]] * subS[[1]],
            {i, 1, Length[mults]}]

multipliers[S11_, restS_] :=
      Map[(#/S11)&, Transpose[restS][[1]]]

LUdecomp2[S_] :=
   Module[{soln = LUdecomp1[S]},
      soln[[1]] - IdentityMatrix[Length[S]] + soln[[2]]]
```

7.6 | Trees

Mathematica expressions can be visualized as *upside-down trees*; for example, f[x, y + 1] could be expressed as

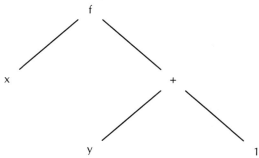

Such structures are called *trees* (they're always drawn upside-down), and they have many uses in programming. In this section, we will discuss a way of representing trees in *Mathematica*, develop some basic functions on trees, and give a well-known application: Huffman encoding trees.

7.6.1 | Binary Trees

First, some terminology: Trees consist of *nodes*, which have *labels* (the symbols f, x, +, y, and 1 in the example above) and some number of *children*, which are themselves nodes (*e.g.*, the nodes labeled x and + are the children of the node labeled f). If a node has no children, it is called a *leaf*, otherwise, it is an *interior node*. The node at the top of the tree is called the *root* of the tree. In the example above, the interior nodes are the ones labeled f and +, and the root is the node labeled f.

More specifically, we will be discussing *binary trees*—trees in which every interior node has two children, called the *left child* and *right child*.

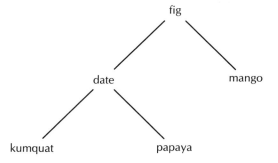

We will be interested in trees whose labels are data values like numbers and strings. The simplest way to represent them in *Mathematica* is to use lists: An interior node is represented by a three-element list containing the node's label and its two children; a leaf node by a one-element list containing the label. For example, the tree above is represented as:

```
{"fig", {"date", {"kumquat"}, {"papaya"}}, {"mango"}}
```

Many—in fact, most—algorithms that operate on trees are recursive. It's natural, because simply *visiting* every node in a tree is a recursive process. For example, suppose we have a tree of strings (like the fruit tree above), and we want to find the alphabetically smallest string in the tree (*i.e.*, the first string in the lexicographic ordering of strings):

```
In[1]:= fruittree =
            {"fig", {"date", {"kumquat"}, {"papaya"}}, {"mango"}};


In[2]:= minInTree[fruittree]

Out[2]= date
```

As usual, we should try not to think about exactly how the function works, but just ask this question: Given the minimum strings in the children of a node, how can we find the minimum for the entire tree? Just pick the minimum among the label of this node and the minima (recursively computed) of its children. The easiest way to find the minimum of a collection of strings is to sort them and take the first element.

```
minInTree[{lab_}] := lab
minInTree[{lab_, lc_, rc_}] :=
   Sort[{lab, minInTree[lc], minInTree[rc]}][[1]]
```

A useful function is the one that determines the *height* of a tree: the distance from the root to the farthest leaf node.

```
height[{lab_}] := 0
height[{lab_, lc_, rc_}] := 1 + Max[height[lc], height[rc]]
```

It would be nice to have a better way to display trees than as lists. In Chapter 10 (page 340), we'll discuss the graphical display of trees, but for now we can at least print them in a nicely indented style.

```
In[3]:= printTree[fruittree]

Out[3]= fig
           date
                kumquat
                papaya
           mango
```

We need an auxiliary function: `printTree[t, k]` prints *t* in indented form, with the entire tree moved over *k* units. To put it another way, it prints *t*, assuming it occurs *k* levels down. We've chosen, arbitrarily, to indent three spaces for each level in the tree.

```
printTree[t_] := printTree[t, 0]

printTree[{lab_}, k_] := printIndented[lab, 3k]
printTree[{lab_, lc_, rc_}, k_] :=
   (printIndented[lab, 3k];
    Map[(printTree[#, k+1])&, {lc, rc}];)
printIndented[x_, spaces_] :=
   Print[Apply[StringJoin, Table["  ", {spaces}]], x]
```

7.6.2 | Huffman Encoding

Computers represent textual information—*i.e.*, lists of characters—as *bit strings*—sequences of 0s and 1s. Especially in the transmission of large amounts of data, it is important to minimize the number of bits used to encode the text.

For simplicity, most of the time strings are represented using *fixed-length* codes—those in which each character is represented by a bit string of the same length. The most common such code, as discussed in Section 3.7, is ASCII. Each character has a number that can be represented in 8 bits; here is part of the table from page 83, giving the 8-bit codes:

ASCII codes		
Character	Decimal	8-bit binary
A	65	01000001
B	66	01000010
E	69	01000101
H	72	01001000
N	78	01001110
O	79	01001111
S	83	01010011
T	84	01010100
(space)	32	00100000

So, for example, the string,

HONEST ABE

is represented as

01001000010011110100111001000101010100011
01010100001000000100000101000010001000101

However, this representation is far from being optimally compact. Better codes are **variable-length codes,** using shorter bit strings for more common characters (just as Morse code uses the shortest code—a single dot—for the most common letter in English: *e*). Given a list of characters and their relative frequencies, the most compact encoding of strings that respect those frequencies is called the **Huffman encoding**. David Huffman showed how to construct this code and represent it using a tree (see [Knuth 1991] or [Sedgewick 1988] for more information). We'll define what Huffman encoding trees are and show how to use them to encode and decode strings, and then show how to construct them.

Simply put, a Huffman encoding tree is a binary tree with characters labeling the leaf nodes. An example is shown in Figure 7.1. Note that the space (b) appears in the tree as an ordinary character, just as it does in the ASCII code.

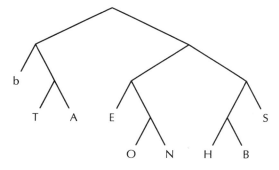

Figure 7.1: A Huffman encoding tree.

Here is how to use the tree to find the code for a character: look for the character in the tree, and record the sequence of branches going from the root to the character. For example, for H, the trip is: right branch, then right, then left, and left again. Recording a 1 for a right branch and a 0 for a left, this gives the code for H: 1100. Here, then, are the codes for all the characters given in this tree:

Char.	Code	Char.	Code	Char.	Code
b	00	E	100	O	1010
A	011	H	1100	S	111
B	1101	N	1011	T	010

Note how the most common characters have shorter codes; for example, the space—which occurs very often—uses only two bits. (Of course, if we included the entire alphabet, our tree would be much bigger, and many letters would have longer codes.)

With this code, the string HONEST ABE is represented by:

11001010101011100111010000111101100

We need to put some more information in our tree. To allow for efficiently finding where a character occurs in the tree, we need to label every interior node with the set of characters labeling leaves below it, as shown in Figure 7.2. Now we can give two programs: one to encode character strings, and one to decode bit strings. The programs we write will assume that Htree contains the tree in Figure 7.2.

```
In[1]:= Htree = {" ABEHONST", {" AT", {" "}, {"AT", {"T"}, {"A"}}},
            {"BEHONS", {"EON", {"E"}, {"ON", {"O"}, {"N"}}},
            {"BHS", {"BH", {"H"}, {"B"}}, {"S"}}}}
```

```
Out[1]= { ABEHONST, { AT, { }, {AT, {T}, {A}}},
         {BEHONS, {EON, {E}, {ON, {O}, {N}}},
         {BHS, {BH, {H}, {B}}, {S}}}}
```

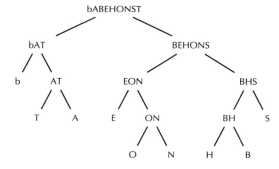

Figure 7.2: A Huffman encoding tree, with interior labels.

We consider encoding character strings first. What we really need is the function to give the bit-string encoding of a single character. Given that function—call it encodeChar—we can easily encode an entire string:

```
In[2]:= encodeString[str_] :=
            Flatten[Map[encodeChar, Characters[str]]]
```

```
In[3]:= encodeString["HONEST ABE"]
```

```
Out[3]= {1, 1, 0, 0, 1, 0, 1, 0, 1, 0, 1, 1, 1, 0, 0, 1,
         1, 1, 0, 1, 0, 0, 0, 0, 1, 1, 1, 1, 0, 1, 1, 0, 0}
```

So how do we encode a single character? The method is essentially recursive: find whether the character occurs in the left or the right subtree, recursively find its code in that subtree, and then prepend a 0 if it was in the left or a 1 if it was in the right. For example, consider H again: we can tell from Htree that H occurs in the right subtree; within that subtree, its code is 100 (right, left, left); since it was in the right subtree, we prepend a 1, to get 1100.

So let encodeChar have two arguments—the character and the Huffman tree.

```
encodeChar[c_, {_, lc_, rc_}] :=
    If[stringMemberQ[First[lc], c],      (* if c in left subtree *)
       Join[{0}, encodeChar[c, lc]],     (* prepend 0 *)
       Join[{1}, encodeChar[c, rc]]]     (* otherwise prepend 1 *)
```

The base case is when we reach a leaf; of course, if the character is in the tree at all—which we are assuming—then it *must* be the label on this leaf, so we don't have to check:

```
encodeChar[_, {_}] := {}
```

Finally, we can give a one-argument version of encodeChar that uses Htree.

```
encodeChar[c_] := encodeChar[c, Htree]
```

Decoding of messages works similarly. We use the list of bits to guide our path down the tree, and when we get to a leaf we 'emit' that character and start over at the root. Again, we will use a function with two arguments: the list of bits, and the tree. There are two cases: when we are at a leaf, we've reached the end of the encoding of a character; otherwise, we choose the left or right subtree, depending upon the next bit in the code.

```
decode[code_, {ch_}] :=        (* leaf node - label is character *)
    StringJoin[ch, decode[code, Htree]]
decode[{0, r___}, {_, lc_, _}] := decode[{r}, lc]
decode[{1, r___}, {_, _, rc_}] := decode[{r}, rc]
decode[{}, _] := ""
```

As usual, we can then give the desired one-argument form.

```
decode[code_] := decode[code, Htree]
```

There is an important point to notice here: In Huffman codes, we always know when a character's code ends. But how? The decode function breaks up the code into characters in some way, but how do we know it's the only possible way?

In fact, a bit of thought will convince you that it must be, because Huffman codes have an interesting property: No character's code can be extended to be the code of another character. For example, no character's code begins with 00, which is the code for space, except space itself; and none begins with 100 except letter "E"s. This property implies that our decoding algorithm finds the *unique* decoding of a string of bits.

Finally, we discuss how Huffman trees are constructed. This is actually very simple—and not really recursive—so we'll describe the method and leave the programming as an exercise.

Keep in mind that the code for a character should be based on a set of *frequencies* of the characters, given at the outset. For example, these might be the frequencies of the characters in our example, based on their occurrences in a large body of English writing (not just our sample phrase).

Characters	Frequency
space	6
E	5
S, T, A	3
H, O, N	2
B	1

So now suppose we are given the list of characters along with their frequencies. For purposes of the algorithm, it's better for us to think that what we've gotten is a list of *trees*, each of which has only a single node, which is labeled by a letter and its frequency.

```
{ {{{b}, 6}}, {{{A}, 3}}, {{{B}, 1}}, {{{E}, 5}}, ...}
```

Still thinking of this as a list of trees, the frequency of each character is called the *weight* of the node containing that character. What we want to do is to combine these single-node trees into larger trees, and keep doing it until they have all been joined into one big tree. So repeatedly perform the following operation on the list of trees.

> Suppose t1 = {{cl1, w1}, ...} and t2 = {{cl2, w2}, ...} are the trees in the list with the lowest weights (*i.e.*, w1 and w2 are as small as possible). Remove them from the list, and replace them by the single tree t = {{Join[cl1, cl2], w1+w2}, t1, t2}.

This operation always reduces the number of trees in the list by one. When there is only one tree in the list, that is the Huffman encoding tree for these characters. Or rather it is *a* Huffman encoding tree: The algorithm does not specify how to choose when there are more than two trees of minimal weight, nor in which order to place those two trees once they're chosen, so there are actually many trees that might result. Huffman proved that they all give equally compact representations of bit strings.

Let's see how this works for our example. To make it easier to read, we'll draw the trees instead of writing them in *Mathematica* list notation.

1. Start with,

    {   b,6   A,3   B,1   E,5   H,2   N,2   O,2   S,3   T,3   }

2. Pick H and B (we could have picked N or O instead of H, but we picked H).

    {   b,6   A,3   BH,3   E,5   N,2   O,2   S,3   T,3   }
                            /\
                         H   B

We've dropped the weights from the H and B nodes, since they won't contribute any more to the algorithm.

3. Now we have to choose N and O (although we can put them in either order).

    {   b,6   A,3   BH,3   E,5   NO,4   S,3   T,3   }
                       /\   /\
                     H   B   O   N

4. We have four trees of weight 3. We (arbitrarily) choose T and A.

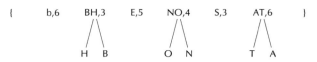

5. Now we join the BH tree with the S tree.

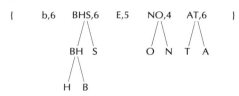

6. Join E with NO,

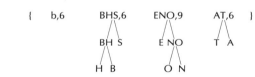

7. And b with AT,

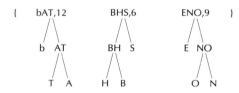

8. BHS with ENO.

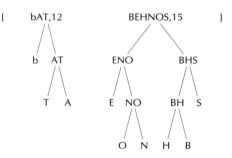

Finally, we join the last two trees, yielding the tree shown in Figure 7.2.

Exercises

1. Suppose you have a tree all of whose labels are numbers. Write a function to sum all the labels.

```
In[1]:= numbertree = {4, {5}, {6, {7}, {9, {10}, {11}}}}
```

```
In[2]:= sumNodes[numbertree]
Out[2]= 52
```

2. Assume now that your tree's labels are all strings. Write a function to concatenate the strings in *depth-first* order. This is the order you get by following the leftmost children of any node as far as possible before visiting their siblings on the right.

```
In[1]:= catNodes[fruittree]
Out[1]= figdatekumquatpapayamango
```

3. A tree is said to be *balanced* if, for every node, the heights of its children differ from one another by no more than 1; that is, the difference in height between the taller child and the shorter is zero or one. (fruittree is balanced, but numbertree from Exercise 1 is not.) Note that the condition must hold at *all* nodes, not just the root. Here is a function to test whether a tree is balanced.

```
balanced[{_}] := True
balanced[{_, lc_, rc_}] :=
   balanced[lc] && balanced[rc] &&
      Abs[height[lc] - height[rc]] <= 1
```

However, this is very expensive because of all the computing of heights of subtrees. For example, it first checks the height of the two children of the root (which involves visiting every node in the tree except the root itself), and then it calls balanced on those two children, which then computes the height of *their* children *for the second time*.

To avoid this extra cost, define a function balancedHeight[*t*] that returns a list of two elements: the first is the height of *t*, and the second is a Boolean value saying whether *t* is balanced. Then you can define balanced by

```
balanced[t_] := balancedHeight[t][[2]]
```

Exercises (cont.)

4. listLevel[n, t] gives a list of all the labels in tree t at level n, where the root is at level 0, its children are at level 1, its grandchildren at level 2, and so on.

```
In[1]:= listLevel[2, numbertree]

Out[1]= {7, 9}
```

Write listLevel.

5. In trees of arbitrary degree, one node can have any (finite) number of children. Represent such a tree by a list containing the label of the root and its children. For example, the following tree

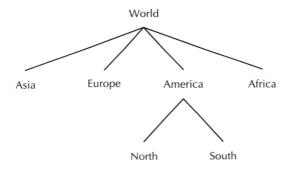

can be represented as

```
{World, {Asia}, {Europe}, {America, {North}, {South}}, {Africa}}
```

Write functions minInTree, height, and printTree for trees of arbitrary degree.

6. Program the function that constructs a Huffman encoding tree, as shown in the last part of this section.

7. Write a more efficient version of encodeString that creates a table of all the encodings of all the characters in a given tree, then applies the table to the list of characters. This table can be represented as a list of rewriting rules, like "a" -> {0, 1, 1}, which can then be applied to the list of characters using ReplaceAll (/.).

Solutions

1.
```
sumNodes[{lab_}] := lab
sumNodes[{lab_, lc_, rc_}] := lab + sumNodes[lc] + sumNodes[rc]
```

2.
```
catNodes[{lab_}] := lab
catNodes[{lab_, lc_, rc_}] :=
    StringJoin[lab, catNodes[lc], catNodes[rc]]
```

3.
```
balanced[t_] := balancedHeight[t][[2]]
balancedHeight[{lab_}] := {0, True}
balancedHeight[{lab_, lc_, rc_}] :=
   Module[{lbh, rbh},
      lbh = balancedHeight[lc];
      If[lbh[[2]], rbh = balancedHeight[rc];
                   If[rbh[[2]] && Abs[lbh[[1]]-rbh[[1]]] <= 1,
                      {Max[lbh[[1]], rbh[[1]]] + 1, True},
                      {0, False}],   (* height doesn't matter *)
                   {0, False}]]   (* height doesn't matter *)
```

4.
```
listLevel[0, t_] := {t[[1]]}
listLevel[n_, {lab_}] := {}
listLevel[n_, {lab_, lc_, rc_}] := Join[listLevel[n-1, lc],
                                        listLevel[n-1, rc]]
```

5.
```
minInTree[{lab_}] := lab
minInTree[{lab_, subtrees__}] :=
    Sort[Join[{lab}, Map[minInTree, {subtrees}]]][[1]]
```

```
In[1]:= height[{lab_}] := 0
        height[{lab_, subtrees__}] := 1 + Apply[Max, Map[height,
        {subtrees}]]
```

```
In[2]:= printTree[t_] := printTree[t, 0]

        printTree[{lab_}, k_] := printIndented[lab, 3k]
        printTree[{lab_, subtrees__}, k_] :=
            (printIndented[lab, 3k]; Map[(printTree[#, k+1])&,
        {subtrees}];)
```

```
In[3]:= printIndented[x_, spaces_] :=
            Print[Apply[StringJoin, Table[" ", {spaces}]], x]
```

6. We've used a slightly different representation for the list of trees than the one shown on page 204. Instead of a node's label containing a list of characters and a number, it contains a string and a number. The only reason for this is that it makes the result come out looking like the tree called Htree (shown on page 201). Note that the algorithm may give different results depending upon how it is programmed, since there are arbitrary choices made at each step. The result of applying our function constructHTree to the initial list of trees shown on page 205 (which we've included here as testlist) is different from Htree.

In our solution, we solve the problem of finding the two trees of smallest weight by keeping the list of trees sorted by weight; then we simply always pick the first two.

```
In[1]:= HTreeSort[trees_] :=
            Sort[trees, (#1[[1, 2]] < #2[[1, 2]])&]


In[2]:= joinHTrees[{{{cl_, wt_}, kids___}}] := {cl, kids}
        joinHTrees[{{{cl1_, wt1_}, kids1___},
                    {{cl2_, wt2_}, kids2___},
                    trees___}] :=
            joinHTrees[
                HTreeSort[{{{StringJoin[cl1, cl2], wt1+wt2},
                            {cl1, kids1}, {cl2, kids2}}, trees}] ]


In[3]:= constructHTree[t_] := joinHTrees[HTreeSort[t]]


In[4]:= htnode[a_,b_] := {{{a,b}}}


In[5]:= testlist =
            Join[htnode[" ",6], htnode["A",3], htnode["B",1],
                 htnode["E",5], htnode["H",2], htnode["N",2],
                 htnode["O",2], htnode["S",3], htnode["T",3]]
Out[5]= {{{ , 6}}, {{A, 3}}, {{B, 1}}, {{E, 5}}, {{H, 2}},
         {{N, 2}}, {{O, 2}}, {{S, 3}}, {{T, 3}}}
```

7. To make the results here comparable to those in the book, we will use Htree from the book as our sample tree.

makeTreeTable[tree] produces a list of rules as described in the problem. encodeString[str, rules] decodes the string according to those rules.

```
In[1]:= Htree =
        {" ABEHONST", {" AT", {" "}, {"AT", {"T"}, {"A"}}},
               {"BEHONS", {"EON", {"E"}, {"ON", {"O"}, {"N"}}},
                      {"BHS", {"BH", {"H"}, {"B"}}, {"S"}}}};

In[2]:= makeTreeTable[prefix_, {ch_}] = {ch->prefix}
        makeTreeTable[prefix_, {_, left_, right_}] :=
            Join[makeTreeTable[Join[prefix, {0}], left],
                makeTreeTable[Join[prefix, {1}], right]]

        makeTreeTable[tree_] := makeTreeTable[{}, tree]

In[3]:= HtreeRules = makeTreeTable[Htree]

Out[3]= {  -> {0, 0}, T -> {0, 1, 0}, A -> {0, 1, 1},
           E -> {1, 0, 0}, O -> {1, 0, 1, 0},
           N -> {1, 0, 1, 1}, H -> {1, 1, 0, 0},
           B -> {1, 1, 0, 1}, S -> {1, 1, 1}}

In[4]:= encodeString[str_, rules_] :=
            Flatten[Characters[str] /. rules]

        encodeString[str_] := encodeString[str, HtreeRules]
```

7.7 Dynamic Programming

In Section 5.5, "dynamic programming" was described as a method in which rewrite rules are added to the global rule base *dynamically*; that is, during the running of a program. A well-known application of this is to speed up the computation of Fibonacci numbers.

The function F defined in Section 7.1 is simple, but quite expensive to execute. For example, here is a table giving the number of additions needed to compute F[n] for various values of n (these are the values FA_n from Exercise 2 on page 171):

n	5	10	15	20	25
F[n]	5	55	610	6765	75025
number of additions	7	88	986	10945	121392

The reason for this excessive cost is easy to see: In the course of computing F[n], there are numbers $m < n$ for which F[m] is computed many times. For instance, F[$n-2$] is computed twice (it is called from F[n] and also from F[$n-1$]), F[$n-3$] three times, and F[$n-4$] five times. This continual recalculation can be eliminated easily by memorizing these values as they're computed—that is, by dynamic programming.

The following definition of function FF is just like the definition of F, but it adds a rule FF[n] = F_n to the global rule base the first time the value is computed. Since *Mathematica* always chooses the most specific rule to apply when rewriting, whenever a future request for FF[n] is made, the new rule will be used instead of the more general rule in the program. Thus, for every n, FF[n] will be *computed* just once; after that, its value will be found in the rule base.

```
In[1]:= FF[0]  := 0
        FF[1]  := 1
        FF[n_] := FF[n] = FF[n-2] + FF[n-1]
```

We can see the change in the trace of FF[4] as compared with that on page 170. Specifically, there is only one evaluation of FF[2] now, since the second evaluation of it is just a use of a global rewrite rule.

```
In[2]:= TracePrint[FF[4], FF[_Integer] | (FF[_] = FF[_] + FF[_])]
          FF[4]
          FF[4] = FF[4 - 2] + FF[4 - 1]
            FF[2]
            FF[2] = FF[2 - 2] + FF[2 - 1]
              FF[0]
              FF[1]
             FF[2]
            FF[3]
            FF[3] = FF[3 - 2] + FF[3 - 1]
              FF[1]
              FF[2]
             FF[3]
           FF[4]

Out[2]= 3
```

Another way to understand what's going on is to look at the global rule base *after* evaluating FF[4].

```
In[3]:= ?FF

Global`FF
FF[0] := 0
FF[1] := 1
FF[2] = 1
FF[3] = 2
FF[4] = 3
FF[n_] := FF[n] = FF[n - 2] + FF[n - 1]
```

The cost of executing this version of F is dramatically lower:

n	5	10	15	20	25
number of additions in FF[n]	4	9	14	19	24

Furthermore, these costs are only for the first time FF[n] is computed; in the future, we can find FF[n] for free, or rather, for the cost of looking it up in the global rule base.

Dynamic programming can be a useful technique, but needs to be used with care. It can entail enormous cost in memory, as the global rule base is expanded to include the new rules.

Exercises

1. Using dynamic programming is one way to speed up F, but another is to use a different algorithm. A much more efficient algorithm than F can be designed, based on these identities.

$$F_{2n} = 2F_{n-1}F_n + F_n^2, \quad \text{for } n \geq 1$$

$$F_{2n+1} = F_{n+1}^2 + F_n^2, \quad \text{for } n \geq 1$$

Program F using these identities.

2. You can still speed up the code for generating Fibonacci numbers by using dynamic programming. Do so, and construct tables, like those in this section, giving the number of additions performed for various n by the two programs you've just written.

3. Calculation of the Collatz numbers c_i^n, from Exercise 3 in section 7.1, can also be speeded up by using dynamic programming. Program the function c this way, and compare its speed to that of your original solution.

Solutions

1.
```
In[1]:= F[0]  := 0
        F[1]  := 1
        F[n_ /; EvenQ[n]] := 2 F[n/2-1] F[n/2] + F[n/2]^2
        F[n_ /; OddQ[n]]  := F[(n-1)/2+1]^2 + F[(n-1)/2]^2
```

2.
```
FF[0] := 0
FF[1] := 1
FF[n_ /; EvenQ[n]] := FF[n] = 2 FF[n/2-1] FF[n/2] + FF[n/2]^2
FF[n_ /; OddQ[n]]  := FF[n] = FF[(n-1)/2+1]^2 + FF[(n-1)/2]^2


In[1]:= Fcount[0]  := 0
        Fcount[1]  := 0
        Fcount[n_ /; EvenQ[n]] :=
                Fcount[n/2-1] + Fcount[n/2] + Fcount[n/2]^2 + 1
        Fcount[n_ /; OddQ[n]]  :=
                Fcount[(n-1)/2+1] + Fcount[(n-1)/2] + 1
```

The number of additions for F is given by,

```
In[2]:= Table[{i, Fcount[i]}, {i, 5, 25, 5}] //TableForm
Out[2]//TableForm= 5     4
                   10    24
                   15    22
                   20    609
                   25    93
```

(By the way, note that this computes only the number of *additions*. However, for this version of Fibonacci, multiplications and divisions should also be counted.) To compute the number of additions when memoizing is used, run the following version of Fcount. It computes answers the same way as the previous Fcount, but then *zeroes out* the value of Fcount [*n*]. This reflects the fact that future calls to Fcount [*n*] will be *free*.

```
In[3]:= Fcount[0]  := 0
        Fcount[1]  := 0
        Fcount[n_ /; EvenQ[n]] :=
           Module[{answer = Fcount[n/2-1] + Fcount[n/2] +
                            Fcount[n/2]^2 + 1},
                Fcount[n] = 0; answer]
        Fcount[n_ /; OddQ[n]]  :=
           Module[{answer = Fcount[(n-1)/2+1] +
                            Fcount[(n-1)/2] + 1},
                Fcount[n] = 0; answer]
```

Number of additions for FF: Note that this cannot be computed simply by generating a table, as above, because the computation of Fcount[5] affects the subsequent computation of Fcount[10], and so on. The correct way to do this is to evaluate Fcount[5], Fcount[10], etc., entering Clear[Fcount] after each evaluation. The result is:

```
Out[3]= 5      3
        10     5
        15     6
        20     7
        25     9
```

3.

```
cc[0, n_] := n
cc[i_, n_] := (cc[i, n] = cc[i-1, n]/2) /; EvenQ[c[i-1, n]]
cc[i_, n_] := (cc[i, n] = 3 cc[i-1, n] + 1) /; OddQ[c[i-1, n]]
```

7.8 Higher-Order Functions and Recursion

As a final wrap-up on recursion, we note that many of the built-in functions discussed in Chapter 4 could be written as user-defined functions using recursion (although they may not run as fast).

A simple example is Map. Since it has the attribute Protected, *Mathematica* won't let us give any new rules for it, so we'll call our version map. map[f, L] applies f to each element of L. This is a simple recursion on the tail of L: if we assume that map[f, Rest[L]] works, map[f, L] is easily obtained from it by adding f[First[L]] to the beginning.

```
In[1]:= map[f_, {}] := {}
        map[f_, {x_, y___}] := Join[{f[x]}, map[f, {y}]]
```

We can quickly check that map does what it was intended to.

```
In[2]:= map[Cos, {1, 2, 3}]

Out[2]= {Cos[1], Cos[2], Cos[3]}
```

Like many of the functions in Chapter 4, this function has a *function* as an argument. This is the first time we've seen *user-defined* higher-order functions.

We will give one more example of a built-in that can be defined using recursion, and leave the rest as exercises.

Consider Nest (page 79): Nest [f, x, n] applies f to x, n times. The recursion is, obviously, on n.

```
In[3]:= nest[f_, x_, 0] := x
        nest[f_, x_, n_] := f[nest[f, x, n-1]]
```

Here is an example of the use of this function.

```
In[4]:= nest[Sin, theta, 4]

Out[4]= Sin[Sin[Sin[Sin[theta]]]]
```

Before leaving this topic, we note that it is sometimes useful to write your own higher-order functions. Consider this problem: You're given a function f whose argument must be an integer in the range $1 \ldots 1000$, and whose result is also in that range, and you need to answer the following question: on average, for a number n_1, how many times can f be applied before it repeats itself? That is, on average, if we form the sequence n_1, $n_2 = f[n_1]$, $n_3 = f[n_2]$, ..., what is the smallest i such that $n_i = n_j$ for some $j < i$? Assume f is so expensive to compute that we prefer to approximate this average by just checking 10 randomly chosen numbers. This technique, known as *random sampling*, is used in many areas where statistical analysis of data is required.

If we had a function repeatCount [n] to answer this question for a particular n, then we might answer the question in this way:

```
Sum[repeatCount[Random[Integer, {1, 1000}]], {10}] / 10
```

So how do we write repeatCount? What we'll do is to define our own higher-order function.

```
In[5]:= repeat[f_, L_, pred_] := L /; pred[Drop[L, -1], Last[L]]
        repeat[f_, L_, pred_] := repeat[f, Append[L, f[Last[L]]], pred]
```

repeat takes an argument list *L*, and repeatedly applies *f* to its last element, and adds that new value to the end, until the predicate pred returns True. repeatCount becomes,

```
In[6]:= repeatCount[f_, n_] := repeat[f, {n}, MemberQ]
```

For example,

```
In[7]:= plus4mod20[x_] := Mod[x+4, 20]
```

```
In[8]:= repeatCount[plus4mod20, 0]
Out[8]= {0, 4, 8, 12, 16, 0}
```

Exercises

1. Write recursive definitions for Fold, FoldList, and NestList.

2. Recall the Collatz sequence from Exercise 3 on page 171. It has long been conjectured that for any $n \geq 1$, the list of numbers c_i^n contains a 1; yet this seemingly elementary fact has so far defeated all attempts at proof. To demonstrate it, define the function tryC[n_] to return the list $c_0^n, c_1^n, \ldots, c_k^n$, such that $c_k^n = 1$ (but no $c_j^n = 1$ for $j < k$). Use repeat. (If the Collatz conjecture is false, there is the danger that your function may compute forever, but don't worry too much about that—the conjecture is known to hold for all $n < 10^{15}$. The interested reader should consult [Lagarias 1985] and [Vardi 1993, Chapter 7].)

3. Recall the notion of a random walk on a two-dimensional lattice from Chapter 3 (see, for example, page 71). Use repeat to define a special kind of random walk, one which continues until it steps onto a location it had previously visited. That is, define landMineWalk as a function of no arguments which produces the list of the locations visited in such a random walk, starting from location (0, 0).

4. One deficiency in the set of built-in operations of *Mathematica* is their inability to deal elegantly with lists of varying length. For example, Plus cannot be applied to lists of unequal length.

Exercises (cont.)

```
In[1]:= {1, 2, 3} + {4, 5, 6, 7, 8}

        Thread::tdlen:
            Objects of unequal length in
                {1, 2, 3} + {4, 5, 6, 7, 8} cannot be combined.

Out[1]= {1, 2, 3} + {4, 5, 6, 7, 8}
```

This problem with built-in operations is why interleave2 couldn't be written easily using them.

You can remedy this situation by defining your own higher-order functions. Define map2 so that map2 [*f*, *g*, *L*, *M*] applies *f* to pairs of elements from *L* and *M*; when it reaches the end of either list, it applies *g* to the remainders of the two lists.

```
In[2]:= map2[Plus, Join, {1, 2, 3}, {4, 5, 6, 7, 8}]

Out[2]= {5, 7, 9, 7, 8}


In[3]:= map2[Plus, ({})&, {1, 2, 3}, {4, 5, 6, 7, 8}]

Out[3]= {5, 7, 9}


In[4]:= map2[{#1, #2}&, Join, {1, 2, 3}, {a, b, c, d, e}]

Out[4]= {{1, a}, {2, b}, {3, c}, d, e}
```

(The latter is almost interleave2[{1, 2, 3}, {a, b, c, d, e}]; it just needs to be flattened.) Define map2.

Solutions

1.
```
fold[f_, x_, {}] := x
fold[f_, x_, {a_, r___}] := fold[f, f[x, a], {r}]


foldList[f_, x_, {}] := {x}
foldList[f_, x_, {a_, r___}] := Join[{x}, foldList[f, f[x, a],
{r}]]


nestList[f_, x_, 0] := {x}
nestList[f_, x_, n_] := Join[{x}, nestList[f, f[x], n-1]]
```

2.
```
repeat[f_, L_, pred_] := L /; pred[Drop[L, -1], Last[L]]
repeat[f_, L_, pred_] := repeat[f, Append[L, f[Last[L]]], pred]

tryC[n_] := repeat[cc[n,#]&, {n}, (#2==1)&]
```

3.
```
landMineWalk[] :=
    repeat[(#+{{0,1},{0,-1},{1,0},{-1,0}}[[Random[Integer,
                                                    {1, 4}]]])&,
        {{0,0}},
        MemberQ[#1, #2]&]
```

4.
```
map2[f_, g_, {}, {}] := {}
map2[f_, g_, L_, {}] := g[L, {}]
map2[f_, g_, {}, M_] := g[{}, M]
map2[f_, g_, {a_, r___}, {b_, s___}] :=
        Join[{f[a, b]}, map2[f, g, {r}, {s}]]
```

7.9 Debugging

Whenever you program, much of your time will be spent in *debugging*—figuring out why your program doesn't work. In this section, we offer a few tips on debugging, and also give some examples of common programming errors to avoid. Though our examples in this section are mainly functions that are defined using recursion, the principles of debugging in *Mathematica*—using its trace facilities and inserting calls to the Print function—apply to the debugging of any *Mathematica* code.

7.9.1 | Tracing Evaluation

We've already seen the two functions Trace and TracePrint. They can be especially useful when you know how to use their second argument. It can have one of several forms: if it is just a symbol, then only those parts of the trace that use rewrite rules for that symbol are shown. If it is a pattern, only those lines of the trace that match the pattern will be printed; an example of this was seen in Section 7.1. If it is a transformation rule, then when the pattern matches a line of the trace, the rule is applied before printing it.

For example, the full TracePrint of F[2] follows. In this section, we will use TracePrint, but the discussion applies equally to Trace:

```
In[1]:= Trace[F[2]]

Out[1]= {F[2], {{2 > 1, True},
            RuleCondition[F[2 - 2] + F[2 - 1], True],
            F[2 - 2] + F[2 - 1]}, F[2 - 2] + F[2 - 1],
         {{2 - 2, -2 + 2, 0}, F[0], 0},
         {{2 - 1, -1 + 2, 1}, F[1], 1}, 0 + 1, 1}
```

Most of this, you'll agree, is not very interesting. We can confine it to only those parts that involve the applications of an F rule by giving F as the second argument to Trace.

```
In[2]:= Trace[F[2], F]

Out[2]= {F[2], F[2 - 2] + F[2 - 1], {F[0], 0}, {F[1], 1}}
```

Perhaps more useful here would be the pattern F[_], which includes all lines in the original trace of the form F[*something*].

```
In[3]:= TracePrint[F[2], F[_]]

Out[3]= 1
```

```
        F[2]
         F[2 - 2]
         F[0]
         F[2 - 1]
         F[1]
```

The example in Section 7.1 uses a pattern which shows all lines of the trace that are either F applied to an integer, or the right-hand side of the recursive rule. Using a transformation rule, we can show just the arguments of the various calls.

```
In[4]:= TracePrint[F[2], F[n_Integer] -> n]

Out[4]= 1
```

```
          2
           0
           1
```

7.9.2 | Printing Variables

The classic debugging method, used in all programming languages, is to insert `Print` statements in the body of a program to show where evaluation is occurring and what the values of variables are at that point. Keep in mind that if *expr* is any expression, the compound expression (`Print[...]`; *expr*) has the same value as *expr*, so it is easy to insert `Print` statements without changing how the program works.

The most common use of `Print` is to show the values of a function's arguments. A rule `f[x_] := ` *expr* can be changed to `f[x_] := (Print[x]; ` *expr*`)` and it will print the value of the argument in each call.

```
In[1]:= F[n_] := (Print[n]; F[n-2] + F[n-1]) /; n > 1


In[2]:= F[4]
Out[2]= 3

        4
        2
        3
        2

```

It may also be useful to know which rule is being applied, if you have `Print` statements in more than one.

```
f[x_, y_] := (Print["Rule 1: ", x, y]; e)
```

7.9.3 | Common Errors

Many of the errors you will see are obvious. For example, one of the most common looks like,

```
Part::partw: Part i of L does not exist.
```

for some integer *i* and list *L*. Here, you are attempting to extract a part of an expression that doesn't have that part; *i.e.*, trying to extract the first element of the empty list. There are a number of variations of this error message, such as the one you get when the subscript is not of the right type. In any case, inserting a `Print` expression should quickly show you what is going on.

Another thing you'll often see is an entire expression returned instead of a value—sometimes the exact same expression you entered.

```
In[1]:= f[n] := Table[i, {i, -n, n}]
```

```
In[2]:= f[10]
Out[2]= f[10]
```

What's the problem? The n in f's argument list is missing the blank. *Mathematica* sees the left-hand side as a pattern that matches the expression f[n] and nothing else—in particular, not f[10]. Of course, when there are no rules to apply to an expression, *Mathematica* is done—it doesn't even know there's an error!

Another very common case where this occurs is when you fail to supply enough arguments to a function,

```
In[3]:= f[{x_, r___}, y_] := If[x<0, {y, f[{r}]}, f[{r}]]
```

```
In[4]:= f[{}, _] := {}
```

```
In[5]:= f[{-5, 4, 17}, -1]
Out[5]= {-1, f[{4, 17}]}
```

or when you supply too many:

```
In[6]:= g[{x_, r___}, y_] := If[x<0, {y, g[r, y]}, g[r, y]]
```

```
In[7]:= g[{}, _] := {}
```

```
In[8]:= g[{-5, 4, 17, 12, 21}, -1]
Out[8]= {-1, g[4, 17, 12, 21, -1]}
```

In the first example, the recursive call to f had just one argument, and there were no rules for this case. In the second, we forgot to put the r in list braces in the recursive call, so g was called with all the elements of r as arguments, giving it too many arguments.

Another very common error is to get your program in a loop—where it seems to go on forever. If this happens when you're doing exercises in this chapter, the chances are that your function is continually making recursive calls and not finishing them. In this case, you will reach *Mathematica*'s limit on the number of recursive calls it allows, which is stored in the variable $RecursionLimit:

```
In[9]:= h[x_] := h[x-1] + h[x+1]


In[10]:= h[0] := 0


In[11]:= h[1]
```

$RecursionLimit::reclim:
    Recursion depth of 256 exceeded.

General::stop:
    Further output of $RecursionLimit::reclim
        will be suppressed during this calculation.

Another possibility is to get the same message, but for $IterationLimit, as we saw on page 187.

In either case, *Mathematica* may not stop the computation, but instead continue to give this message. In this occurs, you will have to terminate the program from the keyboard, as described on page 5. There are times when you may want to increase the recursion limit, which you can do by assigning a larger integer to $RecursionLimit, but usually if you exceed it you're in a loop.

When doing exercises in the next chapter, you may go into a loop without doing recursive calls, in which case the program will just go on forever without printing any error messages, or may print an error message indicating that $IterationLimit is exceeded. Again, terminate the program from the keyboard. You can increase the value of $IterationLimit, but only do that if you're sure there is no error.

8 Iteration

This book is very different from introductory books about conventional languages. For one thing, the entire first part is about built-in operations on *Mathematica*'s data types; by contrast, conventional languages provide fewer and less powerful built-in data types and operations. More significantly, our treatment of programming has centered on the use of functional programming, rewrite rules, and recursion. This is typical of *Mathematica* programming, but not of programming in conventional languages. Rather, in those languages it is far more common to use *iteration*. In this technique, the values of variables, and the contents of arrays, are repeatedly modified until they have the correct values. The method can be used in *Mathematica*, and can sometimes be very useful. This chapter is about how to use iteration in *Mathematica*.

8.1 Newton's Method

Among the most famous of algorithms is Newton's method for finding the roots of a function. This is a classic use of iteration, and will be our first example.

As in some of the examples in the early part of Chapter 7, there is an easier way to do this in *Mathematica*: use the built-in operation FindRoot. Our example here, and for most of this section, is the function $x^2 - 50$, whose root is, of course, the square root of 50.

```
In[1]:= FindRoot[x^2 - 50 == 0, {x, 50}]

Out[1]= {x -> 7.07107}
```

The number 50 in {x, 50} is the initial guess of the root.

So why should you learn to program a root-finder yourself? For the same reason you learned to solve linear systems of equations in Section 7.5: because the built-in operation doesn't always work, and then the only way out is to program it yourself. An example is the function $f(x) = x^{\frac{1}{3}}$:

```
In[2]:= FindRoot[x^(1/3) == 0, {x, .1}]

        FindRoot::cvnwt:
            Newton's method failed to converge to the prescribed
                accuracy after 15 iterations.

Out[2]= {x -> -0.000195312}
```

By learning to program, we'll be able to write our own root-finding functions that can solve this problem. We will discuss this topic further in Section 9.4.1.

8.1.1 | Do Loops

Suppose we are given a function f and can compute its derivative, f'. Then Newton's algorithm works as follows:

- Give an initial estimate of the root, say a_0.

- Keep generating better estimates, a_1, a_2, ..., using the rule,

$$a_{i+1} = a_i - \frac{f(a_i)}{f'(a_i)}$$

until you're done (we'll discuss this later).

The method is illustrated in Figure 8.1. Under the favorable circumstances pictured there, the estimates get closer and closer to the root. We'll discuss in a moment when to stop, but first let's look at an example. For the function $f(x) = x^2 - 50$, the derivative is $f'(x) = 2x$. This specific case is shown in Figure 8.2, with 50 itself as the initial estimate:

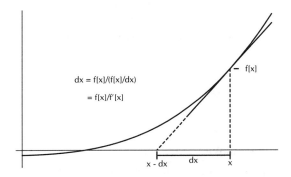

Figure 8.1: Illustration of Newton's method.

```
In[3]:= f[x_] := x^2 - 50


In[4]:= a0 = 50;


In[5]:= a1 = N[a0 - f[a0] / f'[a0]]
Out[5]= 25.5
```

```
In[6]:= a2 = N[a1 - f[a1] / f'[a1]]
```

Out[6]= 13.7304

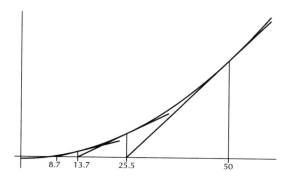

Figure 8.2: Newton's method for $f(x) = x^2 - 50$.

```
In[7]:= a3 = N[a2 - f[a2] / f'[a2]]
```

Out[7]= 8.68597

```
In[8]:= a4 = N[a3 - f[a3] / f'[a3]]
```

Out[8]= 7.22119

```
In[9]:= a5 = N[a4 - f[a4] / f'[a4]]
```

Out[9]= 7.07263

As you can see, these values are getting closer and closer to the real square root of 50, which is approximately 7.07107. (Recall that in *Mathematica*, f'[x] is special syntax for Derivative[1][f][x], which gives the first derivative of f, whenever *Mathematica*'s built-in algorithms can compute it; see page 32.)

We need to discuss how to decide when we're satisfied with the answer we've computed. First, though, note one thing: Wherever we decide to stop, say at a_5, all the previous values we computed are of no interest. So we could have avoided introducing those new names by instead just writing

```
In[10]:= a = 50;
```

```
In[11]:= a = N[a - f[a] / f'[a]]
```

Out[11]= 25.5

```
In[12]:= a = N[a - f[a] / f'[a]]
Out[12]= 13.7304

In[13]:= a = N[a - f[a] / f'[a]]
Out[13]= 8.68597

In[14]:= a = N[a - f[a] / f'[a]]
Out[14]= 7.22119

In[15]:= a = N[a - f[a] / f'[a]]
Out[15]= 7.07263
```

Now, we can put the expression in a function and save some typing.

```
In[16]:= next[x_, fun_] := N[x - fun[x] / fun'[x]]

In[17]:= a = 50;

In[18]:= a = next[a, f]
Out[18]= 25.5

In[19]:= a = next[a, f]
Out[19]= 13.7304

In[20]:= a = next[a, f]
Out[20]= 8.68597

In[21]:= a = next[a, f]
Out[21]= 7.22119

In[22]:= a = next[a, f]
Out[22]= 7.07263
```

To return to the question of when to terminate the computation, one simple answer is: repeat it ten times. Here's how:

```
In[23]:= Do[a = next[a, f], {10}]
```

In general, Do[*expr₁*, {*expr₂*}], evaluates *expr₁ expr₂* times. So, in this case, we can say,

```
In[24]:= a = 50;
         Do[a = next[a, f], {10}]
```

```
In[25]:= a
```
```
Out[25]= 7.07107
```

Note that the Do loop itself yields no value (or rather, it yields the special value Null, which is useless and doesn't get printed). But the important thing is that the Do loop assigns a value to a that is very close to the square root of 50.

The arguments of Do are the same as those of Table (page 61; see also Exercise 3). In particular, they can have the form,

```
Do[expr, {i, imin, imax, di}]
```

which repeats *expr* with variable *i* having values *imin*, *imin* + *di*, and so on, as long as the value of *imax* is not exceeded. (Thus, the loop is repeated a total of $|(imax - imin)/di|$ times.) Furthermore, if *di* is omitted, it is assumed to be 1; and if only *i* and *imax* are given, both *imin* and *di* are assumed to be 1. For example, if we wanted to print each approximation and label it with a number, we could write:

```
In[26]:= a = 50;
```

```
In[27]:= Do[a = next[a, f];
         Print["approximation ", i, ": ", a],
         {i, 1, 10}]
```
```
approximation 1: 25.5
approximation 2: 13.7304
approximation 3: 8.68597
approximation 4: 7.22119
approximation 5: 7.07263
approximation 6: 7.07107
approximation 7: 7.07107
approximation 8: 7.07107
approximation 9: 7.07107
approximation 10: 7.07107
```

8.1.2 | While Loops

The question of when to stop the iterations needs attention. Ten times is okay for $f(x) = x^2 - 50$, but it isn't always. Consider this function:

```
In[28]:= g[x_] := x - Sin[x]
```

It has a root at 0:

```
In[29]:= g[0]
Out[29]= 0
```

However, ten iterations of Newton's algorithm doesn't get us very close to it.

```
In[30]:= a = 1.0;
         Do[a = next[a, g], {10}]
```

```
In[31]:= a
Out[31]= 0.0168228
```

Twenty-five iterations does a bit better.

```
In[32]:= a = 1.0;
         Do[a = next[a, g], {25}]
```

```
In[33]:= a
Out[33]= 0.0000384172
```

In truth, no fixed number of iterations is going to do the trick for all functions. We need to iterate repeatedly until our estimate is close enough to stop. When is that? There are a number of ways to answer that question, none always best, but here's an easy one: when $f(a_i)$ is very close to zero. So, let ϵ be a very small number, and iterate until $|f(a_i)| < \epsilon$.

But how can we write a loop that will stop by testing this condition? The looping construct Do iterates a number of times that is fixed when the loop is begun. We need a new kind of iterative function. It is While, and it has the following form.

```
While[cond, expr]
```

The first argument is the *test* or *condition*, the second the *body*. It works like this: evaluate the test; if it's true, then evaluate the body and then the test again. If it's true again, then again evaluate the body and the test. Continue this way until the test evaluates to `False`. Note that the body may not be evaluated at all (if the test is false the first time), or it may be evaluated once, or a thousand times.

This is just what we want: if the estimate is not yet close enough, compute a new estimate and try again.

```
In[34]:= epsilon = .001;
         a = 50;
```

```
In[35]:= While[Abs[f[a]] > epsilon, a = next[a, f]]
```

```
In[36]:= a
```
```
Out[36]= 7.07107
```

To wrap things up, let's put this all into a function.

```
In[37]:= findRoot[fun_, init_, eps_] :=
             Module[{a = init},
                 While[Abs[fun[a]] > eps,
                     a = N[a - fun[a] / fun'[a]]
                     ];
                 a]
```

```
In[38]:= findRoot[f, 50, .001]
```
```
Out[38]= 7.07107
```

Instead of setting a global variable to the final estimate, this function returns that estimate as its value. (For an explanation of why we introduced the local variable a, see the end of this subsection.)

Let's work with this example a little more. Suppose we'd like to know how many iterations were needed to find the answer. One possibility is to insert a `Print` to show the value of a each time through the loop:

```
In[39]:= findRoot[fun_, init_, eps_] :=
            Module[{a = init},
                While[Abs[fun[a]] > eps,
                    Print["a = ", a];
                    a = N[a - fun[a] / fun'[a]]
                    ];
                a]
```

```
In[40]:= findRoot[f, 50, 0.001]
```

```
Out[40]= 7.07107
```

```
        a = 50
        a = 25.5
        a = 13.7304
        a = 8.68597
        a = 7.22119
        a = 7.07263
```

Counting the lines shows that the function *converged* after six iterations (note that we're seeing the value of a at the *beginning* of each execution of the body).

A better idea would be to have the function actually compute this number and return it as part of its answer.

```
In[41]:= findRoot[fun_, init_, eps_] :=
            Module[{a = init, count = 0},
                While[Abs[fun[a]] > eps,
                    count = count + 1;
                    a = N[a - fun[a] / fun'[a]]
                    ];
                {a, count}]
```

```
In[42]:= findRoot[f, 50, 0.001]
```

```
Out[42]= {7.07107, 6}
```

Here's another question: in all these versions of findRoot, f[a] is computed two times at each iteration, once in the condition and once in the body. In many circumstances, calls to f are very time-consuming, and should be minimized. Can we arrange that f[a] only be computed once in each iteration?

The solution to this is to create a new local variable, funa, which *always* contains the value of fun[a] for the current value of a. We can ensure that it does so by recomputing it whenever a is reassigned:

```
findRoot[fun_, init_, eps_] :=
   Module[{a = init, funa = fun[init]},
        While[Abs[funa] > eps,
             a = N[a - funa / fun'[a]];
             funa = fun[a]
            ];
        a]
```

In all our examples, we used Module to introduce a local variable to which we assigned values in the body of the While. We did this to avoid a common error in the use of iteration: *attempting to assign to a function's argument.* For example, the following version of findRoot doesn't work:

```
In[43]:= findRoot[fun_, x_, eps_] :=
            (While[Abs[fun[x]] > eps,
                x = N[x - fun[x] / fun'[x]]
                ];
            x)
```

```
In[44]:= findRoot[Sin, .1, .01]
```

Set::setraw: Cannot assign to raw object 0.1.

General::stop:
   Further output of Set::setraw
     will be suppressed during this calculation.

```
In[45]:= TracePrint[findRoot[Sin, .1, .01], findRoot]
```

findRoot

While[Abs[Sin[0.1]] > 0.01,

$$0.1 = N[0.1 - \frac{Sin[0.1]}{Sin'[0.1]}]]; \ 0.1$$

What happened can be seen from the trace (of which we've only shown some of the output). The x in the body of findRoot is replaced by the argument .1—which is perfectly normal—leaving an expression of the form 0.1 = *something*—which is not normal. There is a way around this, using the HoldFirst attribute as described on page 245, but introducing local variables is much better style. It is very disconcerting, after all, to call a function and find, when it's done, that your variables have changed their values.

Exercises

1. Compute the square roots of 50 and 60 simultaneously, *i.e.*, with a single Do loop.

2. Compare the use of a Do loop to using the function Nest (page 79). In particular, compute the square root of 50 using Nest; then do Exercise 1 using a *single* Nest.

3. Do is closely related to Table. It is fair to say that Table is to Do as NestList is to Nest. Use Table instead of Do in your solution to Exercise 1. What do you get?

4. Compute Fibonacci numbers iteratively. You will need to have two variables, say this and prev, giving the two most recent Fibonacci numbers, so that after the ith iteration, this and prev have the values F_i and F_{i-1}, respectively.

5. Another termination criterion for root-finding is to stop when $|a_i - a_{i+1}| < \epsilon$; that is, when two successive estimates are very close. The idea is that if we're not getting much improvement, we must be very near the root. The difficulty in programming this is that we need to remember the *two* most recent estimates computed. (It is similar to computing Fibonacci numbers iteratively, as in Exercise 4.) Program findRoot this way.

6. Modify each of the versions of findRoot presented in the text to produce a *list* of all the estimates computed.

    ```
    In[1]:= findRoot[f, 50, 0.001]

    Out[1]= {50, 25.5, 13.7304, 8.68597, 7.22119, 7.07263, 7.07107}
    ```

7. To guard against starting with a poor choice of initial value, modify findRoot to take, as an argument, a *list* of initial values, and simultaneously compute approximations for each until one converges, then return that one.

8. The bisection method is quite useful for finding zeros of functions. If a continuous function $f(x)$ is such that $f(a) < 0$ and $f(b) > 0$ for two real numbers a and b, then as a consequence of the Intermediate Value Theorem of calculus, a zero of f must occur between a and b. If f is now evaluated at the midpoint of a and b, and if $f((a + b)/2) < 0$, then the zero must occur between $(a + b)/2$ and b; if not, then it occurs between a and $(a + b)/2$. This bisection can be repeated until a zero is found to any specified tolerance. Define bisect[f, a, b, *eps*] to compute a root of f, within *eps*, using the bisection method. You should give it two initial values a and b and assume that $f(a) \cdot f(b) < 0$.

Solutions

1. This computes the next approximations of both square roots each time through the loop.

```
In[1]:= next[x_, fun_] := N[x - fun[x] / fun'[x]]
```

```
In[2]:= a = 50;
        b = 60;
        Do[a = next[a, (#^2 - 50)&];
           b = next[b, (#^2 - 60)&],
          {10}]
```

```
In[3]:= Print["square root of 50 = ", a]
        Print["square root of 60 = ", b]
```

2. Notice that the function to compute square root of one number using Nest is very concise, but the one to compute the square root of two numbers simultaneously using Nest is verbose. Since Nest applies just one function to the value computed at the previous step, we need to put the two values into a list and then write a function that extracts each element in the list, applies the two functions to the two elements, and then puts them back into a list.

```
In[1]:= nestSqrt[a_] := Nest[next[#, (#^2-a)&]&, a, 10]

        nestSqrt2[a_, b_] :=
          Module[{},
            nexta[x_] := next[x, (#^2-a)&];
            nextb[x_] := next[x, (#^2-b)&];
            Nest[{nexta[#[[1]]], nextb[#[[2]]]}&,
                {a, b},
                10]]
```

3. We need to place the two expressions that were in the body of the Do into a list. Try copying the body of the Do exactly as above and see what happens.

```
In[1]:= a = 50;
        b = 60;
        Table[{a = next[a, (#^2 - 50)&],
               b = next[b, (#^2 - 60)&]},
              {10}]
```

```
Out[1]= {{25.5, 30.5}, {13.7304, 16.2336}, {8.68597, 9.96482},
         {7.22119, 7.993}, {7.07263, 7.74978},
         {7.07107, 7.74597}, {7.07107, 7.74597},
         {7.07107, 7.74597}, {7.07107, 7.74597},
         {7.07107, 7.74597}}
```

4. Note that this version of the Fibonacci function is much more efficient than the simple recursive version, and is closer to the version that uses dynamic programming.

```
fib[n_] :=
  Module[{prev = 0,    (* fib[0] *)
          this = 1,    (* fib[1] *)
          next},
     (* At the ith step, prev = fib[i-1], this = fib[i] *)
     Do[ next = prev+this;   (* fib[i+1] *)
         prev = this;        (* prev = fib[i] *)
         this = next,        (* this = fib[i+1] *)
       {n}];
     prev]
```

5. The variable b is the current approximation, and the variable a is the previous approximation.

```
In[1]:= findRoot[fun_, init_, eps_] :=
          Module[{a = init, b = fun[init]},
                 While[Abs[b-a] > eps,
                       a = b;
                       b = N[b - fun[b] / fun'[b]]
                      ];
                 b]
```

6. This solution is based on the solution to problem 5.

```
In[1]:= findRootList[fun_, init_, eps_] :=
          Module[{a = init, b, solns = {init}},
                 b = N[a - fun[a] / fun'[a]];
                 While[Abs[b-a] > eps,
                       a = b;
                       b = N[b - fun[b] / fun'[b]];
                       solns = Join[solns, {a}]
                      ];
                 Join[solns, {b}]]
```

7. We go back to the version of findRoot on page 231, and add multiple initial values.

```
In[1]:= findRootList[fun_, inits_, eps_] :=
            Module[{a = inits},
                While[Min[Abs[Map[fun, a]]] > eps,
                    a = Map[N[# - fun[#] / fun'[#]]&, a]
                    ];
                Select[a, (Min[Abs[Map[fun,a]]]==Abs[fun[#]])&]]

In[2]:= findRootList[(#^2-50)&, {25, 50, 75, 100}, .001]

Out[2]= {{25, 50, 75, 100}, {13.5, 25.5, 37.8333, 50.25}}
```

8.

```
In[1]:= bisect[f_, a_, b_, eps_] :=
            Module[{low = Min[a,b],
                    high = Max[a,b],
                    mid = N[(a+b)/2],
                    fofMid = N[f[(a+b)/2]]},
                While[Abs[fofMid] > eps,
                    If[fofMid < 0, low = mid, high = mid];
                    mid = N[(low+high)/2];
                    fofMid = N[f[mid]]];
                mid]

In[2]:= bisect[(#^2-50)&, 0, 50, .001]

Out[2]= 7.07111
```

8.2 Vectors and Matrices

Vectors and matrices are used throughout mathematics, and many fundamental algorithms are expressed in terms of them. We have seen in Chapter 3 that *Mathematica* represents vectors as lists and matrices as nested lists. In Section 7.5, we showed how to solve linear systems recursively.

There is, however, something you will occasionally find in mathematics books that we have not seen: algorithms in which the contents of an array *change* during the computation.

As an example, consider again the problem given in Exercise 6 on page 71. You're given a vector P containing a permutation of the numbers 1 through n, and a vector S of length n, and you need to produce a vector T of length n such that for all k between 1 and n, T[[P[[k]]]] = S[[k]]:

```
In[1]:= permute[{"a", "b", "c", "d"}, {3, 2, 4, 1}]
Out[1]= {d, b, a, c}
```

The functional style solution was described on page 71, and it is certainly fair to describe it as nonobvious. Here is a simpler way to do this:

1. Initialize T to be a list of length n containing all zeros.

2. For each i from 1 to n:

   (a) Let p = P[[i]].

   (b) Set the pth entry of T to S[[i]].

3. Return T as the result.

All of these steps are easy except 2b—we know how to initialize T, how to loop with loop variable i, and how to return T. So the outline of this function is,

```
permute[S_, P_] :=
   Module[{T = Table[0, {Length[S]}], p},
       Do[p = P[[i]];
          set p^th element of T to S[[i]],
          {i, Length[[S]]}];
       T]
```

How can we change the pth element of T? *Mathematica* provides a way to alter T directly: *list component assignment*. It uses the syntax, T[[$expr_1$]] = $expr_2$. Thus, 2b can be accomplished by, T[[p]] = S[[i]], which leads to the program:

```
permute[S_, P_] :=
   Module[{T = Table[0, {Length[S]}], p},
       Do[p = P[[i]];
          T[[p]] = S[[i]],
          {i, Length[S]}];
       T]
```

8.2.1 | List Component Assignment

The capability to alter elements of lists merits detailed consideration. The general syntax for modifying a list is:

name[[*integer-valued-expression*]] = *expression*

The *name* must be the name of a list. The *integer-valued-expression* must evaluate to a legal subscript, that is, a number whose absolute value is less than or equal to the length of the list. The assignment returns the value of *expression* (as assignments always do), but has the effect of changing the list to which *name* is bound.

```
In[2]:= L = {0, 1, 2, 3, 4}
Out[2]= {0, 1, 2, 3, 4}

In[3]:= L[[1]] = 10
Out[3]= 10

In[4]:= L
Out[4]= {10, 1, 2, 3, 4}
```

Components of nested lists can be modified using:

name[[*int-expr$_1$*, *int-expr$_2$*]] = *expression*

where *int-expr$_1$* and *int-expr$_2$* are expressions that evaluate to integers. *int-expr$_1$* chooses the sublist of *name*, and *int-expr$_2$* the element of that sublist:

```
In[5]:= A = {{1, 2, 3}, {4, 5, 6}}

In[6]:= A[[2, 3]] = 20
Out[6]= 20

In[7]:= A
Out[7]= {{1, 2, 3}, {4, 5, 20}}
```

However, note that assigning one array variable to another one makes a copy of the first, so that component assignments to either one don't affect the other.

```
In[8]:= B = A

Out[8]= {{1, 2, 3}, {4, 5, 20}}

In[9]:= B[[1, 2]] = 30

Out[9]= 30

In[10]:= B

Out[10]= {{1, 30, 3}, {4, 5, 20}}

In[11]:= A

Out[11]= {{1, 2, 3}, {4, 5, 20}}

In[12]:= A[[2, 1]] = 40

Out[12]= 40

In[13]:= B

Out[13]= {{1, 30, 3}, {4, 5, 20}}
```

8.2.2 | Finding prime numbers

One of the oldest algorithms in the history of computing is the Sieve of Eratosthenes. Named for the famous Greek astronomer Eratosthenes (*ca.* 276–*ca.* 194 B.C.), this method is used to find all prime numbers below a given number n. The great feature of this algorithm is that it finds prime numbers without doing any divisions—an operation that took considerable skill and concentration before the introduction of the Arabic numeral system. In fact, its only operations are addition and component assignment. Here is how it works:

1. Let `primes` be a list containing all the integers between 1 and n.

2. Let p = 2.

3. Repeat these steps until p = n + 1:

    (a) Starting at position 2p, *cross out* every p[th] value in `primes`. That is, assign 1 to `primes` at positions 2p, 3p, ...

    (b) Keep incrementing p by 1, until `primes[[p]]` is not 1 (or until p = n + 1).

4. The non-1s in `primes` are all the prime numbers less than or equal to n.

For a better idea of how this works, here is how the variables look after five iterations for $n = 12$; by an *iteration* we mean a complete execution of steps 3a and 3b.

initial state: primes = {1, 2, 3, 4, 5, 6, 7, 8, 9, 10, 11, 12}, p=2
                             ↑

1 iteration: primes = {1, 2, 3, 1, 5, 1, 7, 1, 9, 1, 11, 1}, p=3
                                ↑

2 iterations: primes = {1, 2, 3, 1, 5, 1, 7, 1, 1, 1, 11, 1}, p=5
                                  ↑

3 iterations: primes = {1, 2, 3, 1, 5, 1, 7, 1, 1, 1, 11, 1}, p=7
                                   ↑

4 iterations: primes = {1, 2, 3, 1, 5, 1, 7, 1, 1, 1, 11, 1}, p=11
                                      ↑

5 iterations: primes = {1, 2, 3, 1, 5, 1, 7, 1, 1, 1, 11, 1}, p=13

(Note that the program can be improved in two ways: (1) the *crossing out* operation can begin at p^2, since $2p, 3p, \ldots, (p-1)p$ must already have been crossed out; and (2) by the same token, the iterations can stop as soon as p is greater than $\sqrt{n}$. We'll leave these optimizations as an exercise.)

Only step 3a involves component assignments; all other steps are straightforward:

```
primes[n_] := Module[{primes = Range[n], p = 2},
   While[p != n + 1,
         cross out every p^th value in primes;
         p = p + 1;
         While[p != n + 1 && primes[[p]] == 1, p = p + 1]
         ];
   Select[primes, (# != 1)&]]
```

As for 3a, this is just a simple loop with a component assignment.

```
Do[primes[[i]] = 1, {i, 2p, n, p}]
```

We can use the `Print` function to see how our program follows the algorithm:

```
In[14]:= primes2[n_] := Module[{primes = Table[i, {i,1,n}], p=2},
            Print["Initially: primes = ", primes, ", p = ", p];
            While[p!= n+1,
                 Do[primes[[i]] = 1, {i, 2p, n, p}];
                 p = p + 1;
                 While[p != n+1 && primes[[p]]==1, p = p+1];
                 Print["In loop: primes = ", primes, ", p = ", p];
                 ];
            Select[primes, (# != 1)&]]
```

```
In[15]:= primes2[12]
```

```
Initially: primes = {1, 2, 3, 4, 5, 6, 7, 8, 9, 10, 11, 12}, p = 2
In loop: primes = {1, 2, 3, 1, 5, 1, 7, 1, 9, 1, 11, 1}, p = 3
In loop: primes = {1, 2, 3, 1, 5, 1, 7, 1, 1, 1, 11, 1}, p = 5
In loop: primes = {1, 2, 3, 1, 5, 1, 7, 1, 1, 1, 11, 1}, p = 7
In loop: primes = {1, 2, 3, 1, 5, 1, 7, 1, 1, 1, 11, 1}, p = 11
In loop: primes = {1, 2, 3, 1, 5, 1, 7, 1, 1, 1, 11, 1}, p = 13
```

```
Out[15]= {2, 3, 5, 7, 11}
```

Exercises

1. Write the following functions (that return vectors or matrices), by creating a new list and filling in the elements, as we did for permute.

   (a) reverse[V], where V is a vector.

   (b) transpose[A], where A is a matrix.

   (c) rotateRight[V, n], where V is a vector and n is a (positive or negative) integer.

   (d) rotateRows, which could be defined this way:

   ```
   rotateRows[A_] := Map[(rotateRight[A[[#]], #-1])&,
                         Range[1, Length[A]]]
   ```

   That is, it rotates the ith row of A $i - 1$ places to the right.

   (e) rotateRowsByS, which could be defined this way:

   ```
   rotateRowsByS[A_, S_] /; Length[A] == Length[S] :=
       Map[(rotateRight[A[[#]], S[[#]]])&,
           Range[1, Length[A]]]
   ```

   That is, it rotates the ith row of A by the amount S[[i]].

   (f) compress[L, B], where L and B are lists of equal length, and B contains only Boolean values (False and True), selects out of L those elements corresponding to True's in B. For example, the result of compress[{a, b, c, d, e}, {True, True, False, False, True}] should be {a, b, e}. To know what size list to create, you will first need to count the True's in B.

Exercises (cont.)

(g) Modify `sieve` by replacing the final call to `Select` with code that creates a list of appropriate size and then fills it from `primes`.

Solutions

1.a Create a table `rv` of zeros, then use a `Do` loop to set `rv[[i]]` to `V[[n-i]]`, where `n` is the length of `V`.

```
reverse[V_] :=
  Module[{rV = Table[0, {Length[V]}]},
     Do[ rV[[i]] = V[[Length[V]-i+1]],
      {i, 1, Length[V]}];
     rV]
```

1.b Similarly, create a matrix `tA` of zeros, and use a doubly-nested loop to set `tA[[i, j]]` to `A[[j, i]]`.

```
transpose[A_] :=
  Module[{tA = Table[Table[0, {Length[A]}],
                     {Length[A[[1]]]}]},
         Do[tA[[i, j]] = A[[j, i]],
          {i, 1, Length[A[[1]]]},
          {j, 1, Length[A]}];
         tA]
```

1.c The key to this problem is to use the `Mod` operator to compute the target address for any item from V. That is, the element $V[i]$ must move to, *roughly speaking*, position $(n + i)$ mod $length[V]$. The *roughly speaking* is due to the fact that the `Mod` operator returns values in the range $0, \ldots, length[V]-1$, whereas vectors are indexed by values $1, \ldots, length[V]$. This causes a little trickiness in this problem.

```
rotateRight[V_, n_] :=
  Module[{rV = Table[0, {Length[V]}]},
         Do[rV[[1+Mod[n+i-1, Length[V]]]] = V[[i]],
          {i, 1, Length[V]}];
         rV]
```

1.d   Iterate over the rows of A, setting row i to the result of calling `rotateRight`.

```
rotateRows[A_] :=
  Module[{tA = Table[0, {Length[A]}]},
    Do[ tA[[i]] = rotateRight[A[[i]], i],
      {i, 1, Length[A]}];
    tA]
```

1.e

```
rotateRowsByS[A_, S_] :=
  Module[{tA = Table[0, {Length[A]}]},
      Do[tA[[i]] = rotateRight[A[[i]], S[[i]]],
        {i, 1, Length[A]}];
      tA]
```

1.f   Create a list cL of correct length, then iterate over L and B, moving L[[i]] to cL whenever B[[i]] is True. The position in cL that receives this value is not necessarily i; we use the variable last to keep track of the next position in cL that will receive a value from L.

```
compress[L_, B_] :=
  Module[{cL = Table[0, {Count[B, True]}],
          last = 1},
        Do[If[B[[i]],
            cL[[last]] = L[[i]];  last = last+1,
            NULL (* do nothing *)],
          {i, 1, Length[B]}];
        cL]
```

1.g   This is just a matter of taking the code for primes shown in the book and replacing the last line by code that is very similar to the compress function just given.

```
primes[n_] :=
Module[{primes = Table[i, {i, 1, n}], p = 2},
    While[p!= n+1,
          Do[primes[[i]] = 1, {i, 2p, n, p}];
          p = p + 1;
          While[ p != n+1 && primes[[p]]==1, p = p+1];
          ];
      Module[{justprimes = Table[0, {n-Count[primes, 1]}],
              last = 1},
          Do[If[primes[[i]] != 1,
            justprimes[[last]] = primes[[i]]; last = last+1,
            NULL (* do nothing *)],
            {i, 1, Length[primes]}];
          justprimes]]
```

8.3 Passing Arrays to Functions

In our next algorithm—Gaussian elimination using iteration—we will pass an array into a function as an argument, and have the function modify the array. Passing an array to a function is no problem, *unless* the function is going to modify the array. In that case, some extra work is required. To see why, consider this example:

```
In[1]:= sqrElement[L_, i_] := L[[i]] = L[[i]]^2
```

sqrElement should square L[[i]]. But let's try it:

```
In[2]:= x = {0, 10, 20};
```

```
In[3]:= sqrElement[x, 3]
        Part::setps:
            {0, 10, 20} in assignment of part is not a symbol.

Out[3]= 400
```

```
In[4]:= TracePrint[sqrElement[x, 3], sqrElement]
        Part::setps:
            {0, 10, 20} in assignment of part is not a symbol.

Out[4]= 400
            sqrElement
            sqrElement[{0, 10, 20}, 3]
                                                2
            {0, 10, 20}[[3]] = {0, 10, 20}[[3]]
```

This is the same kind of error as we had on page 233. It will be a problem whenever we attempt to pass a list into a function and use component assignment. We could solve it as we did there, by introducing a local variable, but that solution is not a very good one here, because it would mean copying the array every time the function is called—an expensive proposition. You can get around it by giving sqrElement the HoldFirst attribute. This tells *Mathematica* not to evaluate the first argument to sqrElement. That way, the body of sqrElement rewrites to x[[3]] = x[[3]]^2, which is legal.

```
In[5]:= SetAttributes[sqrElement, HoldFirst]
```

In[6]:= **sqrElement[x, 3]**

Out[6]= 400

In[7]:= **x**

Out[7]= {0, 10, 400}

Of course, when we do this, we must also be sure to pass a list *name* into the function, or we'll have the same problem.

In[8]:= **sqrElement[{5, 10, 15, 20}, 2]**

       Part::setps:
          {5, 10, 15, 20} in assignment of part is not a
            symbol.

Out[8]= 100

Finally, note that HoldFirst causes only the *first* argument to be left unevaluated. So in each of our functions that use component assignment on an argument, that argument will have to be the function's first. If you need to have two array arguments that get modified, use the attribute HoldAll.

Exercises

1. Define swap[V, i, j] so that it swaps elements i and j of vector V. For example, if v = {a, b, c, d}, then after evaluating swap[v, 2, 4], v will be {a, d, c, b}.

2. Define reverseInPlace[V] to reverse the elements of a list by modifying the list instead of creating a new list. You should use swap to switch elements.

Solutions

1.

In[1]:= **SetAttributes[swap, HoldFirst]**

```
swap[V_, i_, j_] :=
   Module[{temp = V[[i]]},
      V[[i]] = V[[j]];
      V[[j]] = temp;
      Null]
```

2.

```
In[1]:= SetAttributes[reverseInPlace, HoldFirst]

reverseInPlace[V_] :=
    Module[{l = Length[V]},
        Do[swap[V, i, l-i+1], {i, 1, l/2}]]
```

8.4 Gaussian Elimination Revisited

Recall the method of Gaussian elimination from Section 7.5. Despite its natural recursive statement, Gaussian elimination is easily programmed in iterative style as well. The key observation is that after the elimination phase, the original matrix is not needed, so we are justified in *changing* it to the transformed matrix. In other words, we can transform it *in place*.

The overall structure of the iterative version is:

```
SetAttributes[{solveI, elimInPlace}, HoldFirst]

solveI[S_] :=
    Module[{xs = {}, n = Length[S]},
        Do[elimInPlace[S, i], {i, 1, n-1}];
        Do[PrependTo[xs, soln[S, i, xs]], {i, n, 1, -1}];
        xs]
```

We've called the new function `solveI` (I for "iterative") to distinguish it from the recursive version, called `solve`.

The function `elimInPlace` *modifies* its argument so that the ith row is eliminated. Thus, after the call `elimInPlace[S, 1]`, the matrix produced by `elimx1[S]` (page 193), which represented the system S', occupies positions `S[[2, 2]]`,..., `S[[n, n + 1]]`; `S[[1, 1]]`, ..., `S[[1, n + 1]]` are unchanged, as are `S[[2, 1]]`, ..., `S[[n, 1]]`. So, if `S` starts out like this:

$$
\begin{pmatrix}
a_{11} & a_{12} & \cdots & a_{1n} & b_1 \\
a_{21} & a_{22} & \cdots & a_{2n} & b_2 \\
\vdots & \vdots & \ddots & \vdots & \vdots \\
a_{n1} & a_{n2} & \cdots & a_{nn} & b_n
\end{pmatrix}
$$

then after the call elimInPlace[S, 1], it looks like this:

$$\begin{pmatrix} a_{11} & a_{12} & \cdots & a_{1n} & b_1 \\ a_{21} & d'_{22} & \cdots & d'_{2n} & b'_2 \\ \vdots & \vdots & \ddots & \vdots & \vdots \\ a_{n1} & d'_{n2} & \cdots & d'_{nn} & b'_n \end{pmatrix}$$

where

$$\begin{pmatrix} d'_{22} & \cdots & d'_{2n} & b'_2 \\ \vdots & \ddots & \vdots & \vdots \\ d'_{n2} & \cdots & d'_{nn} & b'_n \end{pmatrix}$$

is the matrix returned by elimx1[S].

Thus, after calling the expressions elimInPlace[S, 1], elimInPlace[S, 2], ..., elimInPlace[S, n1], S is:

$$\begin{pmatrix} a_{11} & & \cdots & & a_{1n} & b_1 \\ & d'_{22} & & \cdots & a'_{2n} & b'_2 \\ & & d''_{33} & \cdots & d''_{3n} & b''_3 \\ & & & \ddots & \vdots & \vdots \\ & & & & a_{nn}^{(n-1)} & b_n^{(n-1)} \end{pmatrix}$$

the other positions being irrelevant. We can now solve for $x_1, \ldots, x_n$ by back substitution:

$$x_n \quad = \quad \text{S[[}n, n+1\text{]]/S[[}n, n\text{]]}$$
$$x_{n-1} \quad = \quad (\text{S[[}n\text{-}1, n+1\text{]]} - x_n\text{S[[}n\text{-}1, n\text{]]})/\text{S[[}n\text{-}1, n\text{-}1\text{]]}$$

and so on. Here is the program:

```
In[1]:= elimInPlace[S_, i_] :=
          Module[{m, n = Length[S]},
              Do[m = S[[j, i]] / S[[i, i]];
                Do[S[[j, k]] = S[[j, k]] - m*S[[i, k]],
                  {k, i+1, n+1}],
                {j, i+1, n}]]
```

We can follow this by printing S after each call to elimInPlace:

```
In[2]:= solveI[S_] :=
            Module[{xs = {}, n = Length[S]},
                Do[elimInPlace[S, i];
                    Print[MatrixForm[S]];
                    Print[" \n"],
                   {i, 1, n-1}];
                Do[PrependTo[xs, soln[S, i, xs]], {i, n, 1, -1}];
                xs]


In[3]:= m = {{2, 3, 5, 7}, {3, 5, 7, 11}, {5, 7, 11, 13}};


In[4]:= solveI[m]

Out[4]= {-6, 3, 2}
```

$$
\begin{array}{cccc}
2 & 3 & 5 & 7 \\[4pt]
3 & \dfrac{1}{2} & -\left(\dfrac{1}{2}\right) & \dfrac{1}{2} \\[8pt]
5 & -\left(\dfrac{1}{2}\right) & -\left(\dfrac{3}{2}\right) & -\left(\dfrac{9}{2}\right)
\end{array}
$$

$$
\begin{array}{cccc}
2 & 3 & 5 & 7 \\[4pt]
3 & \dfrac{1}{2} & -\left(\dfrac{1}{2}\right) & \dfrac{1}{2} \\[8pt]
5 & -\left(\dfrac{1}{2}\right) & -2 & -4
\end{array}
$$

Finally, the function soln[S, i, xs] computes the value of x_i given that xs is a list of the values $\{x_{i+1}, \ldots, x_n\}$. The computation here is exactly like the one at the end of solve (page 193):

```
soln[S_, i_, xs_] :=
    Module[{r = 0, n = Length[S]},
        Do[r = r + S[[i, k]] * xs[[k-i]], {k, i+1, n}];
        (S[[i, n+1]] - r) / S[[i, i]]]
```

The algorithm being essentially the same as before, it can handle the same matrices and, in particular, it still fails for the Hilbert matrix. In the next chapter, the algorithm itself will be changed to handle some cases that it currently has trouble with.

Exercises

1. Look at Exercise 1 on page 194. The problem described there is still a problem in solveI, and can still be solved by the pivoting method described there. Implement this method in solveI.

2. Look at Exercise 3 on page 195, describing LU-decomposition. The matrix S after all the calls to elimInPlace (that is, the first *n* columns of S) is exactly the matrix described in part (b) of that problem. Make the (very minor) changes to solveI to produce function LUdecompI.

3. The problem discussed in Exercise 1 on page 194 is also a problem for LU-decomposition. It can be solved using pivoting, but we didn't discuss it in Section 7.5 because incorporating pivoting into the recursive versions of LUdecomp is tricky. It is, however, fairly easy to incorporate it into LUdecompI. The idea is this: when you are working on the ith step—that is, the submatrix of S whose upper-left corner is at S_{ii}—and you decide to swap rows i and j, swap *the entire rows*, not just the parts that are within the submatrix. Implement this.

Solutions

1. The basic idea is simple: before calling elimInPlace, call pivot, which will swap rows if necessary.

```
In[1]:= SetAttributes[{elimInPlace, solveI, pivot}, HoldFirst]

solveI[S_] :=
    Module[{xs = {}, n = Length[S]},
        Do[pivot[S, i];
            elimInPlace[S, i],
            {i, 1, n-1}];
        Do[PrependTo[xs, soln[S, i, xs]],
            {i, n, 1, -1}];
        xs]
```

elimInPlace and soln do not change:

```
In[2]:= elimInPlace[S_, i_] :=
        Module[{m, n = Length[S]},
            Do[m = S[[j, i]] / S[[i, i]];
                Do[S[[j, k]] = S[[j, k]] - m*S[[i, k]],
                    {k, i+1, n+1}],
                {j, i+1, n}]]
```

```
In[3]:= soln[S_, i_, xs_] :=
           Module[{r = 0, n = Length[S]},
               Do[r = r + S[[i, k]] * xs[[k-i]],
               {k, i+1, n}];
               (S[[i, n+1]] - r) / S[[i, i]]]
```

pivot is similar to pivot written for Exercise 1 on page 194, except that it swaps the rows *in place* instead of returning an entire new array. (Note that since rows are the elements of an array, they can be swapped using swap from the previous exercise set.) However, what makes it trickier than the pivot from that exercise is that the argument is the entire array, together with an index *i* indicating what part of the array we are interested in. Thus, pivotrow below gives the pivot row, considering only rows from *i* to *n*. So if pivotrow = 1, it really means that *i* itself is the pivot row (*i.e.*, no pivoting is needed); if pivotrow > 1, then row *i* should be swapped with row i + pivotrow − 1.

```
In[4]:= pivot[Q_, i_] := Module[{p, pivotrow, Qsuffix, ST1},
           Qsuffix=Drop[Q,i-1];  (* look only in remaining rows *)
           ST1=Transpose[Qsuffix][[i]]; (* ith elements of rows *)
           p = Position[ST1, x_/; x!= 0];
           If[p == {},
               Print["Matrix is singular"]; Q,
               pivotrow = p[[1]][[1]];
               If[pivotrow != 1, swap[Q, i, i+pivotrow-1]]
           ]]
```

```
In[5]:= m = {{1, 1, 1, 1}, {1, 1, 2, 2}, {1, 2, 2, 1}};
```

```
In[6]:= solveI[m]
```

```
Out[6]=
       Power::infy: Infinite expression  1/0  encountered.

       {Indeterminate, Indeterminate, Indeterminate}
```

2. We return to the nonpivoting version of solveI for this problem. The change in ludecompI from solveI is simply to omit the final part of computing the solution vector. However, the changes to elimInPlace are more significant.

```
In[1]:= SetAttributes[{elimInPlace, ludecompI}, HoldFirst]

ludecompI[S_] :=
    Module[{xs = {}, n = Length[S]},
        Do[elimInPlace[S, i], {i, 1, n-1}];
        S]
```

For `elimInPlace`, note first that *S* is now a square matrix instead of a rectangular one, since it does not include the vector *B*. Also, we had previously used the multipliers only to eliminate rows, but now we need to save them in the array. These are the two changes with respect to the version of `elimInPlace` from Problem 1.

```
In[2]:= elimInPlace[S_, i_] :=
    Module[{m, n = Length[S]},
        Do[m = S[[j, i]] / S[[i, i]];
           S[[j, i]] = m;
           Do[S[[j, k]] = S[[j, k]] - m*S[[i, k]],
              {k, i+1, n}],
           {j, i+1, n}]]
```

3.

```
In[1]:= SetAttributes[{elimInPlace, ludecompI}, HoldFirst]

ludecompI[S_] := Module[{xs = {}, n = Length[S]},
    Do[pivot[S, i];
       elimInPlace[S, i],
       {i, 1, n-1}];
    S]
```

For `elimInPlace`, note first that *S* is now a square matrix instead of a rectangular one, since it does not include the vector *B*. Also, we had previously used the multipliers only to eliminate rows, but now we need to need to save them in the array. These are the two changes with respect to the version of `elimInPlace` from Problem 1.

```
In[2]:= elimInPlace[S_, i_] := Module[{m, n = Length[S]},
    Do[m = S[[j, i]] / S[[i, i]];
       S[[j, i]] = m;
       Do[S[[j, k]] = S[[j, k]] - m*S[[i, k]],
          {k, i+1, n}],
       {j, i+1, n}]]
```

```
In[3]:= pivot[Q_, i_] := Module[{p, pivotrow, Qsuffix, ST1},
          Qsuffix=Drop[Q,i-1];  (* look only in remaining rows *)
          ST1=Transpose[Qsuffix][[i]]; (* ith elements of rows *)
          p = Position[ST1, x_/; x!= 0];
          If[p == {},
             Print["Matrix is singular"]; Q,
             pivotrow = p[[1]][[1]];
             If[pivotrow != 1, swap[Q, i, i+pivotrow-1]]]]
```

```
In[4]:= m = {{2, 3, 5}, {3, 5, 7}, {5, 7, 11}};
```

```
In[5]:= ludecompI[m]
```

$$Out[5]= \{\{2, 3, 5\}, \{\frac{3}{8}, -(\frac{23}{8}), -(\frac{49}{8})\}, \{\frac{5}{8}, -(\frac{47}{161}), -(\frac{262}{161})\}\}$$

```
In[6]:= m = {{1, 1, 1}, {1, 1, 2}, {1, 2, 2}};
```

```
In[7]:= ludecompI[m]
```

$$Out[7]= \{\{1, 1, 1\}, \{1, -1, 0\}, \{1, 0, 0\}\}$$

9 | Numerics

Of the many data types that are available in *Mathematica*—numbers, strings, symbols, lists—numbers are perhaps the most familiar. *Mathematica* recognizes many different kinds of numbers and has the ability to operate with numbers in a variety of ways. Most importantly, it can operate on numbers of any size and to any degree of precision. In this chapter we will explore some computational aspects of numbers and look into the issues that must be addressed when using numbers in programs.

9.1 Types of Numbers

9.1.1 | Integers and Rationals

There are four kinds of numbers represented in *Mathematica*—integers, rationals, reals, and complex. Integers are considered to be exact and are represented without a decimal point; rational numbers are quotients of integers and are also considered to be exact.

```
In[1]:= 3/27
```

$$Out[1]= \frac{1}{9}$$

As can be seen in the above example, *Mathematica* simplifies rational numbers to lowest terms and leaves them as exact numbers.

The Head function can be used to identify the type of number you are working with. FullForm is used to show how *Mathematica* represents objects internally.

```
In[2]:= {Head[1/9], FullForm[1/9]}

Out[2]= {Rational, Rational[1, 9]}
```

In stark contrast to programming languages such as C or Pascal that typically restrict computations with integers to 16- or 32-bits,[1] *Mathematica* allows you to compute with integers and rational numbers of arbitrary size. Related to this is the fact that all integers (and rational numbers) are treated as *exact*. You can see this by examining the `Precision` of any integer or rational number.

```
In[3]:= {Precision[7], Precision[1/9]}

Out[3]= {Infinity, Infinity}
```

The *precision* of numbers in *Mathematica* depends upon the type of number you are computing with. In general, the precision of an approximate real number x is the number of significant digits in x. For real numbers, this notion has to do with the total number of decimal digits needed to specify that number.

```
In[4]:= Precision[92.2226669991111111111122]

Out[4]= 24
```

Integers and rational numbers, on the other hand, are exact and so have more precision than any approximate number. Representing a number with infinite precision is another way of saying that it is exact. This allows *Mathematica* to operate on such a number differently than if the number were only approximate. But in fact, more is true. As far as *Mathematica* is concerned, all integers are not created equal. If two numbers are to be added, 3 + 6 for example, *Mathematica* checks to see if the numbers can be added as *machine integers*. A machine integer is an integer whose magnitude is small enough to fit into your machine's natural word size, and to be operated on by the machine's instructions. *Word size* means the number of bits used to represent integers. For example, the most common word size is 32 bits, which can represent integers in the range from $-2,147,483,648$ to $2,147,483,647$. Arithmetic operations on integers within this range can be performed using the machine's own instructions, whereas operations on integers out of that range must be done by programs, and so are much less efficient.

If the two numbers to be added are machine integers and *Mathematica* can determine that their sum is a machine integer, then the addition is performed at this low level.

If, on the other hand, the two integers to be added are large and either the integers themselves or their sum is larger than the size of a machine integer, then

---

[1] This restricts integers to a magnitude of 2^{16} (in the case of 16-bit integers), or to a magnitude of 2^{32} in the case of 32-bit integers.

Mathematica performs the arithmetic using special algorithms. Integers in this range are referred to as *extended precision integers*.

```
In[5]:= 2^256 + 2^1024
Out[5]= 17976931348623159077293051907890247336179769789423065
7\
       2734300811577326758055009631327084773224075360211201\
       1387987139335765878976881441662249284743063947412437\
       7767893424865485276302219601246094119453082952085005\
       7688381506823424628815897051997781434325869214956932\
       7420609321723060412028034429294033753735377152
```

`Rational` numbers are treated somewhat similarly to `Integer` numbers in *Mathematica* since the rational number a/b can be thought of as a pair of integers and in fact, as we saw earlier, it is represented as a pair (`Rational[a, b]`). In this way, algorithms for exact rational arithmetic will use integer arithmetic (either machine or extended) to perform many of the necessary computations.

This representation of rational numbers as a pair of integers has one more consequence. If you try to pattern match with rational numbers, you should be aware of their internal representation. For example, trying to pattern match with a_/b_ will not work:

```
In[6]:= 3/4 /. x_/y_ -> {x, y}
Out[6]= 3
        -
        4
```

But pattern matching instead with `Rational` works fine:

```
In[7]:= 3/4 /. Rational[x_, y_] -> {x, y}
Out[7]= {3, 4}
```

9.1.2 | Real Numbers

Real numbers (often referred to as "floating point numbers") contain decimal points, and although they can contain any number of digits, are not considered exact.

```
In[8]:= {Head[1.61803], Precision[1.61803]}
Out[8]= {Real, 16}
```

In a manner similar to how integers are treated, *Mathematica* uses different internal algorithms to do arithmetic on real numbers depending upon whether you are using very high precision reals or not. Real numbers that can be computed at the hardware level of the machine are referred to as *fixed precision reals*. Typically, these have 16 decimal digits of precision. The number of digits that each machine uses for fixed-precision real numbers is given by the system variable $MachinePrecision:

```
In[9]:= $MachinePrecision

Out[9]= 16
```

Arithmetic operations on real numbers are performed whenever possible using machine-precision (fixed) reals.

```
In[10]:= Precision[1.23]

Out[10]= 16
```

Although this last result may seem odd at first, it is a consequence of how *Mathematica* represents real numbers internally. A Precision of 16 (on a computer with $MachinePrecision of 16) indicates that the number 1.23 is viewed as a machine-precision real number which will allow *Mathematica* to perform arithmetic with it using the efficient machine-precision arithmetic routines. *Mathematica* views the number 1.23 as a machine precision real by effectively padding with zeros out to 16 significant digits. The only way to know for sure is to test the number as follows:

```
In[11]:= MachineNumberQ[1.23]

Out[11]= True
```

You can see more clearly how *Mathematica* computes Accuracy by looking at the following example:

```
In[12]:= {Accuracy[1.23], Accuracy[12.3], Accuracy[123.]}

Out[12]= {16, 15, 14}
```

Each addition of a digit to the left of the decimal point has the effect of reducing the number of significant digits to the right of the decimal point by 1.

One fact to keep in mind when working with machine-precision numbers is that any computations of expressions containing machine-precision numbers will be done at the machine precision.

```
In[13]:= 2.0^100
```

$$Out[13]= 1.26765 \ 10^{30}$$

```
In[14]:= Precision[%]
```

$$Out[14]= 16$$

Note: By default, only 6 digits of a machine number are displayed. Do not assume that typing in what is displayed will result in the same value.

```
In[15]:= pi = N[Pi]
```

$$Out[15]= 3.14159$$

```
In[16]:= pi - %
```

$$Out[16]= 0.$$

```
In[17]:= pi - 3.14159
```

$$Out[17]= 2.65359 \ 10^{-6}$$

Although there is a limit to the magnitude of the machine precision numbers on any given computer,

```
In[18]:= {$MaxMachineNumber, $MinMachineNumber}
```

$$Out[18]= \{1.79769 \ 10^{308}, \ 1.11254 \ 10^{-308}\}$$

numbers outside of this range can still be handled. Real numbers larger than machine-precision reals are referred to as *multiple precision reals* and arithmetic on such numbers is called *multiple precision* arithmetic or *variable precision* floating point arithmetic. So for example, on a machine whose $MachinePrecision is 16, computations involving real numbers with greater than 16 significant digits will be performed using multiple precision algorithms. (This last statement is not quite correct. In Section 9.1.4, we will see how *Mathematica* operates on expressions containing numbers of different precision.)

Built-in constants are not treated as real numbers by *Mathematica*.

```
In[19]:= {Head[Pi], NumberQ[Pi]}
```

$$Out[19]= \{Symbol, \ False\}$$

This last point has particular importance when you try to use the built-in constants Pi or E, for example, in functions that expect a real number argument.

```
In[20]:= Random[Real, {-Pi, Pi}]

        Random::randn:
            Range specification {-Pi, Pi} in
             Random[Real, {-Pi, Pi}]
               is not a valid number or pair of numbers.

Out[20]= Random[Real, {-Pi, Pi}]
```

If you are using versions of *Mathematica* after Ver. 2.2, then you will be able to use mathematical constants more like ordinary numbers. For example,

```
In[21](Version 3.0):= Random[Real, {E, Pi}]

Out[21](Version 3.0)= 3.12176


In[22](Version 3.0):= Rationalize[Pi, .0001]
```
$$Out[22](Version\ 3.0) = \frac{355}{113}$$

This behavior is available to you because mathematical constants now have an additional attribute that essentially alerts *Mathematica* to the fact that these are numbers.

```
In[23](Version 3.0):= Map[NumericQ, {Pi, E, EulerGamma}]

Out[23](Version 3.0)= {True, True, True}
```

When *Mathematica* recognizes that a quantity has this attribute, it converts the symbol to a real number, using what it perceives to be necessary precision.

```
In[24](Version 3.0):= E^Pi > Pi^E

Out[24](Version 3.0)= True
```

In all versions though, exact results can be obtained by using such symbols in their exact form, together with transformation rules for operating with such quantities. Notice the contrast between trigonometric expressions involving an approximation to π and those that use the symbolic expression.

In[25]:= **1/Sin[N[Pi]]**

Out[25]= 8.16589 10^{15}

In[26]:= **1/Sin[Pi]**

Out[26]= ComplexInfinity

9.1.3 | Complex Numbers

Complex numbers are of the form *a + bi*, where *a* and *b* are any numbers—integer, rational, or real. *Mathematica* represents $\sqrt{-1}$ by the symbol I.

Mathematica views complex numbers as a distinct data type, different from integers or real numbers.

In[1]:= **z = 3 + 4 I**

Out[1]= 3 + 4 I

In[2]:= **Head[z]**

Out[2]= Complex

You can add and subtract complex numbers.

In[3]:= **z + (2 - I)**

Out[3]= 5 + 3 I

You can find the real and imaginary parts of any complex number.

In[4]:= **Re[z]**

Out[4]= 3

In[5]:= **Im[z]**

Out[5]= 4

The conjugate and absolute value can also be computed.

In[6]:= **Conjugate[z]**

Out[6]= 3 - 4 I

In[7]:= **Abs[z]**

Out[7]= 5

Recall that the absolute value of any number is its distance to the origin in the complex plane. Its phase angle is given by the argument.

In[8]:= **Arg[4 I]**

Out[8]= $\dfrac{Pi}{2}$

Each of these properties of complex numbers can be visualized geometrically, as shown in Figure 9.1.

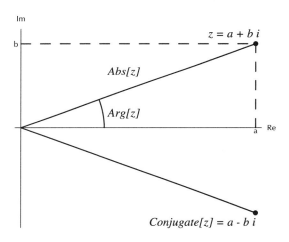

Figure 9.1: Geometric representation of complex numbers in the plane.

For purposes of pattern matching, complex numbers are quite similar to rational numbers. x_ + I y_ will not match with complex numbers. A complex number a + b I is treated as a single object for many operations, and is stored as Complex[a, b]. So to match a complex number z, use Complex[x_, y_] (or z_Complex and Re[z] and Im[z]) on the right-hand side of any rule you define.

9.1.4 | Computing with Different Number Types

When doing computations with numbers, *Mathematica* tries to work with the most general type of number in the expression at hand. For example, when adding two rational numbers, the sum is a rational number, unless of course it can be reduced to an integer.

```
In[1]:= 34/21 + 2/11
```

$$Out[1]= \frac{416}{231}$$

```
In[2]:= 3/4 + 9/4
```

Out[2]= 3

But if one of the terms is a real number, then all computations are done in real-number arithmetic.

```
In[3]:= Precision[10^100 + 1.3]
```

Out[3]= 16

One point to keep in mind is that when a Symbol is present in the expression to be computed, *Mathematica* does not convert the symbol to a machine number. This ability to perform *symbolic* computations is an extremely important feature that separates *Mathematica* from most other computer languages.

```
In[4]:= 1 + 2.1 + Pi
```

Out[4]= 3.1 + Pi

```
In[5]:= Expand[(a + b - 7.1)(c - d + EulerGamma)]
```

```
Out[5]= -7.1 c + a c + b c + 7.1 d - a d - b d -
           7.1 EulerGamma + a EulerGamma + b EulerGamma
```

When two extended precision approximate numbers are multiplied, the precision of the result will be the minimum of the precision of the two factors.

```
In[6]:= Precision[N[Sqrt[2], 50] * N[Sqrt[3], 80]]
```

Out[6]= 50

In fact, whenever two numbers are multiplied, the precision of the product will be the minimum of the precision of the factors, even if one factor is a machine precision real and the other factor is a high precision real number.

```
In[7]:= a = N[2];
```

```
In[8]:= b = N[2^99, 30];
```

```
In[9]:= {Precision[a], Precision[b], Precision[a b]}

Out[9]= {16, 30, 16}
```

For addition of real numbers, it is their *accuracy* that counts most. In *Mathematica*, Accuracy gives the number of significant digits to the right of the decimal point. In essence, Accuracy[x] measures the absolute error in the number x.

```
In[10]:= {Accuracy[1.23], Accuracy[12.5]}

Out[10]= {16, 15}
```

For machine-precision numbers, adding a digit to the left of the decimal point essentially removes one digit from the right of the decimal point. These numbers have a fixed number of digits. This is not the case though, for extended-precision numbers, where all the digits to the right of the decimal can be considered significant.

```
In[11]:= Accuracy[123.4444444444444444444444444444]

Out[11]= 28
```

```
In[12]:= Accuracy[12321.4444444444444444444444444444]

Out[12]= 28
```

In an analogous manner to the use of Precision with multiplication, the Accuracy of an addition will be the minimum of the accuracies of the summands.

```
In[13]:= Accuracy[1.23 + 12.3]

Out[13]= 15
```

This last point can lead to some unexpected results if you are not careful.

```
In[14]:= 1.0 + 10^(-25)
Out[14]= 1.
```

```
In[15]:= Accuracy[%]

Out[15]= 16
```

The number 1.0 is a machine number, so this computation was performed using machine accuracy, hence the 1 in the 25th decimal place to the right was lost in the rounding operation.

9.1.5 | Digits and Number Bases

A list of the digits of a number can be obtained with the commands `Integer-GerDigits` or `RealDigits`.

```
In[1]:= x = IntegerDigits[1293]
Out[1]= {1, 2, 9, 3}
```

You could square each of these digits,

```
In[2]:= x^2
Out[2]= {1, 4, 81, 9}
```

and add them together.

```
In[3]:= Apply[Plus, x^2]
Out[3]= 95
```

The exercises contain a problem on the sums of *cubes* of digits that will make use of this construction.

Converting from one number base to another is useful for many applications in programming. For example, some algorithms that run slowly in base 10 arithmetic can be sped up considerably by performing arithmetic operations in base 2^{32} or some other base that is a power of 2.

Numbers in base 10 can be displayed in other bases by means of the `BaseForm` function. For example, the following shows 18 in base 2:

```
In[4]:= BaseForm[18, 2]
Out[4]//BaseForm= 10010₂
```

The operator $b\hat{\ }\hat{\ }n$ takes the number n in base n and converts it to base 10.

```
In[5]:= 2^^10010
Out[5]= 18
```

For numbers in bases larger than 10, the letters of the alphabet are used. For example, here are the numbers 1 through 20 in base 16:

In[6]:= **Table[BaseForm[j, 16], {j, 1, 20}]**

Out[6]= {1_{16}, 2_{16}, 3_{16}, 4_{16}, 5_{16}, 6_{16}, 7_{16}, 8_{16}, 9_{16}, a_{16},

b_{16}, c_{16}, d_{16}, e_{16}, f_{16}, 10_{16}, 11_{16}, 12_{16}, 13_{16},

14_{16} }

Numbers other than integers can be represented in bases different from 10. Here are the first few digits of π in base 2:

In[7]:= **BaseForm[N[Pi, 5], 2]**

Out[7]//BaseForm= 11.0010010001_{2}

Recall that *Mathematica* is only displaying six significant *decimal* digits while storing quite a few more. In the exercises, you are asked to convert the base 2 representation back to base 10. You will need the digits from the base 2 representation which are obtained with the RealDigits[*number, base*] function.

In[8]:= **RealDigits[N[Pi], 2]**

Out[8]= {{1, 1, 0, 0, 1, 0, 0, 1, 0, 0, 0, 0, 1, 1, 1, 1, 1,
1, 0, 1, 1, 0, 1, 0, 1, 0, 1, 0, 0, 0, 1, 0, 0, 0,
1, 0, 0, 0, 0, 1, 0, 1, 1, 0, 1, 0, 0, 0, 1, 1, 0,
0, 0}, 2}

The 2 in this last result indicates where the binary point is placed and can be stripped off this list by wrapping the First function around the expression RealDigits[N[Pi], 2].

You are not restricted to integral bases such as in the previous examples. The base can be any real number greater than 1. For example,

In[9]:= **RealDigits[N[Pi], N[GoldenRatio]]**

Out[9]= {{1, 0, 0, 0, 1, 0, 0, 1, 0, 1, 0, 1, 0, 0, 1, 0, 0,
0, 1, 0, 1, 0, 1, 0, 1, 0, 0, 0, 0, 0, 1, 0, 1, 0,
0, 1, 0, 0, 0, 0, 1, 0, 0, 1, 0, 1, 0, 0, 0, 1, 0,
0, 0, 0, 0, 1, 0, 1, 0, 1, 0, 1, 0, 1, 0, 1, 0, 0,
0, 0, 0, 0, 1, 0, 0, 0}, 3}

Noninteger bases are used in certain branches of number theory.

Exercises

1. Define a function `complexToPolar` that converts complex numbers to their polar representations. Then, convert the numbers $3 + 3i$ and $e^{\pi i/3}$ to polar form.

2. How close is the number $e^{\pi\sqrt{163}}$ to an integer? Use `N`, but be careful about the `Precision` of your computations.

3. Using the built-in `Fold` function, write a function `convert[list, base]` that accepts a list of digits in any base (less than 20) and converts it to a base 10 number. For example, 1101_2 is 13 in base 10, so your function should handle this as follows:

   ```
   In[1]:= convert[{1, 1, 0, 1}, 2]
   Out[1]= 13
   ```

4. Write a function `sumsOfCubes[n]` that takes a positive integer argument n and computes the sums of cubes of the digits of n. (This exercise and the next three exercises are excerpted from an article by Allan Hayes [Hayes 1992])

5. Use `NestList` to iterate this process of summing cubes of digits. That is, generate a list, starting with an initial integer, say 4, of the successive sums of cubes of digits. For example, starting with 4, the list should look like: {4, 64, 280, 520, 133, ...}. Note, $64 = 4^3$, $280 = 6^3 + 4^3$, etc. Extend the list for at least fifteen values and make an observation about any patterns you notice. Experiment with other starting values.

6. Prove the following statements:

   (a) If n has more than four digits, then `sumsOfCubes[n]` has fewer digits than n.

   (b) If n has four digits or less, then `sumsOfCubes[n]` has four digits or less.

   (c) If n has four digits or less, then `sumsOfCubes[n]` $\leq 4 \cdot 9^3$.

   (d) If n is less than 2916, then `sumsOfCubes[n]` is less than 2916.

7. Write a function `sumsOfPowers[n, p]` that computes the sums of pth powers of n.

8. Binary shifts arise in the study of computer algorithms because they often allow one to speed up calculations by operating in base 2 or in bases that are powers of 2. Try to discover what a binary shift does by performing the following shift on 24 (base 10). First get the integer digits of 24 in base 2.

Exercises (cont.)

```
In[1]:= IntegerDigits[24, 2]

Out[1]= {1, 1, 0, 0, 0}
```

Then, do a binary shift, one place to the right.

```
In[2]:= RotateRight[%]

Out[2]= {0, 1, 1, 0, 0}
```

Finally, convert back to base 10.

```
In[3]:= 2^^01100

Out[3]= 12
```

Experiment with other numbers (including both odd and even integers) and make some conjectures.

9. The survivor[n] function from Chapter 4 (page 115) can be programmed using binary shifts. This can be done by rotating the base 2 digits of the number n by one unit to the left and then converting this rotated list back to base 10. For example, if $n = 10$, the base 2 representation is 1010_2; the binary shift gives 0101_2; converting this number back to base 10 gives 5, which is the output to survivor[5]. Program the a new survivor function using the binary shift.

Solutions

1. This function gives the polar form as a list consisting of the magnitude and the polar angle.

```
In[1]:= complexToPolar[z_] := {Abs[z], Arg[z]}
```

Here are the computations for the examples in the text.

```
In[2]:= complexToPolar[3 + 3 I]

                       Pi
Out[2]= {3 Sqrt[2], ──}
                        4
```

```
In[3]:= complexToPolar[E^(Pi I/3)]

                 Pi
Out[3]= {1, ──}
                  3
```

3. This function uses a default value of 2 for the base. (You should replace `Fold` with `FoldList` to more clearly see what this function is doing.)

```
In[1]:= convert[digits_List, base_:2] :=
            Fold[(base #1 + #2)&, 0, digits]
```

Here are the digits for 9 in base 2:

```
In[2]:= IntegerDigits[9, 2]
Out[2]= {1, 0, 0, 1}
```

This converts them back to the base 10 representation.

```
In[3]:= convert[%]
Out[3]= 9
```

This does the same for the number 129 in base 16.

```
In[4]:= IntegerDigits[129, 16]
Out[4]= {8, 1}
```

```
In[5]:= convert[%, 16]
Out[5]= 129
```

This function is essentially an implementation of Horner's method for fast polynomial multiplication.

```
In[6]:= convert[{a,b,c,d,e}, x]
Out[6]= e + x (d + x (c + x (b + a x)))
```

```
In[7]:= Expand[%]
```

$$Out[7]= e + d\,x + c\,x^2 + b\,x^3 + a\,x^4$$

5. Here is the `sumsOfCubes` function.

```
In[1]:= sumsOfCubes[n_Integer] := Apply[Plus, IntegerDigits[n]^3]
```

Here is the function that performs the iteration.

```
In[2]:= sumsOfSums[n_Integer, iter_]:=
            NestList[sumsOfCubes, n, iter]
```

We see that the number 4 enters into a cycle.

```
In[3]:= sumsOfSums[4, 12]
```
```
Out[3]= {4, 64, 280, 520, 133, 55, 250, 133, 55, 250, 133, 55,
         250}
```

In fact, it appears as if many initial values enter cycles.

```
In[4]:= sumsOfSums[32, 12]
```
```
Out[4]= {32, 35, 152, 134, 92, 737, 713, 371, 371, 371, 371,
         371, 371}
```

```
In[5]:= sumsOfSums[7, 12]
```
```
Out[5]= {7, 343, 118, 514, 190, 730, 370, 370, 370, 370, 370,
         370, 370}
```

```
In[6]:= sumsOfSums[372, 12]
```
```
Out[6]= {372, 378, 882, 1032, 36, 243, 99, 1458, 702, 351,
         153, 153, 153}
```

7. The function sumsOfPowers is a straightforward generalization of the previous cases.

```
In[1]:= sumsOfPowers[n_, p_] := Apply[Plus, IntegerDigits[n]^p]
```

```
In[2]:= sumsOfPowers[123, 5]
```
```
Out[2]= 276
```

9. Using the number 100 as an example, let's first put it in base 2,

```
In[1]:= BaseForm[100, 2]
```
```
Out[1]//BaseForm= 1100100
                         2
```

and then get a list of its digits.

```
In[2]:= IntegerDigits[100, 2]
Out[2]= {1, 1, 0, 0, 1, 0, 0}
```

This performs a binary shift of one unit,

```
In[3]:= l = RotateLeft[IntegerDigits[100, 2], 1]
Out[3]= {1, 0, 0, 1, 0, 0, 1}
```

and this converts back from base 2 to base 10 (using the convert function from Exercise 3).

```
In[4]:= convert[l, 2]
Out[4]= 73
```

Now we can put all of this code together to make the survivor function.

```
In[5]:= survivor[n_]:= Module[{p},
            p = RotateLeft[IntegerDigits[n, 2], 1];
            Fold[(2 #1 + #2)&, 0, p]]

In[6]:= survivor[100]
Out[6]= 73
```

You could of course do the same thing without the symbol p, but it is just a bit less readable.

```
In[7]:= survivor2[n_Integer]:= Fold[(2 #1 + #2)&, 0,
                RotateLeft[IntegerDigits[n, 2], 1]]

In[8]:= survivor2[100]
Out[8]= 73
```

9.2 Random Numbers

Statistical work and numerical experimentation often require random numbers to test hypotheses. We can generate a variety of random numbers in *Mathematica* by means of the Random function.

Here is a random real number between 0 and 1:

```
In[1]:= Random[]
```

```
Out[1]= 0.0560708
```

This generates a random integer in the range {0, 100}:

```
In[2]:= Random[Integer, {0, 100}]
```

```
Out[2]= 34
```

A good random number generator will distribute random numbers evenly over many trials. For example, this generates a list of 1000 integers between 0 and 9,

```
In[3]:= numbers = Table[Random[Integer, {0,9}], {1000}];
```

Here is a plot of the frequency with which each of the digits 0–9 occur.

```
In[4]:= (* Load definition of BarChart *)
        Needs["Graphics`Graphics`"]
```

```
In[5]:= (* Load definition of Frequencies *)
        Needs["Statistics`DataManipulation`"]
```

```
In[6]:= BarChart[Frequencies[numbers]]
```

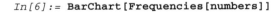

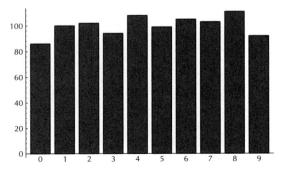

```
Out[6]= -Graphics-
```

We see each of the numbers 0–9 occur *roughly* 1/10 of the time. You wouldn't want these numbers to occur *exactly* 1/10 of the time, as there would be no randomness in this. In fact, for a uniform distribution of the numbers 0–9, any sequence of one thousand digits is equally as likely to occur as any other sequence of one thousand digits. A sequence of 1000 numbers that contains exactly 100 occurrences of the digit 0 followed by 100 occurrences of the digit 1, followed by 100 occurrences of the digit 2, . . ., is no more likely than the sequence that contains one thousand 7s, for example.

As a mathematical example of a typical use of random numbers, suppose you were searching for a formula for the sum of the squares of the first n integers: $\sum_{i=1}^{n} i^2$. You might compute a few sums first.

```
In[7]:= Sum[i^2, {i, 1, 3}]

Out[7]= 14
```

This command adds up the squares of i as i takes on the values from 1 to 3; that is, $1^2 + 2^2 + 3^2$. Here is a table of the first five sums of squares of the positive integers.

```
In[8]:= Table[Sum[i^2, {i, 1, n}], {n, 5}]

Out[8]= {1, 5, 14, 30, 55}
```

Suppose that after a bit of work you suspect that the general term is given by the formula $n(2n + 1)(n + 1)/6$. Before setting about proving such a fact, you could check it using random numbers.

```
In[9]:= numb = Random[Integer, {1, 1000}]

Out[9]= 299
```

```
In[10]:= Sum[i^2, {i, 1, numb}]

Out[10]= 8955050
```

This just computed the $\sum_{i=1}^{299} i^2$. Here is the closed form using the random number.

```
In[11]:= (2 numb + 1)(numb + 1)numb / 6

Out[11]= 8955050
```

You certainly would want more evidence than this that your conjecture is true. You could check 10 random integers in the range from 0 to 1000.

```
In[12]:= Table[(Sum[i^2, {i,x}] == (2x + 1)(x + 1)x/6 ) /.
            x -> Random[Integer, {1, 1000}], {10}]
```

```
Out[12]= {True, True, True, True, True, True, True, True, True,
            True}
```

Reasonably assured that your formula is correct, you could now set out to prove it (by mathematical induction).

Exercises

1. Simulate the throwing of two dice by defining a function `rollEm` that, when executed, displays two integers between 1 and 6.

2. Create a random two-dimensional image using `ListDensityPlot`.

3. A surprisingly simple pseudo-random number algorithm is the *linear congruential method*. It is quite easy to implement and has been studied extensively. Sequences of random numbers are generated by a formula such as:

$$x_{n+1} = x_n b + 1 \quad (\text{mod } m)$$

where x_0 is the *seed*, b is the *multiplier*, and m is the *modulus*. Recall that 7 mod 5 is the remainder upon dividing 7 by 5. This is represented in *Mathematica* as

```
In[1]:= Mod[7, 5]
```

```
Out[1]= 2
```

Implement the linear congruential method and test it with a variety of numbers m and b. If you find that the generator gets in a loop easily, try a large value for the modulus m. See [Knuth 1981] for a full treatment of random number generating algorithms.

4. Write a function `quadCong[a, b, c, m, x0]` that implements a *quadratic congruential method* where a, b, and c are the parameters, m is the modulus, and x0 is the starting value. The iteration is given by

$$x_{n+1} = (ax_n^2 + bx_n + c) \quad (\text{mod } m)$$

5. (a) Numerous tests are available for determining the effective *randomness* of a sequence. One of the more fundamental tests is known as the χ^2 (chi-square) test. It tests to see how evenly spread out the numbers appear in

Exercises (cont.)

the sequence and uses their frequency of occurrence. If n is the upper bound of a sequence of m positive numbers, then in a well distributed random sequence, we would expect about m/n copies of each number. To take into account the actual frequency with which each number occurs, the χ^2 test is implemented by the following formula:

$$\chi^2 = \frac{\sum_{1 \leq i \leq n}(f_i - m/n)^2}{m/n}$$

where f_i is the number of copies of i in the sequence. If the χ^2 statistic is close to n, then the numbers are reasonably random. In particular, we will consider the sequence sufficiently random, if the statistic is within $2\sqrt{n}$ of n.

Write a function chiSquare[list] that takes a sequence of numbers and returns the χ^2 statistic. You will find the built-in Count function helpful for calculating the frequencies.

(b) Determine the χ^2 statistic for a sequence of 1000 integers generated with the linear congruential method with $m = 381$, $b = 15$, and a starting value of 0.

6. John von Neumann, considered by many to be the "father of computer science", suggested a random number generator known as the *middle-square* method. Starting with a ten-digit integer, square the initial integer and then extract its middle ten digits to get the next number in the sequence. For example, starting with 1234567890, squaring it produces 1524157875019052100. The middle digits are 1578750190, so the sequence starts out 1234567890, 1578750190, 4521624250, ... Implement a middle square random number generator and then test it on a 1000-number sequence using the χ^2 test. Was the "father of computer science" a good random number generator?

Solutions

1. This function has a straightforward implementation. Each die can be viewed as a random integer between 1 and 6.

```
In[1]:= rollEm :=
          {Random[Integer,{1,6}], Random[Integer,{1,6}]}
```

```
In[2]:= rollEm

Out[2]= {2, 4}
```

Here are five rolls in a row.

```
In[3]:= Table[rollEm, {5}]

Out[3]= {{1, 4}, {5, 6}, {2, 4}, {5, 5}, {4, 1}}
```

3. Here is the linear congruential generator.

```
In[1]:= linearCong[x_] := Mod[b x + 1, m]
```

With modulus 100 and multiplier 15, this generator quickly gets into a cycle.

```
In[2]:= m = 100;
        b = 15;


In[3]:= NestList[linearCong, 5, 10]

Out[3]= {5, 76, 41, 16, 41, 16, 41, 16, 41, 16, 41}
```

With a larger modulus and multiplier, it *appears* as if this generator is doing better.

```
In[4]:= m = 381;
        b = 15;
```

Here are the first 60 terms starting with a seed of 0.

```
In[5]:= data = NestList[linearCong, 0, 60]

Out[5]= {0, 1, 16, 241, 187, 139, 181, 49, 355, 373, 262,
         121, 292, 190, 184, 94, 268, 211, 118, 247, 277,
         346, 238, 142, 226, 343, 193, 229, 7, 106, 67,
         244, 232, 52, 19, 286, 100, 358, 37, 175, 340,
         148, 316, 169, 250, 322, 259, 76, 379, 352,
         328, 349, 283, 55, 64, 199, 319, 214, 163,
         160, 115}
```

Sometimes it is hard to see if your generator is getting into a rut. Graphical analysis can help by allowing you to see patterns over larger domains. Here is a ListPlot of this sequence taken out to 5000 terms.

```
In[6]:= m = 381;
        b = 15;
```

```
In[7]:= ListPlot[NestList[linearCong, 0.0, 5000]]
```

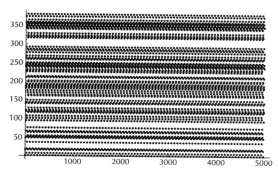

```
Out[7]= -Graphics-
```

It appears as if certain numbers are repeating. Looking at the plot of the Fourier data shows peaks at certain frequencies, indicating a periodic nature to the data.

```
In[8]:= ListPlot[Abs[Fourier[NestList[linearCong, 0.0, 5000]]]]
```

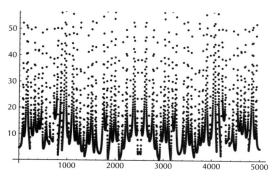

```
Out[8]= -Graphics-
```

Using a much larger modulus and multiplier (chosen carefully), you can keep your generator from getting in such short loops.

```
In[9]:= m = 2^16;
         b = 27421;
         ListPlot[data=NestList[linearCong, 0.0, 5000]]
```

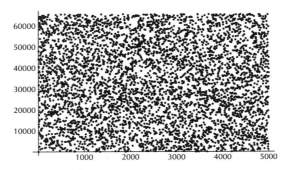

```
Out[9]= -Graphics-
```

```
In[10]:= ListPlot[Abs[Fourier[ data ]]]
```

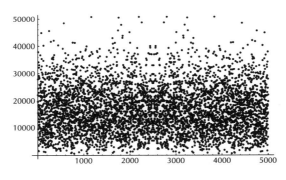

```
Out[10]= -Graphics-
```

5.a This implements the chi-square test.

```
In[1]:= chiSquare[l_List] :=
           Module[{m = Length[l], n = Max[l]},
               Sum[(Count[l,i] - m/n)^2, {i,1,n}] / (m/n)]
```

5.b Here is some data generated using the linear congruential generator with small modulus and multiplier.

```
In[2]:= m = 381;
         b = 15;
         data = NestList[linearCong, 0, 1000];
```

```
In[3]:= chiSquare[data]

        5018521
Out[3]= ————————
         1001
```

```
In[4]:= N[%]

Out[4]= 5013.51
```

Notice that the statistic is quite far from $2\sqrt{n}$ or n. This is a particularly pathological sequence. You can see a cycle of length 63.

```
In[5]:= m = 381;
        b = 15;
        NestList[linearCong, 0, 100]

Out[5]= {0, 1, 16, 241, 187, 139, 181, 49, 355, 373, 262,
         121, 292, 190, 184, 94, 268, 211, 118, 247,
         277, 346, 238, 142, 226, 343, 193, 229, 7, 106,
         67, 244, 232, 52, 19, 286, 100, 358, 37, 175,
         340, 148, 316, 169, 250, 322, 259, 76, 379,
         352, 328, 349, 283, 55, 64, 199, 319, 214, 163,
         160, 115, 202, 364, 127, 1, 16, 241, 187, 139,
         181, 49, 355, 373, 262, 121, 292, 190, 184, 94,
         268, 211, 118, 247, 277, 346, 238, 142, 226,
         343, 193, 229, 7, 106, 67, 244, 232, 52, 19,
         286, 100, 358}
```

Here are those positions that contain the number 1.

```
In[6]:= Position[%, 1]

Out[6]= {{2}, {65}}
```

9.3 Precision and Accuracy

When doing exact arithmetic—multiplying two integers, for example—*Mathematica* first checks that both numbers are in fact integers (actually, *machine integers*). If they are small and do not overflow the machine's arithmetic registers, then it goes ahead and multiplies them at the hardware level. If they are large (on most machines, integers are 32 bits long), then *Mathematica* goes to its extended precision algorithms and multiplies the integers there. In either case, all work done is *exact*.

When doing computations on inexact numbers, *Mathematica* uses two different types of arithmetic, depending upon the precision of the numbers involved. *Fixed precision* floating point arithmetic is used whenever the numbers can be handled in the machine's hardware routines. Sometimes, this arithmetic is referred to as *machine precision* arithmetic. In the previous section, we gave the following example:

```
In[1]:= {Precision[1.23], Accuracy[1.23]}

Out[1]= {16, 16}
```

Precision[*numb*] gives the total number of significant digits in the number *numb*, and Accuracy[*numb*] gives the number of significant digits to the right of the decimal point. So in our example above, we see that *Mathematica* has converted 1.23 to a machine floating point number and will use machine arithmetic on it whenever possible. The Accuracy of 16 in this example indicates that there are implicit trailing zeros in this number.

```
In[2]:= n = 12345.6789101112


In[3]:= {Precision[n], Accuracy[n]}

Out[3]= {16, 12}
```

When using N[*expr*], *Mathematica* uses machine-precision numbers as the default.

```
In[4]:= {N[Pi], Precision[N[Pi]]}

Out[4]= {3.14159, 16}
```

There are at least two situations where this observation needs to be kept in mind. When using N it is quite easy to forget just what this function does and does not do. Consider the following example:

```
In[5]:= x1 = N[Pi, 2]

Out[5]= 3.1


In[6]:= Sin[3.1]

Out[6]= 0.0415807
```

```
In[7]:= Sin[x1]
```

$$Out[7]= 1.22461\ 10^{-16}$$

This seemingly strange behavior—the fact that x1 does not appear to be equal to 3.1—can be explained by looking at the internal representation of x1:

```
In[8]:= InputForm[x1]
```

```
Out[8]//InputForm= 3.141592653589793
```

The command N[Pi, 2] causes *Mathematica* to first convert Pi to a machine precision number (on a NeXT, 16 significant digits), and then to display only two digits. Any computations with this number occur using the machine-precision.

Mathematica allows the user to adjust the precision of numbers with the Set-Precision command. It is to be noted though, that this command will not make an inexact number more exact. An example will point up some of the issues that arise with using this command.

```
In[9]:= a = SetPrecision[1/3, 30]
```

```
Out[9]= 0.333333333333333333333333333333
```

When SetPrecision is used with exact numbers, such as integers and rational numbers, it creates a few more bits than were asked for—thirty in this case. You can see this by increasing the precision.

```
In[10]:= b = SetPrecision[a, 50]
```

```
Out[10]= 0.33333333333333333333333333333333333333235375470765
```

When the number a was first created, the extended precision number was represented as a finite number of *binary* bits, followed by infinitely many (implicit) trailing zeros. Increasing the precision of this number uncovered the decimal digits which are not zeros. We can see this by converting 1/3 to a binary representation and then taking a finite number of the binary digits to convert back to base 10.

```
In[11]:= RealDigits[N[1/3], 2]
```

```
Out[11]= {{1, 0, 1, 0, 1, 0, 1, 0, 1, 0, 1, 0, 1, 0, 1, 0, 1,
           0, 1, 0, 1, 0, 1, 0, 1, 0, 1, 0, 1, 0, 1, 0,
           1, 0, 1, 0, 1, 0, 1, 0, 1, 0, 1, 0, 1, 0, 1,
           0, 1}, -1}
```

```
In[12]:= 2^^.010101
```

```
Out[12]= 0.33
```

```
In[13]:= FullForm[%]
```

```
Out[13]//FullForm= 0.328125
```

The other issue that you should be concerned with when working with inexact numbers is how precision and accuracy are affected by performing computations with these types of numbers. The magnification of error due to *roundoff* is discussed next.

9.3.1 | Roundoff Errors

Here is a machine-precision approximation to $\sqrt{2}$ raised to a large power.

```
In[14]:= N[Sqrt[2]]^200
```

$$Out[14]= 1.26765 \; 10^{30}$$

Working with approximations necessarily introduces some error. Comparing the machine-precision result with the exact result gives a measure of how the error is magnified.

```
In[15]:= % - Sqrt[2]^200
```

$$Out[15]= 1.74514 \; 10^{16}$$

That is an error of over 17 thousand trillion! This loss of accuracy is typically referred to as *roundoff error*. You can see how this loss gets progressively worse by repeating the above example for larger and larger exponents.

```
In[16]:= Table[N[Sqrt[2]]^j - Sqrt[2]^j, {j, 100, 1000, 100}]
```

$$Out[16]= \{7.75, \; 1.74514 \; 10^{16}, \; 2.9156 \; 10^{31}, \; 4.38879 \; 10^{46},$$
$$6.18671 \; 10^{61}, \; 8.36779 \; 10^{76}, \; 1.1 \; 10^{92}, \; 1.4105 \; 10^{107},$$
$$1.78821 \; 10^{122}, \; 2.23866 \; 10^{137}\}$$

```
In[17]:= Map[Accuracy, %]
```

```
Out[17]= {15, 0, -16, -31, -46, -61, -76, -91, -106, -121}
```

Recall that Accuracy[*n*] gives the number of significant digits to the right of the decimal point in *n*. The negative values indicate that the significant digits are to the *left* of the decimal point.

Exercises

1. Explain the output of the following:

```
Table[Cos[N[Pi/2, j]], {j, 1, 10}]
```

In particular, why are the ten elements of the output all identical?

2. Explain why the following computation produces an unexpected result (*i.e.*, why the value 0.00000000000001 is not returned).

```
In[1]:= 1.0 - 0.9999999999999
```

$$Out[1]= 1.00031\ 10^{-13}$$

3. Explain why *Mathematica* comes up with the following output:

```
In[1]:= (10^13 - 10^13) + 1.0/3.0
```

```
Out[1]= 0.333984
```

Consider the Attributes of the Plus function. How can you write the sum so that the correct value of 0.3333333333333333 is returned? You might want to use the Evaluate function in this exercise. (This example appeared in [Jacobson 1992].)

Solutions

1. First note the exact result.

```
In[1]:= Cos[Pi/2]
```

```
Out[1]= 0
```

```
In[2]:= Table[Cos[N[Pi/2, j]], {j, 1, 10}]

Out[2]= {6.12303 10   , 6.12303 10   , 6.12303 10   ,
                    -17            -17            -17

           6.12303 10   , 6.12303 10   , 6.12303 10   ,
                    -17            -17            -17

           6.12303 10   , 6.12303 10   , 6.12303 10   ,
                    -17            -17            -17

           6.12303 10   }
                    -17
```

It is instructive to look at the Precision of each result.

```
In[3]:= Precision[Cos[N[Pi/2, 10]]]

Out[3]= 16
```

So, each of these numbers is forced to be a machine-precision real. If you increased the precision past machine-precision, you would see different results.

```
In[4]:= Table[Cos[N[Pi/2, j]], {j, 10, 50, 10}]

Out[4]= {6.12303 10   , 0. 10   , 0. 10   , 0. 10   , 0. 10   }
                    -17       -32       -40       -50       -61
```

3. Here is the computation.

```
In[1]:= (10^13 - 10^13) + 1.0/3.0

Out[1]= 0.333984

In[2]:= Attributes[Plus]

Out[2]= {Flat, Listable, OneIdentity, Orderless, Protected}

In[3]:= FullForm[Hold[(10^13 - 10^13) + 1.0/3.0]]

Out[3]= Hold[Plus[Power[10, 13], Times[-1, Power[10, 13]],
            Times[1., Power[3., -1]]]]
```

Tracing the computation shows what is happening in *Mathematica*'s evaluation procedure. The next to last line below shows that *Mathematica* reorders the numbers placing 0.333333 in the middle of the two exact numbers. At this stage, *Mathematica* performs a floating point computation at machine precision using default accuracy on the first two numbers -10000000000000 and 0.333333.

$In[4]:=$ **Trace[(10^13 - 10^13) + 1.0/3.0]**

$Out[4]=$ {{10^{13}, 10000000000000},

{{10^{13}, 10000000000000}, -1 10000000000000,

-10000000000000}, {{$\frac{1}{3.}$, 0.333333}, 1. 0.333333,

0.333333 1., 0.333333},

10000000000000 - 10000000000000 + 0.333333,

-10000000000000 + 0.333333 + 10000000000000,

0.333984}

-10000000000000 + 0.333333

-1. 10^{13}

$In[5]:=$ **FullForm[%]**

$Out[5]=$ -9.99999999999967 10^{12}

Notice the decreased accuracy due to adding these two numbers (one with many digits to the right of the decimal and the other with many digits to the left of the decimal).

$In[6]:=$ **{Precision[%], Accuracy[%]}**

$Out[6]=$ {16, 3}

9.4 Numerical Computations

9.4.1 | Newton's Method Revisited

In Section 8.1.2 we wrote a program to implement Newton's method for finding roots of equations:

```
findRoot[fun_, init_, eps_] :=
   Module[{a = init, funa = fun[init]},
        While[Abs[funa] > eps,
             a = N[a - funa / fun'[a]];
             funa = fun[a]
             ];
        a]
```

Although findRoot will be sufficient for many types of functions, it can still give inaccurate results in certain situations. In this section we will explore the types of numerical problems that can arise in root finding and see how to avoid them.

As mentioned earlier, when an expression contains machine precision real numbers, computations with that expression will generally be performed at this low precision. If we start with an extended precision real number as the initial guess in Newton's method, we want to be assured that all calculations will be performed at this higher level of precision. This can be accomplished by first checking the precision of the initial guess and then increasing its precision so that all later calculations have this higher precision. By how much should we increase the precision? We will increase it by 12 digits, although a user may increase by any convenient amount. Here then is Newton's method using increased precision:

```
In[1]:= newton[fun_, init_]:=
          Module[{p = Precision[init], fi = fun[init]},
               x = SetPrecision[init, p + 12];
               While[Accuracy[fi] - Precision[fi] < p,
                    x = x - fun[x]/fun'[x];
                    fi = fun[x]
                    ];
               N[x, p]
          ]
```

This function differs from findRoot in a number of important ways. First, it determines the precision of the initial guess init and assigns this to the variable p. It then bumps up the precision of this number by 12 digits (x = SetPrecision[init, p + 12]). The iteration is continued until the number of leading zeros to the right of the decimal point in the result is at least as great as the Precision of the initial guess (While[Accuracy[fi] - Precision[fi] <p). Finally, the output is printed using the initial precision of the input (N[x, p]).

Let's use this function to calculate $\sqrt{2}$:

```
In[2]:= f[x_] := x^2 - 2
```

```
In[3]:= newton[f, 1.0]
```
```
Out[3]= 1.414213562373095
```

Notice that the output is at machine precision. If our input was an extended machine-precision number, the output from newton would reflect this.

```
In[4]:= x0 = N[Sqrt[3], 40];
```

```
In[5]:= newton[f, x0]
```
```
Out[5]= 1.4142135623730950488016887242096980785697
```

There are still a number of problems that can arise with our implementation of Newton's method. First is the possibility that the derivative of the function we are working with might be equal to zero. This will produce a division by zero error.

Another type of difficulty that can arise in root finding occurs when the derivative of the function in question is either difficult or impossible to compute. As a very simple example, consider the function $|x + 3|$ which has a root at $x = -3$. Both Find-Root and newton will fail with this function since a symbolic derivative cannot be computed. One way around such problems is to use a numerical derivative (as opposed to an analytic derivative). The *Secant method* approximates $f'(x_k)$ using the difference quotient:

$$\frac{f(x_k) - f(x_{k-1})}{x_k - x_{k-1}}$$

Although this will require two initial values to start, it has the advantage of not having to compute symbolic derivatives.

```
In[6]:= secant[f_, a_, b_]:=
           Module[{x1 = a, x2 = b},
                 While[Abs[ f[x2] ] > 10^(-10),
                       df = (f[x2] - f[x1])/(x2 - x1);
                       {x1, x2} = {x2, x2 - f[x2]/df}
                      ];
                 x2]


In[7]:= f[x_] := Abs[x + 3]


In[8]:= secant[f, -3.1, -1.8]

Out[8]= -3.
```

In the exercises, the reader is asked to refine this program and to compare it to previous methods.

9.4.2 | Gaussian Elimination Revisited Again

When solving the linear system $Ax = b$ by numerical techniques, some error is often introduced by using machine numbers as opposed to exact numbers. For many matrices A, there is little propagation of roundoff error. But for some matrices the error tends to magnify in a startling way and can lead to very incorrect results. Such matrices are called *ill-conditioned* and in this section, we will identify ill-conditioned matrices and discuss what to do about them when doing numerical linear algebra.

In Section 7.5, we used Gaussian elimination to solve the system $Ax = b$. In the exercises at the end of that section, we gave a very brief discussion of the conditions under which the method might fail. In this section we will give a more detailed treatment of the potential pitfalls with Gaussian elimination.

Since the method of Gaussian elimination is essentially list manipulation involving additions, subtractions, multiplications, and divisions, clearly one avenue of failure would be if we were to divide by zero. We formed what are commonly called *multipliers* (z, in the example below) as follows:

```
subtractE1[E1_, Ei_] :=
   Module[{z = Ei[[1]]/E1[[1]]},
         Module[{newE1 = z * Rest[E1]},
               Rest[Ei] - newE1]]
```

If the element E1[[1]] were ever equal to zero, the method would fail. Recall the example from the exercises at the end of Section 7.5:

```
In[1]:= m = {{0, 3}, {3, 0}};
        b = {5, 6};


In[2]:= m . {2, 5/3}

Out[2]= {5, 6}
```

The system m.x = b clearly has a solution x = {2, 5/3}. Unfortunately, the solve command we developed earlier in Section 7.5 will fail on this linear system:

```
In[3]:= solve[m, b]

                                         1
Out[3]= Power::infy: Infinite expression - encountered.
                                         0
        First::first:
            {} has a length of zero and no first element.
        Rest::norest:
            Cannot take Rest of expression {} with length zero.
        Part::partw: Part 1 of {} does not exist.

        General::stop:
            Further output of Part::partw
                will be suppressed during this calculation.
```

It is pretty clear that our solve command has not been written to take this situation into account. The problem can be remedied as suggested in Exercise 1 in Section 7.5, by interchanging rows (equations) so that the zero element is not in this *pivoting* position. Interchanging rows is equivalent to swapping equations, so this will not change the solution of the system in any way.

However, there is another problem that can arise when solving systems containing finite precision coefficients. The imprecision of the coefficients tends to become magnified in performing the necessary arithmetic. We can see this more clearly with an example.

Suppose we were using six-digit rounded arithmetic on the following system:

$$\begin{bmatrix} 0.000001 & 1.0 \\ 1.0 & 1.0 \end{bmatrix} \begin{bmatrix} x \\ y \end{bmatrix} = \begin{bmatrix} 1.0 \\ 2.0 \end{bmatrix}$$

The augmented matrix would look like:

$$\begin{bmatrix} 0.000001 & 1.0 & 1.0 \\ 1.0 & 1.0 & 2.0 \end{bmatrix}$$

Gaussian elimination would start solving this system by multiplying the first row by 10^6 (which contains seven digits) and subtracting from the second row. But six-digit rounded arithmetic would then produce:

$$\begin{bmatrix} 0.000001 & 1.0 & 1.0 \\ 0.0 & -1000000. & -1000000. \end{bmatrix}$$

Dividing the second row by $-1000000.$ gives the solution for y:

$$\begin{bmatrix} 0.000001 & 1.0 & 1.0 \\ 0.0 & 1. & 1. \end{bmatrix}$$

The second part of the back substitution gives the solution for x; that is, $((-1. \times 1.0)/0.000001)$:

$$\begin{bmatrix} 1.00000 & 0.0 & 0.0 \\ 0.0 & 1. & 1. \end{bmatrix}$$

This *solution* {x, y} = {0, 1}, is in fact, not the least bit close to the correct answer. A much more accurate solution is given by the ordered pair {x, y} = {1.000001000001, 0.999998999999}.

What has gone wrong? In general, accuracy is lost when the magnitude of the pivoting position is small compared to the remaining coefficients in that column. Pivoting can be used not only to avoid a zero element (when the matrix is nonsingular[2]), but also to minimize the potential for round-off errors. It does this by selecting the element from the remaining rows (equations) that is the maximum in absolute value. This will make the multiplier small and will have the effect of reducing possible roundoff errors. The following code selects this pivot and reorders the rows of the system accordingly:

```
In[4]:= pivot[S_] := Module[{p, ST1},
          ST1 = Abs[Transpose[S][[1]]];
          p = Position[ST1, Max[ST1]][[1, 1]];
          Join[{S[[p]]}, Delete[S, p]]]
```

Now the original `solve` function can be rewritten to pivot on this nonzero element. The new function is called `solvePP` (for "partial pivot"):

---

[2]A square matrix A is said to be *nonsingular* if it has an inverse; *i.e.*, if there exists a matrix B such that $AB = I$.

```
In[5]:= solvePP[m_, b_] :=
        Module[{elimx1, pivot, subtractE1, solve},

        elimx1[T_] := Map[subtractE1[T[[1]], #]&, Rest[T]];

        pivot[Q_] := Module[{p, ST1},
          ST1 = Abs[Transpose[Q][[1]]];
          p = Position[ST1, Max[ST1]][[1, 1]];
          Join[{Q[[p]]}, Delete[Q, p]]];

        subtractE1[E1_, Ei_] :=
          Module[{w = Ei[[1]]/E1[[1]]},
            Module[{newE1 = w * Rest[E1]},
              Rest[Ei] - newE1]];

        solve[{{a11_, b1_}}] := {b1/a11};
        solve[S_] :=
          Module[{S1 = pivot[S]},
            Module[{E1 = First[S1],
                    x2toxn = solve[elimx1[S1]]},
              Module[{b1 = Last[E1],
                      a11 = First[E1],
                      a12toa1n = Drop[Rest[E1], -1]},
                Join[{(b1 - a12toa1n . x2toxn)/a11},
                  x2toxn]]]];

        solve[Transpose[Append[Transpose[m], b]]]
        ]
```

Note that this function differs from the `solve` function given in Section 7.5 in that the user now gives the function the matrix m and column vector b, and `solvePP` will form the augmented matrix in the call to `solve` on the last line above.

We can quickly see how partial pivoting solves our first problem of division by zero. Solving the system given earlier with this new function now gives the correct result:

```
In[6]:= m = {{0, 3}, {3, 0}};
        b = {5, 6};


In[7]:= solvePP[m, b]

                5
Out[7]= {2, - }
                3
```

The problem with roundoff error can best be seen by constructing a matrix that would tend to produce quite large intermediate results relative to its original elements. One such class of matrices are referred to as *ill-conditioned* matrices, a complete study of which is outside the scope of this book. (The reader is encouraged to consult [Skeel and Keiper 1993] or [Burden and Faires 1989] for a comprehensive discussion of ill-conditioning.)

A set of classically ill-conditioned matrices are the *Hilbert matrices* which arise in numerical analysis in the solution of what are known as *orthogonal polynomials*. Recall the definition of the nth degree Hilbert matrix that we gave in Section 7.5:

```
In[8]:= HilbertMatrix[n_] := Table[1/(i + j - 1), {i, n}, {j, n}]
```

```
In[9]:= HilbertMatrix[3] //TableForm
```

$$Out[9]//TableForm= \begin{array}{ccc} 1 & \dfrac{1}{2} & \dfrac{1}{3} \\[2ex] \dfrac{1}{2} & \dfrac{1}{3} & \dfrac{1}{4} \\[2ex] \dfrac{1}{3} & \dfrac{1}{4} & \dfrac{1}{5} \end{array}$$

We will use the Hilbert matrices, but instead of working with exact arithmetic, we will work with floating point numbers.

```
In[10]:= NHilbertMatrix[n_] := Table[1.0/N[i + j - 1], {i, n}, {j, n}]
```

```
In[11]:= NHilbertMatrix[3] //TableForm
```

```
Out[11]//TableForm= 1.         0.5        0.333333
                     0.5        0.333333   0.25
                     0.333333   0.25       0.2
```

To compare the simple solver `solve` and the partial pivoting solver `solvePP` along with the built-in `LinearSolve`, we first construct a 50×50 Hilbert matrix and a random 50×1 column vector (and of course, suppress the display of the 2500 elements of the matrix and 50 elements of the column vector):

```
In[12]:= h50 = NHilbertMatrix[50];
```

```
In[13]:= b50 = Table[Random[], {50}];
```

Now let's use each of these three methods to find the solution vector x of the system h50.x = b50. We also give a measure of the total error involved in each case by computing the difference between h50.x and b50:

```
In[14]:= xLS = LinearSolve[h50, b50];

LinearSolve::luc:
    Warning: Result for LinearSolve
        of badly conditioned matrix {<<50>>}
        may contain significant numerical errors.


In[15]:= xGE = solve[h50, b50];


In[16]:= totalerrorGE = Apply[Plus, Abs[h50.xGE - b50]]
Out[16]= 208.6


In[17]:= xPP = solvePP[h50, b50];


In[18]:= totalerrorPP = Apply[Plus, Abs[h50.xPP - b50]]
Out[18]= 89.9402
```

Notice that LinearSolve is smart enough to recognize that the system is ill-conditioned. It is also nice enough to warn us.

It is no surprise that our initial implementation of Gaussian elimination (solve) had a greater total error than the built-in LinearSolve. As we mentioned above, the Hilbert matrices are very ill-conditioned and so we would expect that round-off error would be more significant without pivoting. (It should be noted that results will vary from machine to machine and from session to session. Each evaluation of b50 above will produce a different column vector.)

The importance of these numbers is that they tell us that there is a significant increase in error in using Gaussian elimination without pivoting. We have to be a bit careful in reading too much into them though. Quite a bit of roundoff error is present in these results. You should check that this is in fact the case by running the examples with smaller Hilbert matrices. The exercises outline a method to help reduce such potential roundoff error.

Exercises

1. Modify the function `newton` so that it prints the number of iterations required to compute the root.

2. Modify the function `secant` so that the user may indicate the degree of precision to which the root is computed.

3. Although the function $e^x - x - 1$ has a root at $x = 0$, it appears to be resistant to root finding. Use `newton`, `secant`, and the built-in `FindRoot` to try and compute the root. Then use the option `DampingFactor->2` on `FindRoot` and see if you can improve on the root finding capability of `FindRoot`.

4. Write a functional implementation of Newton's method. Your function should accept as arguments the name of a function and an initial guess. It should maintain the precision of the inputs and it should output the root at the precision of the initial guess, and the number of iterations required to compute the root. You should consider using the built-in functions `FixedPoint` or `Nest`.

5. Some functions tend to cause root finding methods to converge rather slowly. For example, the function $f(x) = \sin(x) - x$ requires over ten iterations of Newton's method with an initial guess of $x_0 = 0.1$ to get three-place accuracy. Implement the following acceleration of Newton's method and determine how many iterations of the function $f(x) = \sin(x) - x$, starting with $x_0 = 0.1$, are necessary for six-place accuracy.

$$\text{accelNewton}(x) = \frac{f(x)f'(x)}{[f'(x)]^2 - f(x)f''(x)}$$

This accelerated method is particularly useful for functions with multiple roots, such as are described in the previous section on Newton's method.

6. The *norm* of a matrix gives some measure of the size of that matrix. The norm of a matrix A is indicated by $\|A\|$. There are numerous matrix norms, but all share certain properties. For $n \times n$ matrices A and B,

   (i) $\|A\| \geq 0$,

   (ii) $\|A\| = 0$ if and only if A is the zero matrix,

   (iii) $\|cA\| = |c|\|A\|$ for any scalar c

   (iv) $\|A + B\| \leq \|A\| + \|B\|$

   (v) $\|AB\| \leq \|A\|\|B\|$

Exercises (cont.)

One particularly useful norm is the L_∞ norm, sometimes referred to as the *max norm*. For a vector $\vec{x} = (x_1, x_2, \ldots, x_n)$, this is defined as:

$$\|\vec{x}\|_\infty = \max_{1 \le i \le n} |x_i|$$

The corresponding matrix norm is defined similarly. Hence, for a matrix $A = a_{ij}$, we have

$$\|A\|_\infty = \max_{1 \le i \le n} \sum_{j=1}^{n} |a_{ij}|$$

This computes the sum of the absolute values of the elements in each row, and then takes the maximum of these sums. That is, the $\| \cdot \|_\infty$ norm is the max of the L_∞ norms of the rows.

Write a function `normInfinity` which takes a square matrix as an argument and outputs its $\| \cdot \|_\infty$ norm.

7. If a matrix A is nonsingular (*i.e.*, is invertible), then its *condition number* $c(A)$ is defined as $\|A\| \cdot \|A^{-1}\|$. A matrix is called *well-conditioned* if its condition number is close to 1 (the condition number of the identity matrix). A matrix is called *ill-conditioned* if its condition number is significantly larger than 1.

Write a function `conditionNumber[m]` that uses the `normInfinity` you defined in the previous exercise as an auxiliary function, and outputs the condition number of *m*.

Use your function `conditionNumber` to compute the condition number of the first ten Hilbert matrices.

8. An additional technique for solving linear systems of equations is known as *scaled pivoting*. Assuming that no column of a matrix *m* contains all zeros (in which case there would be no unique solution), then for each row, a scale factor is determined by selecting the element that is the largest in absolute value. That is, in row i, the scale factor is defined as $s_i = \max_{j=1,2,\ldots,n} |a_{ij}|$. Now a row interchange is determined by finding the first integer k such that

$$\frac{|a_{ki}|}{s_k} = \max_{j=1,2,\ldots,n} \frac{a_{ji}}{s_j}$$

Once such a k is found, then the ith row and the kth row are interchanged. The scaling itself is only done for comparison purposes so no additional roundoff error is introduced by the scaling factor.

Write a function `solveSPP` that implements scaled partial pivoting using the above description.

Solutions

1. Here is the original function as given in the text:

```
In[1]:= newton[fun_, init_]:=
         Module[{p = Precision[init], fi = fun[init]},
               x = SetPrecision[init, p + 12];
               While[Accuracy[fi] - Precision[fi] < p,
                     x = x - fun[x]/fun'[x];
                     fi = fun[x]
                     ];
               N[x, p]]
```

This first initializes a counter (n), and then prints out the result together with the number of iterations through the While loop.

```
In[2]:= newton[fun_, init_]:=
         Module[{p=Precision[init], fi=fun[init], n=0},
               x = SetPrecision[init, p + 12];
               While[n++;
                     Accuracy[fi] - Precision[fi] < p,
                     x = x - fun[x]/fun'[x];
                     fi = fun[x]
                     ];
               Print["The zero is ", N[x,p], " after ",
                     n, " iterations"]]
```

Here is an example of a function with a root at $\pi/2$.

```
In[3]:= f[x_] := Cos[x]
```

```
In[4]:= newton[f, 1.5]
```

```
The zero is 1.570796326794897 after 4 iterations
```

3. First we define the function.

```
In[1]:= f[x_] := E^x - x - 1
```

Here are some attempts at finding the zero, starting with an initial value that is quite close to the actual root.

```
In[2]:= newton[f, 0.0000001]
```

```
                              -8
The zero is 2.5000000625 10    after 3 iterations
```

```
In[3]:= secant[f, -.1, .1]
```

```
Out[3]= 0.0000107544
```

```
In[4]:= FindRoot[f[x]==0, {x, 0.0000001}]
```

$$Out[4]= \{x \to 5.11502 \ 10^{-8}\}$$

Setting the `DampingFactor` to a value of 2 increases the accuracy of the result.

```
In[5]:= FindRoot[f[x]==0, {x, 0.0000001},
              DampingFactor -> 2]
```

$$Out[5]= \{x \to 2.30038 \ 10^{-9}\}$$

6. Here is a three-dimensional vector.

```
In[1]:= v = {1, -3, 2};
```

This computes the $\| \cdot \|_\infty$ norm of the vector.

```
In[2]:= normInfinity[v_?VectorQ] := Max[Abs[v]]
```

```
In[3]:= normInfinity[v]
```

```
Out[3]= 3
```

Here is a 3×3 matrix.

```
In[4]:= m = {{1,2,3}, {1,0,2}, {2,-3,2}}
```

```
Out[4]= {{1, 2, 3}, {1, 0, 2}, {2, -3, 2}}
```

Here then, is the matrix norm.

```
In[5]:= normInfinity[m_?MatrixQ]:=
            normInfinity[Apply[Plus, Abs[Transpose[m]]]]
```

```
In[6]:= normInfinity[m]
```

```
Out[6]= 7
```

Notice how we *overloaded* the definition of the function normInfinity so that it would act differently depending upon what type of argument it was given. This is a particularly powerful feature of *Mathematica*. The expression _?MatrixQ on the left-hand side of the definition causes the function normInfinity to use the definition on the right-hand side *only if* the argument is in fact a matrix (*i.e.*, it passes the MatrixQ test). If that argument is a vector (*i.e.*, it passes the VectorQ test), then the previous definition is used.

Using the definition of HilbertMatrix from earlier in this chapter, here is its norm.

```
In[7]:= normInfinity[ HilbertMatrix[10] ]
```

$$Out[7]= \frac{7381}{2520}$$

7. Here is the function to compute the condition number of a matrix (using the l_∞ norm).

```
In[1]:= conditionNumber[m_?MatrixQ] :=
            normInfinity[m] normInfinity[Inverse[m]]
```

```
In[2]:= conditionNumber[HilbertMatrix[3]]
```

```
Out[2]= 748
```

Here then, are the condition numbers of the first ten Hilbert matrices.

```
In[3]:= Map[conditionNumber[HilbertMatrix[#]]&, Range[10]]
```

$$Out[3]= \{1, 27, 748, 28375, 943656, 29070279, \frac{1970389773}{2},$$

$$33872791095, \frac{2199309082685}{2}, 35357439251992\}$$

```
In[4]:= N[%]
```

$$Out[4]= \{1., 27., 748., 28375., 943656., 2.90703 \times 10^7,$$

$$9.85195 \times 10^8, 3.38728 \times 10^{10}, 1.09965 \times 10^{12},$$

$$3.53574 \times 10^{13}\}$$

10 | Graphics Programming

Graphics programming is a relatively new phenomenon. Hardware and software tools are now available to visualize data and functions and also to simulate very complex systems in a graphical manner. In this chapter we will discuss how to construct graphical images using *Mathematica*, and how to write programs that solve problems that are graphical in nature.

10.1 Graphics Primitives

All *Mathematica* graphics are constructed from objects called *graphics primitives*. These primitive elements (Point, Line, Polygon, Circle, etc.) are used by built-in functions such as Plot. Although it is quite straightforward to create images using *Mathematica's* built-in functions, we frequently find ourselves having to create a graphic image for which no *Mathematica* function exists. This is analogous to the situation in programming where we often have to write a specialized procedure to solve a particular problem. We use the basic building blocks and put them together according to the rules governing the structure of the language and the nature of the problem at hand. In this section we will look at the building blocks of graphics programming and at how we put them together to make graphics.

10.1.1 | Two-Dimensional Graphics Primitives

Graphics created with functions such as Plot and ListPlot, for example, are constructed of lines connecting points, with options governing the display. We can get some insight into this process by looking at the internal representation of a plot. Here is a plot of the Sin function.

```
In[1]:= sinplot = Plot[Sin[x], {x,0,2Pi}]
```

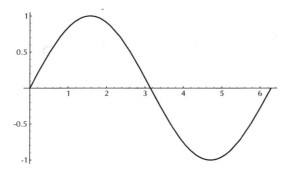

Out[1]= -Graphics-

Mathematica constructs plots by piecing together various graphics elements. The InputForm function displays the expression that we could have entered manually to get the same plot. We use Short to display an abbreviated 14-line listing of that expression. (Note: The formatted output from Short will vary slightly from one computer to another.)

```
In[2]:= Short[InputForm[sinplot], 14]

Out[2]//Short= Graphics[{{Line[{{0., 0.},
                {0.2617993877991494, 0.2588190451025207},
                {0.5235987755982988, 0.5},
                {0.7853981633974483, 0.7071067811865475},
                {0.916297857297023, 0.7933533402912352},
                {1.047197551196598, 0.866025403784439},
                {1.178097245096172, 0.923879532511287},
                <<69>>,
                {5.759586531581287, -0.5000000000000013},
                {6.021385919380436, -0.2588190451025224},
                {6.2831853071795, -(2.4492931757744*10^-16)}}]}
            }, {<<22>>}]
```

This graphic consists of a series of points connected by lines of a certain thickness. There are 79 points that are sampled to make this plot—the 10 displayed here together with 69 more indicated by the notation <<69>>. The <<22>> on the bottom indicates options (omitted from this display), which are discussed a little later. We will examine these graphics elements by constructing a graphic using *only* primitive elements. In a later section we will look into how the built-in functions such as Plot construct graphics out of the primitive elements.

In Section 9.1 on page 262, we displayed a graphic that demonstrated some of the properties of complex numbers. Let's show how this graphic was created, using *Mathematica*'s primitive elements.

The primitives we will work with here are `Point`, `Line`, `Circle`, and `Text`. In this particular example we will have no need for the remaining two-dimensional graphics primitives, `Rectangle`, `Polygon`, `Disk`, or `Raster` (see Table 10.1 at the end of this section for a complete listing).

The graphic we will create will contain the following:

- points in the plane at a complex number $a + bi$ and its conjugate $a - bi$

- lines drawn from the origin to each of these points

- an arc, indicating the polar angle of the complex number

- dashed lines indicating the real and imaginary values

- labels for each of the above elements

- a set of axes in the coordinate plane

First we choose a point in the first quadrant and then construct a line from the origin to this point.

```
In[3]:= z = 8 + 3I;
```

`Line[{{`x_1`,` y_1`},` `{`x_2`,` y_2`}}]` is a graphics primitive that creates a line from the point whose coordinates are (x_1, y_1) to the point (x_2, y_2).

```
In[4]:= line1 = Line[{{0, 0}, {Re[z], Im[z]}}];
```

Let's also create a point in the plane:

```
In[5]:= point1 = {PointSize[.02], Point[{Re[z], Im[z]}]};
```

We have added the graphics directive `PointSize` here so that our displayed point will be reasonably large. A *graphics directive* works by changing only those objects within its scope. In this case, that scope is delineated by the curly braces. The form for directives is {*directive, primitive*}. Additional primitives can also be placed in the scope of any directive, as in,

{ *dir, prim$_1$, prim$_2$, . . ., prim$_n$*}

where the directive *dir* will affect each of the primitives *prim*$_i$ occurring within its scope. You can place as many primitives as you like within the scope of each directive.

To display what we have created so far, we first wrap the Graphics function around the points and lines to turn them into *graphics objects*. Then we display the list of objects with the Show function:

```
In[6]:= Show[Graphics[{line1, point1}]]
```

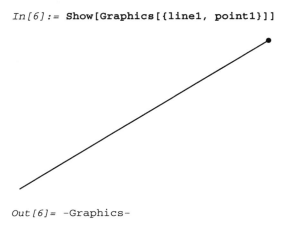

```
Out[6]= -Graphics-
```

Admittedly not too exciting, but it is a start. We can add additional graphics elements indicating the conjugate and a set of axes.

```
In[7]:= cz = Conjugate[z];
        line2 = Line[{{0, 0}, {Re[cz], Im[cz]}}];
        point2 = {PointSize[.02], Point[{Re[cz], Im[cz]}]};
```

```
In[8]:= Show[Graphics[{line1, point1, line2, point2}, Axes -> True]]
```

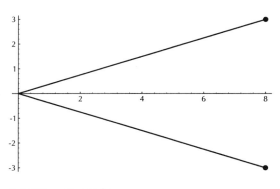

```
Out[8]= -Graphics-
```

Here we have used the Axes option. Options differ from directives in that they affect the entire graphic. Options to functions are placed after any required arguments and are separated by commas. Since Axes is an option to the Graphics function, it is placed after the graphics elements line1, point1, ...

Next, let's create dashed lines indicating the real and imaginary components of our complex number. We use the Dashing directive with Line to get the desired effect.

```
In[9]:= hline = {Dashing[{0.04, 0.04}],
              Line[{{0, Im[z]}, {Re[z], Im[z]}}]};
```

```
In[10]:= vline = {Dashing[{0.04, 0.04}],
              Line[{{Re[z], 0}, {Re[z], Im[z]}}]};
```

Since we were using this graphic to display an arbitrary complex number, we are not interested in the units on the axes, so we suppress the default value and add our own with the Ticks option. Ticks -> {{{Re[z], "a"}}, {{Im[z], "b"}}} places tick marks at Re[z] on the horizontal axis and at Im[z] on the vertical axis and labels them a and b, respectively. In addition, let's add labels on the axes and make the AspectRatio more natural (see page 46 for a description of the AspectRatio option).

```
In[11]:= Show[Graphics[{line1, point1, line2, point2, hline, vline},
              Axes -> True, AxesLabel -> {Re, Im},
              Ticks -> {{{Re[z], "a"}}, {{Im[z], "b"}}},
              AspectRatio -> Automatic]];
```

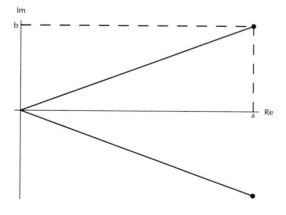

Out[11]= -Graphics-

We now wish to put labels at the two complex numbers and along the line representing the length `Abs[z]`. We will use another graphics element, `Text`, to place text where we need it.

`Text[`*expr*`, {`*x, y*`}]` will create a text object of the expression *expr* and center it at (x, y). So to create $z = a + bi$ as a piece of text centered at a point a little bit above and to the left of z, we use

```
Text["z = a + b i", {Re[z] - 0.75, Im[z] + 0.35}]
```

We are going to add one further element to this graphic object. We would like this text to use a different font and a different size than the default of **Courier**, 10 point. Using `FontForm` we can specify any available font and size. In this example we use the **Times-Italic** font[1] at 9 points.

```
Text[FontForm["z = a + b i", {"Times-Italic", 9}],
    {Re[z] - 0.75, Im[z] + 0.35}]
```

Here then are the labels for the complex number, its conjugate, and the length given by the absolute value of the complex number.

```
In[12]:= text1 = Text[FontForm["z = a + b i", {"Times-Italic", 9}],
                {Re[z] - .75, Im[z] + .35}];


In[13]:= text2 = Text[FontForm["Abs[z]", {"Times-Italic", 9}],
                {3.5, 2}];


In[14]:= text3 = Text[FontForm["Conjugate[z] = a - b i",
                    {"Times-Italic", 9}],
                {Re[cz] - 1.4, Im[cz] - .35}];
```

---

[1] Names of fonts will vary on different computers. Users should check their *Mathematica* documentation for font naming conventions.

```
In[15]:= Show[Graphics[{line1, line2, point1, point2,
                hline, vline, text1, text2, text3},
            Axes -> True,
            AxesLabel -> {Re,Im},
            Ticks -> {{{Re[z], "a"}}, {{Im[z], "b"}}},
            AspectRatio -> Automatic]]
```

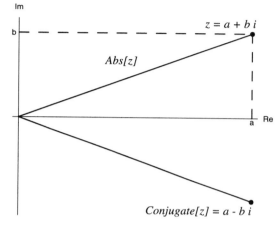

```
Out[15]= -Graphics-
```

Lastly, we need to add the arc representing the polar angle and label it. The arc can be generated with another graphic primitive. Circle[{x, y}, r, {a, b}] will draw an arc of a circle centered at (x, y), of radius r, counterclockwise from an angle of a radians to an angle of b radians. The arc that we are interested in will have a radius smaller than Abs[z] and will be drawn from the real (horizontal) axis to the line connecting the origin and z. Here is the code for the arc and its label, as well as the graphic containing all of the above elements.

```
In[16]:= arc = Circle[{0, 0}, Abs[z]/3, {0, Arg[z]}];
```

```
In[17]:= text4 = Text[FontForm["Arg[z]", {"Times-Italic", 9}],
                {3.5, .5}];
```

```
In[18]:= Show[Graphics[
              {line1, line2, point1, point2, hline, vline,
               text1, text2, text3, text4, arc},
            Axes -> True, AxesLabel -> {Re, Im},
            Ticks -> {{{Re[z], "a"}}, {{Im[z], "b"}}},
            AspectRatio -> Automatic]]
```

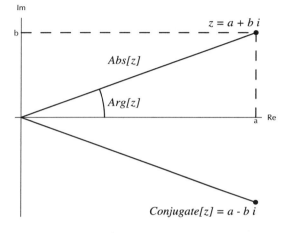

```
Out[18]= -Graphics-
```

The following table lists the graphics primitives that we have used in addition to the remaining two-dimensional elements that are available. Note that three-dimensional versions of Point, Line, Polygon, and Text are also available for constructing three-dimensional graphics. You should consult the next section and the reference manual [Wolfram 1991] for details.

Graphics element	Usage
Point[{x, y}]	a point at position x, y
Line[{{x_1, y_1}, {x_2, y_2}, ...}]	a line through the points {x_i, y_i}
Rectangle[{{$xmin, ymin$}, {$xmax, ymax$}}]	a filled rectangle
Polygon[{{x_1, y_1}, {x_2, y_2}, ...]}	a filled polygon
Circle[{x, y}, r, {$theta_1, theta_2$}]	a circular arc of radius r
Disk[{x, y}, r]	filled disk of radius r
Raster[{{$x_{11}, x_{12}, ...$}, {$x_{21}, x_{22}, ...$}, ...}]	a rectangular array of gray levels
Text[$expr$, {x, y}]	text centered at x, y

Table 10.1: Graphics primitives.

10.1.2 | Three-Dimensional Graphics Primitives

As an example of the use of graphics primitives in three dimensions, we will construct a *random walk* in space. We could just create triples of points using the Random function and then connect them, but this is not really a random walk. We would like each point to be obtained from the previous point in a somewhat random manner.

From a given point, we will obtain the next point by first generating a random number and obtaining a *step* from it as follows:

- the *x*-coordinate of the step is the cosine of this random number;

- the *y*-coordinate of the step is the sine of this random number;

- the *z*-coordinate of the step is a number between −1 and 1 that will be generated by the given random number.

First we generate a triple according to the above specification.

```
In[1]:= Map[{Cos[#], Sin[#], #/N[Pi] - 1}&,
           {Random[Real, {0, N[2Pi]}]}]

Out[1]= {{0.93858, 0.34506, -0.887858}}
```

Note that we are using a random number between 0 and 2π as this gives the *x* and *y* coordinates the ability to move in 360 degrees. To get the *z* coordinate between −1 and 1, we have to scale the random number and translate it (#/N[Pi] - 1).

Now to get a walk, we want to add each step to the location computed by adding all of the previous steps starting at (0, 0, 0). Here is the function that will create a list of *n* points according to the specifications listed above:

```
In[2]:= randomWalk3D[n_Integer]:=
           FoldList[Plus,{0, 0, 0},
               Map[{Cos[#], Sin[#], #/N[Pi] - 1}&,
                   Table[Random[Real, {0, N[2Pi]}], {n}]]]
```

We now wish to connect these points in order with a line and display them in three dimensions. The Line primitive will connect points in order and Graphics3D will create a three-dimensional graphics object that can then be displayed with Show.

```
In[3]:= showRandomWalk3D[n_Integer]:=
           Show[Graphics3D[Line[randomWalk3D[n]]]]
```

Finally, here is a 1000-step walk in three dimensions.

In[4]:= **showRandomWalk3D[1000]**

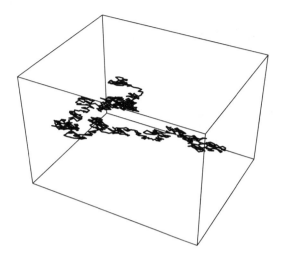

Out[4]= -Graphics3D-

The exercises in this and the next section contain some problems that expand upon this random walk, asking you to color the walk and to normalize the length of each step.

Exercises

1. Create a graphic that consists of points randomly placed in the unit square.

2. Create a graphic that contains one each of a circle, a triangle, and a rectangle. Your graphic should include an identifying label for each object.

3. Create a three-dimensional graphic containing 6 Cuboid graphics primitives randomly placed in the unit cube.

4. Write a function showRandomWalkPL that displays a three-dimensional random walk, with points connected by lines.

5. Modify the function showRandomWalk3D so that each step has unit length.

6. Load the package Graphics`Polyhedra` and then display each of the solids defined in the package, including Tetrahedron, Octahedron, Icosahedron, Cuboid, and the Dodecahedron.

Solutions

1. This creates a list of *n* coordinates in the unit square.

```
In[1]:= coords[n_] = Table[Random[], {n}, {2}]
```

This displays them.

```
In[2]:= Show[Graphics[Map[Point,coords[10]]]]
```

```
Out[2]= -Graphics-
```

This makes the points a bit larger and colors them red.

```
In[3]:= Show[Graphics[
            {RGBColor[1,0,0], PointSize[.03],
             Map[Point, coords[10]]}]]
```

```
Out[3]= -Graphics-
```

2. Here is the circle graphic primitive together with a text label.

```
In[1]:= c = Circle[{0,0}, 1];
```

```
In[2]:= ctext = Text[FontForm["Circle", {"Times-Italic",12}],
                 {Cos[5Pi/4] + .25, Sin[5Pi/4]}];
```

This generates the graphics primitive for the triangle and its text label.

```
In[3]:= t = Line[{{-1,0},{0,1},{1,0},{-1,0}}];
```

```
In[4]:= ttext = Text[FontForm["Triangle",{"Times-Italic",12}],
                 {0, 0+.05}];
```

Here is the rectangle and label.

```
In[5]:= r = Line[{{-1,-1}, {-1,1}, {2,1}, {2,-1}, {-1,-1}}];
```

```
In[6]:= rtext = Text[FontForm["Rectangle", {"Times-Italic",12}],
                 {1.5, -1+.05}];
```

Finally, this displays each of these graphics elements all together.

```
In[7]:= Show[Graphics[{c,t,r,ctext,ttext,rtext}],
            AspectRatio -> Automatic]
```

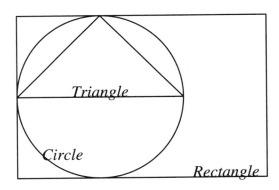

```
Out[7]= -Graphics-
```

3. First, we need to create the cuboid graphic object. Cuboid takes a list of three numbers as the coordinates of lower-left corner. This maps the object across two such lists:

```
In[1]:= Map[Cuboid[#]&, Table[Random[], {2}, {3}]]

Out[1]= {Cuboid[{0.502832, 0.839976, 0.943519}],
            Cuboid[{0.992707, 0.842744, 0.763264}]}
```

Here is a list of six cuboids and the resulting graphic. Notice the large amount of overlap of the cubes. You can reduce the large overlap by specifying minimum *and* maximum values of the cuboid.

```
In[2]:= cubes = Map[Cuboid[#]&, Table[Random[],{6},{3}]];
```

```
In[3]:= Show[Graphics3D[cubes]]
```

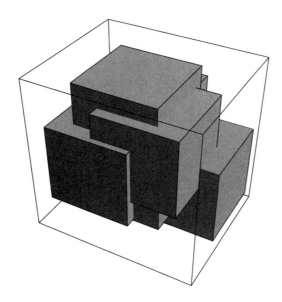

```
Out[3]= -Graphics3D-
```

5. Here is the function to create a three-dimensional random walk, scaling the steps so that each is of unit length.

```
In[1]:= randomWalk3D[n_Integer]:=
            Module[{m = Random[Integer, {1,5}]},
                FoldList[Plus, {0,0,0},
                    Map[{Cos[#]/m, Sin[#]/m, Sqrt[m^2-1]/m}&,
                        Table[Random[Real, {0,N[2Pi]}], {n}]]
            ]]
```

Here are two steps from the origin.

```
In[2]:= r3 = randomWalk3D[2]
```

$$Out[2]= \{\{0, 0, 0\}, \{-0.16676, -0.110413, \frac{2\,Sqrt[6]}{5}\},$$

$$\{-0.364998, -0.136904, \frac{4\,Sqrt[6]}{5}\}\}$$

This function computes the distance between any two vectors *a* and *b*.

```
In[3]:= distance[{a_, b_}]:=
            Sqrt[Apply[Plus, (b-a)^2]]
```

As can be seen, the distance between the second and third steps is 1.

```
In[4]:= distance[{r3[[2]], r3[[3]]}]
```

```
Out[4]= 1.
```

6. This loads the package containing the definitions for the polyhedra.

```
In[1]:= Needs["Graphics`Polyhedra`"]
```

It is often helpful to get a list of the functions defined in a recently loaded package.

```
In[2]:= Names["Graphics`Polyhedra`*"]
```

```
Out[2]= {Cube, Dodecahedron, Faces, Geodesate,
            GreatDodecahedron, GreatIcosahedron,
            GreatStellatedDodecahedron, Hexahedron, Icosahedron,
            Octahedron, OpenTruncate, Polyhedra, Polyhedron,
            SmallStellatedDodecahedron, Stellate, Tetrahedron,
            Truncate, Vertices}
```

Here are the solids displayed in an array.

```
In[3]:= solids = Map[Graphics3D,
            {Cube[], Dodecahedron[], GreatDodecahedron[],
             GreatIcosahedron[], GreatStellatedDodecahedron[],
             Icosahedron[], Octahedron[], Tetrahedron[],
             SmallStellatedDodecahedron[]}];
```

```
In[4]:= gr = {Map[solids[[#]]&, Range[5]],
            Map[solids[[#]]&, Range[6,9]]};
```

```
In[5]:= Show[GraphicsArray[ gr ]]
```

```
Out[5]= -GraphicsArray-
```

10.2 Graphics Directives and Options

In the previous section we saw how to construct a graphics object from primitive elements. We used directives (and options) to change various elements of the graphic. Options work in a global sense, by changing the way the entire graphic object is displayed. To change only certain elements of the graphic, we use graphics directives.

Graphics directives work by changing only certain objects within part of a graphics expression. Those objects that the directive operates on are included in the *scope* of the directive. We can make this clear with two examples. In the first example, we apply a directive to only some of the graphic elements.

```
In[1]:= p1 = Point[{0, 0}];
        p2 = Point[{1, 0}];
        p3 = Point[{.5, .5}];
```

```
In[2]:= Show[Graphics[{{PointSize[.025], p1, p2}, p3}]]
```

```
Out[2]= -Graphics-
```

Note that the directive `PointSize` only affected the points in its scope (p1 and p2). The elements in its scope are delineated by the braces. Since p3 is not in the scope of this `PointSize` directive, its point size is determined by the default value for graphics objects.

As a second example, we consider an algebraic problem that appeared in [Porta, Davis and Uhl 1994]:

Find the positive numbers r such that the system

$$(x-1)^2 + (y-1)^2 = 2$$
$$(x+3)^2 + (y-4)^2 = r^2$$

has exactly one solution in x and y. Once you have found the right number r, then plot the resulting circles in true scale on the same axes.

We leave the algebraic solution to the reader. It is given by

$$r = \frac{10 \pm 2^{3/2}}{2}$$

To display the solution, we will plot the first circle with solid lines and the two solutions with dashed lines together in one graphic. Here is the first circle centered at $(1, 1)$.

```
In[3]:= c = Circle[{1, 1}, Sqrt[2]];
```

```
In[4]:= Show[Graphics[c, Axes -> Automatic,
                       AspectRatio -> Automatic]]
```

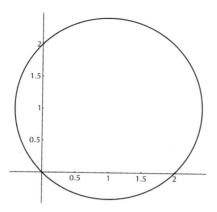

```
Out[4]= -Graphics-
```

Notice that we have used the Axes and AspectRatio options because we want these commands to apply to the entire graphic.

Here are the circles that represent the solution to the problem.

```
In[5]:= r1 = (10 + 2^(3/2))/2;
        r2 = (10 - 2^(3/2))/2;
```

```
In[6]:= Show[Graphics[{c, Circle[{-3, 4}, r1], Circle[{-3, 4}, r2]},
           Axes -> Automatic, AspectRatio -> Automatic]]
```

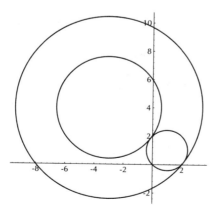

```
Out[6]= -Graphics-
```

We wanted to display the solutions (two circles) using dashed lines. The graphics directive `Dashing[{x, y}]` directs all subsequent lines to be plotted as dashed, alternating the dash x units and the space y units. We use it as a graphics directive on the two circles `c1` and `c2`. The important point to note here is that each of the circles inherits only those directives in whose scope they appear.

```
In[7]:= dashc1 = {Dashing[{.025, .025}], c1};
        dashc2 = {Dashing[{.05, .05}], c2};
```

```
In[8]:= Show[Graphics[{c, dashc1, dashc2},
            Axes -> Automatic, AspectRatio -> Automatic]]
```

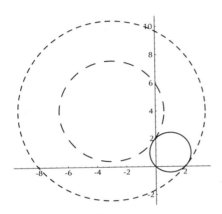

```
Out[8]= -Graphics-
```

Directive	Usage
`AbsoluteDashing[{`d_1`, `d_2`, ... }]`	dashed line segments using absolute units
`AbsoluteThickness[`d`]`	lines of thickness d measured in absolute units
`CMYKColor[{`c`, `m`, `y`, `b`}]`	cyan, magenta, yellow, and black of four color process printing
`Dashing[{`d_1`, `d_2`, ... }]`	dashed line segments of length d_1, d_2, ...
`GrayLevel[`d`]`	gray between 0 (black) and 1 white
`Hue[`h`, `s`, `b`]`	color with hue, saturation, and brightness between 0 and 1
`PointSize[`r`]`	points of radius r given as a fraction of the width of the entire plot
`RGBColor[`r`, `g`, `b`]`	color with red, green, and blue components between 0 and 1
`Thickness[`d`]`	lines of thickness d given as a fraction of the width of the entire plot

Table 10.2: *Mathematica* graphics directives.

Exercises

1. Create a primitive color wheel by coloring successive sectors of a `Disk` according to the `Hue` directive (or `GrayLevel` for those using black and white systems).

2. Modify the graphic in the `showRandomWalk3D` function from the previous section so that the points are colored according to their height along the vertical axis. Use the graphics directive `Hue` or the directive `GrayLevel`.

3. Create a graphic that consists of 500 points placed randomly in the unit square. The points should be of random radii between .01 and .1 units, and colored randomly according to a `Hue` function.

4. Modify the function `showRandomWalk3D` given in the previous section by coloring each step according to its total length.

Solutions

1. The color wheel can be obtained by mapping the `Hue` directive over successive sectors of a disk. Note that the argument to `Hue` must be scaled so that it falls within the range 0 to 1.

```
In[1]:= colorWheel[n_]:=
          Show[Graphics[
            Map[{Hue[#/(2Pi-n)], Disk[{0,0}, 1, {#,#+n}]}&,
              Range[0, 2Pi-n, n]],
            AspectRatio -> Automatic]]
```

Here is a color wheel created from 256 separate sectors (hues).

```
In[2]:= colorWheel[Pi/256]
```

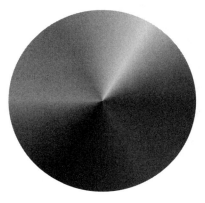

```
Out[2]= -Graphics-
```

3.   First we create the `Point` graphics primitive randomly placed in the unit square.

> `In[3]:= randomcoords := Point[{Random[],Random[]}];`

This creates the point sizes according to the specification given in the statement of the problem.

> `In[4]:= randomsize := PointSize[Random[Real, {.01, .1}]]`

This will assign a random color to each primitive.

> `In[5]:= randomcolor := Hue[Random[]]`

Here then are 500 points. (You may find it instructive to look at just one of these points.)

> `In[6]:= pts =`
> `        Table[{randomcolor, randomsize, randomcoords}, {500}];`

And here is the graphic.

> `In[7]:= Show[Graphics[pts, PlotRange -> All]]`

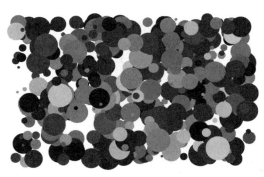

> `Out[7]= -Graphics-`

10.3 Built-In Graphics Functions

Mathematica has a wide range of tools for producing two- and three-dimensional graphics. As we will see, these functions build up lists of primitive elements using any directives necessary, and then add on options before displaying the object. In this section we will look at the structure of *Mathematica's* built-in graphics functions and we will discuss some of the pitfalls to watch out for in constructing graphics.

10.3.1 | The Structure of Built-In Graphics

When *Mathematica* creates a graphic using a function such as `Plot`, it evaluates the function to plot at several points determined by user-supplied information as well as by internal algorithms. For example, here is a plot of `Sin[x]` over the interval from 0 to 2π.

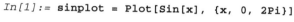

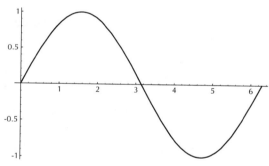

Out[1]= -Graphics-

Notice that the `Output` cell displays the word `Graphics`. This is one type of *graphics object*.[2] It is produced by functions such as `Plot`, `ListPlot`, and `ParametricPlot`. Another type of graphics object—`DensityGraphics`—is created by the `Density-Plot` and `ListDensityPlot` functions.

In[2]:= **sampledata = Table[Random[], {50}, {50}];**

---

[2]More specifically, the output -Graphics- is an abbreviation for a large expression that contains all the information needed to redraw the plot.

In[3]:= **ListDensityPlot[sampledata]**

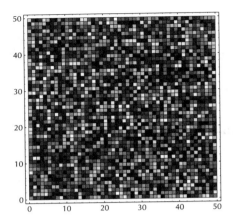

Out[3]= -DensityGraphics-

The other kinds of graphics objects are ContourGraphics, SurfaceGraphics, Graphics3D, and Sound. These can be generated by ContourPlot, Plot3D, ParametricPlot3D, and Play, respectively.

When *Mathematica* constructs a graphic such as the plot of Sin[x] above, it first checks that the syntax of the Plot command is correct. Once the expression is parsed correctly, *Mathematica* uses its internal algorithms to start computing the values of Sin[x] at several values over the user-supplied interval (0 to 2π in this case). Various options determine how the plot is constructed and displayed. For example, the algorithms internal to the Plot command tell it to first break up the interval into 24 evenly spaced subintervals.[3]

In[4]:= **Options[Plot, PlotPoints]**

Out[4]= {PlotPoints -> 25}

Mathematica then computes Sin[x] at the first few values and then uses an adaptive algorithm to determine whether or not to sample more points. If the function has a big second derivative (large curvature) between two sets of points, then *Mathematica* samples more points. If it is not changing too much, then *Mathematica* views the function as linear in this local region and connects these few points with a straight line. It then moves on to the next set of points and repeats the process. Finally, it displays these lines together with any axes and labels.

---

[3] A total of 25 points are needed to divide an interval into 24 subintervals.

To get a bit more insight into just how *Mathematica* determines when to connect consecutive points with a line and when to further subdivide an interval, consider the MaxBend and PlotDivision options to the Plot function.

```
In[5]:= Options[Plot, {MaxBend, PlotDivision}]
```

```
Out[5]= {MaxBend -> 10., PlotDivision -> 20.}
```

These default values tell *Mathematica* to connect points so long as the angle between successive pairs does not exceed 10 degrees. If the angle is greater than 10 degrees, then the function is changing rapidly, and hence more points should be sampled. Plot will subdivide by no more than a factor of 20 (the default value of PlotDivision). Notice what happens if we specify only four points equally spaced in the plotting interval, a MaxBend of 45 degrees and PlotDivision of 1; that is, allow large slope changes and do not subdivide.

```
In[6]:= Plot[Sin[x], {x, 0, 2Pi},
            PlotPoints -> 4, MaxBend -> 45,
            PlotDivision -> 1]
```

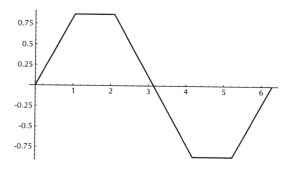

```
Out[6]= -Graphics-
```

10.3.2 | Graphics Anomalies

As mentioned above, if the algorithms used to construct a Plot determine that the angle between successive pairs of points is less than MaxBend, no further subdividing takes place, the points are connected by a line, and the algorithm goes on to the next set of points. This procedure can cause some undesirable behavior if one is not careful. For example, consider the plot of Cos[x] over the interval from 0 to 48Pi.

In[7]:= **Plot[Cos[x], {x, 0, 2Pi (24)}]**

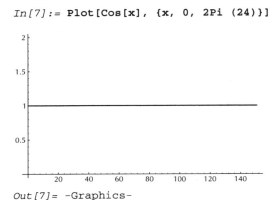

Out[7]= -Graphics-

This is typical of what are called *sampling errors* although it is not really an error on the part of *Mathematica*. *Mathematica* has split the interval [0, 2Pi (24)] into 24 *evenly spaced* subintervals and compute Cos[x] at each of the 25 points that determined this partition. This function oscillates 24 times on the given interval. The reader should look at the InputForm of this plot to see that at each of the sampled points (multiples of 2 Pi), the function has evaluated to 1. Believing that the function was not changing, *Mathematica* decided that it was unnecessary to subdivide the interval and sample more points, so it just connected these points with lines.

Plot3D, on the other hand, does not use an adaptive algorithm. It uses the default value for PlotPoints in both the *x* and *y* directions to construct the graphic.

In[8]:= **Options[Plot3D, PlotPoints]**

Out[8]= {PlotPoints -> 15}

In[9]:= **Plot3D[Sin[x y], {x,0,3Pi/2}, {y,0,3Pi/2}]**

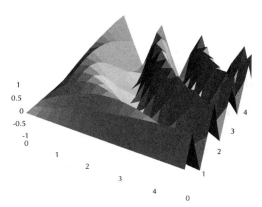

Out[9]= -SurfaceGraphics-

You can actually *count* the 15 places across the front and side where *Mathematica* computes the values of Sin[x y]. Even though we can get a pretty good idea of what this function is doing using the default value of PlotPoints, functions that are changing very rapidly or functions that have singularities will render poorly with so few points sampled.

In[10]:= **Plot3D[Sin[x y], {x,0,2Pi}, {y,0,2Pi}]**

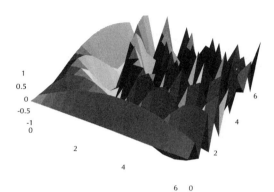

Out[10]= -SurfaceGraphics-

As *x* and *y* increase in value, this function changes at faster and faster rates. The default value of PlotPoints cannot pick up the changes between the sampled points and so we get a very choppy graphic. Increasing the sampled points gives the smoothed image. (Warning: The following graphic takes quite a bit of time and computer memory.)

In[11]:= **Plot3D[Sin[x y], {x,0,2Pi}, {y,0,2Pi}, PlotPoints -> 80]**

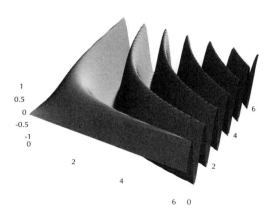

Out[11]= -SurfaceGraphics-

Sometimes a function will not smooth adequately by increasing the number of points sampled in every direction. Such difficulties are usually not related to sampling, but to intrinsic properties of the function in question. The following plot has singularities along the curve $xy = \pi/4$.

In[12]:= `Plot3D[1/(1 - ArcTan[x y]), {x,0,4}, {y,0,4}]`

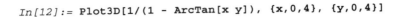

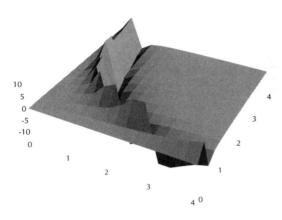

Out[12]= `-SurfaceGraphics-`

Increasing the value of `PlotPoints` here will have little effect on the representation of the singularities.

In[13]:= `atanplot = Plot3D[1/(1 - ArcTan[x y]), {x,0,4}, {y,0,4},`
`PlotPoints -> 35]`

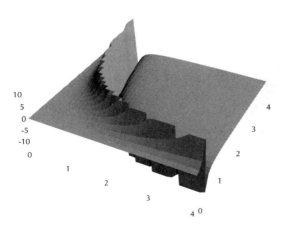

Out[13]= `-SurfaceGraphics-`

10.3.3 | Options for Built-In Graphics Functions

As mentioned above, once *Mathematica* has computed the list of lines and possible points, it has to display these graphics elements. It does this with the Show function. We can use Show to display any graphics we create, or to display previously computed graphics with different options. For example, this changes the ViewPoint of the previous plot.

In[14]:= **Show[atanplot, ViewPoint -> {2.8, 0, 2}]**

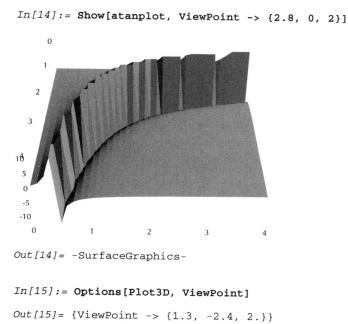

Out[14]= -SurfaceGraphics-

In[15]:= **Options[Plot3D, ViewPoint]**

Out[15]= {ViewPoint -> {1.3, -2.4, 2.}}

The default value of ViewPoint is a point in space that is 1.3 units to the right, −2.4 units in front of, and 2.0 units above the point at the center of the bounding box of the surface. In other words, the coordinates of the ViewPoint are relative to the center of the box containing the plotted object.

We have already seen how to use options for various functions such as View-Point for the Plot3D function above. Each of the functions that produce graphics objects has options for changing the default behavior of the function. Options for graphics functions are specified after the required arguments of the function being used (just like all *Mathematica* functions). For example, to see a list of all available options to Plot3D, we use the Options function.

In[16]:= **Options[Plot3D]**

Out[16]= {AmbientLight -> GrayLevel[0], AspectRatio -> Automatic,
　　　　Axes -> True, AxesEdge -> Automatic, AxesLabel -> None,
　　　　AxesStyle -> Automatic, Background -> Automatic,
　　　　Boxed -> True, BoxRatios -> {1, 1, 0.4},
　　　　BoxStyle -> Automatic, ClipFill -> Automatic,
　　　　ColorFunction -> Automatic, ColorOutput -> Automatic,
　　　　Compiled -> True, DefaultColor -> Automatic,
　　　　Epilog -> {}, FaceGrids -> None, HiddenSurface -> True,
　　　　Lighting -> True, LightSources ->
　　　　　{{{1., 0., 1.}, RGBColor[1, 0, 0]},
　　　　　　{{1., 1., 1.}, RGBColor[0, 1, 0]},
　　　　　　{{0., 1., 1.}, RGBColor[0, 0, 1]}}, Mesh -> True,
　　　　MeshStyle -> Automatic, PlotLabel -> None,
　　　　PlotPoints -> 15, PlotRange -> Automatic,
　　　　PlotRegion -> Automatic, Plot3Matrix -> Automatic,
　　　　Prolog -> {}, Shading -> True, SphericalRegion -> False,
　　　　Ticks -> Automatic, ViewCenter -> Automatic,
　　　　ViewPoint -> {1.3, -2.4, 2.}, ViewVertical -> {0., 0., 1.}
　　　　DefaultFont :> $DefaultFont,
　　　　DisplayFunction :> $DisplayFunction}

Any of the default options can be overridden by specifying different values in the Plot3D command or by passing them to Show. To remove the box and the axes from around the plot, for example, we would enter

In[17]:= **Show[atanplot, Axes -> False, Boxed -> False]**

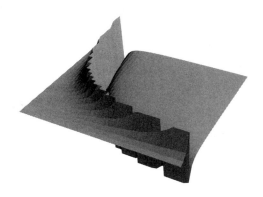

Out[17]= -SurfaceGraphics-

When generating graphics, it is often helpful to list all of the options for the function that you are working with and to adjust the options to suit your needs. The reference manual [Wolfram 1991] contains complete information about these options.

Exercises

1. Plot the function $\cos(x) + \cos(3x)$ over the interval $[-4\pi, 4\pi]$.

   (a) Modify the `AspectRatio` of the above plot by making the height two times the width.

   (b) Label the axes x and y using the `AxesLabel` option.

   (c) Change the tick marks on the plot so that only multiples of π are displayed.

   (d) Give the plot a title using the `PlotLabel` option.

2. Make a plot of the surface $f(x, y) = \sin(xy)$ using `Plot3D`. Plot both x and y over the interval $(0, 3\pi/2)$. Increase the value of `PlotPoints` to get a smooth plot.

   (a) Change the `ViewPoint` option of `Plot3D` so that the surface is viewed from directly overhead.

   (b) Add the option `ColorFunction -> GrayLevel` to the previous plot.

   (c) Make a `DensityPlot` of the same function over the same interval removing the `Mesh` and increasing the `PlotPoints`. What similarities and differences do you observe between the `DensityPlot` and the surface produced with `Plot3D`?

3. Create a graphic of the `Sin` function over the interval $(0, 2\pi)$ that displays vertical lines at each point calculated by the `Plot` function to produce its plot.

4. Plot the function $x + \sin(x)$ together with its first five Taylor polynomials. The Taylor polynomial for a function `f[x]` about $x = 0$ of degree n is produced by `Series[f[x], {x, 0, n}]`. You should wrap the `Normal` function around the `Series` object to convert it to a regular expression that can be plotted. For example, this plots the fifth-degree Taylor polynomial of $x + \sin(x)$ about $x = 0$ over the interval $(-5, 5)$.

   ```
   In[1]:= f[x_] = Normal[Series[x + Sin[x], {x, 0, 5}]]
   ```

Exercises (cont.)

In[2]:= **Plot[f[x], {x, -5, 5}]**

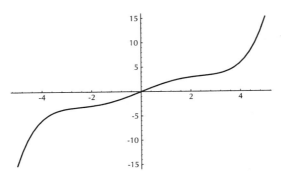

Out[2]= -Graphics-

5. Plot the function Sin[x] over the interval $[-2\pi, 2\pi]$ and then reverse the *x*-and *y*-coordinates of each point by means of a transformation rule.

6. Plot the function Sin[x y] with *x* and *y* taking on values from 0 to $3\pi/2$. Then use a transformation rule to perform a *shear* by sing the graphic in the *x*-direction by a factor of four.

7. Create a function rotatePlot[p, theta] that takes a plot p and rotates it about the origin by an angle theta. For example, to rotate a plot of the sine function, first create the plot,

In[3]:= **plot1 = Plot[Sin[x], {x,0,2Pi}]**

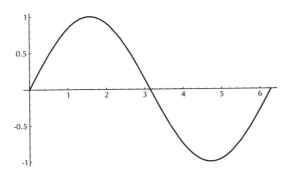

Out[3]= -Graphics-

and then perform the rotation of π radians.

Exercises (cont.)

In[4]:= **rotatePlot[plot1, Pi]**

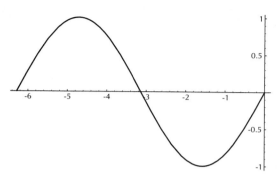

Out[4]= -Graphics-

8. Create a function `rotatePlot3D[phi, theta, psi]` that will rotate a `Graphics3D` object about the origin by the angles `phi`, `theta`, and `psi` in the *x*, *y*, and *z* directions, respectively.

Solutions

1. Here is the plot of the function over the interval $[-4\pi, 4\pi]$.

In[1]:= **p1 = Plot[Cos[x] + Cos[3x], {x,-4Pi,4Pi}]**

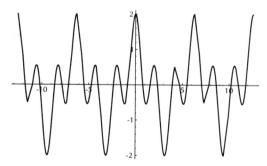

Out[1]= -Graphics-

This changes the AspectRatio so that the height is two times the width.

In[2]:= **Show[p1, AspectRatio -> 2]**

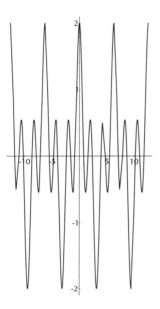

Out[2]= -Graphics-

This adds labels to the axes.

In[3]:= **Show[p1,**
         AxesLabel -> {FontForm["x",{"Times-Italic",14}],
                       FontForm["y",{"Times-Italic",14}]}]

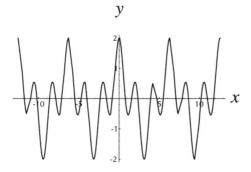

Out[3]= -Graphics-

Here are ticks given explicitly by the `Ticks` option.

```
In[4]:= Show[p1,
            Ticks -> {Range[-4Pi,4Pi,Pi], Automatic},
            DefaultFont -> {"Helvetica", 9}]
```

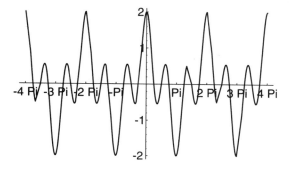

```
Out[4]= -Graphics-
```

Here is the plot with a title.

```
In[5]:= Show[p1, PlotLabel -> FontForm["A wiggly plot",
                                        {"Times-Bold",20}]]
```

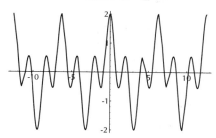

```
Out[5]= -Graphics-
```

3.   Here is a plot of the sine function.

In[1]:= **sinplot = Plot[Sin[x], {x,0,2Pi}]**

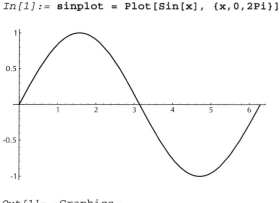

Out[1]= -Graphics-

This solution is essentially that given in *Exploring Mathematics with Mathematica* [Gray and Glynn 1991]. Extracting the points from which *Mathematica* constructs the plot is accomplished by the Nest statement. The Line primitive is then mapped across those points in such a way as to create lines from the points on the graph to points on the *x*-axis with the same *x*-coordinate.

In[2]:= **Show[sinplot,**
                Graphics[
                        {Thickness[.001],
                        Map[Line[{{#[[1]], 0}, #}]&,
                            Nest[First, sinplot, 4]]}]]

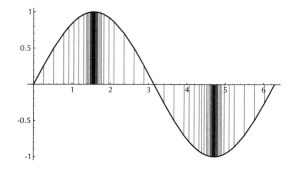

Out[2]= -Graphics-

5. Here is the plot of the sine function.

In[1]:= **splot = Plot[Sin[x], {x,-2Pi,2Pi}];**

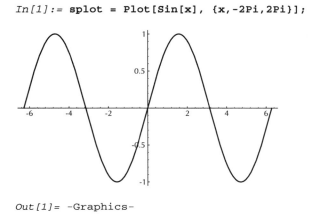

Out[1]= -Graphics-

This replacement rule interchanges each ordered pair of numbers.

In[2]:= **Show[splot /. {x_?NumberQ, y_?NumberQ} -> {y,x}]**

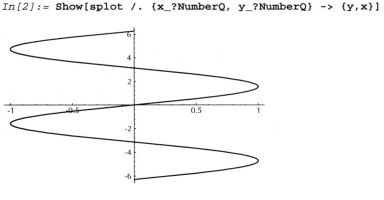

Out[2]= -Graphics-

7. Using the standard rotation matrix, each point is taken to its image under the rotation transformation. Notice that this function first checks that its first argument is in fact a graphic object. This is accomplished via pattern matching.

```
In[1]:= rotatePlot[p_Graphics, theta_] :=
            Show[p /. {x_?NumberQ, y_?NumberQ} ->
                {x, y} . {{Cos[theta], Sin[theta]},
                    {-Sin[theta], Cos[theta]}}]
```

In[2]:= `rotatePlot[plot1, Pi]`

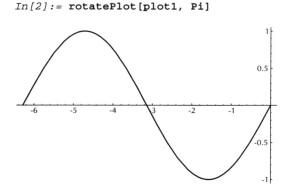

Out[2]= `-Graphics-`

10.4 Graphics Programming

Up until this point, we have looked at the tools that are available to construct relatively simple graphics in *Mathematica*. This has allowed us to create images by using the graphics building blocks. In this section we consider problems whose solution requires geometric insight as we construct our programs. We'll begin with a purely geometric problem on simple closed paths and then move on to construct graphics from programming work we did in Chapter 7—the display of binary trees.

10.4.1 | Simple Closed Paths

As our first example of a programming problem that involves the use of graphics, we will solve a very simplified variation of what are known as *traveling salesman problems*. A *closed path* is one that travels to every point and returns to the original point. The traveling salesman problem asks for the *shortest* closed path that connects an arbitrary set of points.

The traveling salesman problem is one of great theoretical, as well as practical, importance. Airline routing and telephone cable wiring over large regions are examples of problems that could benefit from a solution to the traveling salesman problem.

From a theoretical point of view, the traveling salesman problem is part of a large class of problems that are known as *NP-complete* problems. These are problems that can be solved in polynomial time using nondeterministic algorithms. A *nondeterministic algorithm* has the ability to *choose* among many options when faced with numerous choices, and then to verify that the solution is correct. The outstanding problem in computer science at present is known as the $P = NP$ problem. This equation says that any problem that can be solved by a nondeterministic algorithm in polynomial time (NP) can be solved by a deterministic algorithm in polynomial time (P). It is widely believed that $P \neq NP$, and considerable effort has gone into solving this problem. The interested reader should consult [Lawler et al 1985] or [Skiena 1990].

Our focus will be on a solvable problem that is a substantial simplification of the traveling salesman problem. We will find a *simple closed path* (a closed path that does not intersect itself) through a set of *n* points.

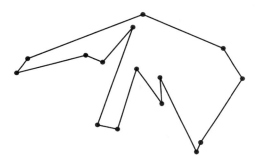

We will demonstrate a graphical solution to the problem by working with a small value of *n* and then generalizing to arbitrary values of *n*. Let us first create a set of ten pairs of points (*n* = 10) in the unit square.

```
In[1]:= coords = Table[Random[], {10}, {2}]

Out[1]= {{0.30547, 0.388389}, {0.884128, 0.467877},
          {0.184384, 0.345779}, {0.572902, 0.733683},
          {0.132478, 0.109042}, {0.842936, 0.384924},
          {0.72443, 0.18245}, {0.590387, 0.49156},
          {0.0783677, 0.332895}, {0.920595, 0.253071}}
```

```
In[2]:= points = Map[Point, coords]

Out[2]= {Point[{0.30547, 0.388389}], Point[{0.884128, 0.467877}],
          Point[{0.184384, 0.345779}], Point[{0.572902, 0.733683}],
          Point[{0.132478, 0.109042}], Point[{0.842936, 0.384924}],
          Point[{0.72443, 0.18245}], Point[{0.590387, 0.49156}],
          Point[{0.0783677, 0.332895}], Point[{0.920595, 0.253071}]}
```

Here we have created a table of ten pairs of numbers (the coordinates of our points in the plane), and then created graphics primitives by mapping `Point` over each pair. We can show the points alone,

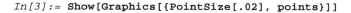

In[3]:= **Show[Graphics[{PointSize[.02], points}]]**

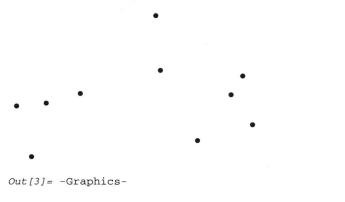

Out[3]= -Graphics-

or connected by lines.

In[4]:= **lines = Line[coords]**

Out[4]= Line[{{0.30547, 0.388389}, {0.884128, 0.467877},
            {0.184384, 0.345779}, {0.572902, 0.733683},
            {0.132478, 0.109042}, {0.842936, 0.384924},
            {0.72443, 0.18245}, {0.590387, 0.49156},
            {0.0783677, 0.332895}, {0.920595, 0.253071}}]

In[5]:= **Show[Graphics[{PointSize[.02], points, lines}]]**

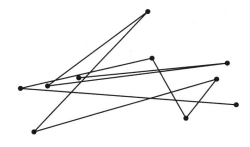

Out[5]= -Graphics-

At this stage, we notice that there are two problems. First, the path is not closed; that is, the last point visited is not the point we started from. The `Line` primitive connects `pt1` to `pt2` to `pt3`, etc., in the sequence that the points are presented to it. So we need to connect the last point to the first point to close the path. This can be accomplished by appending the first point to the end of the list of coordinates, and then evaluating the graphics code again.

```
In[6]:= path = Line[coords /. {a_,b__} -> {a,b,a}];
```

```
In[7]:= Show[Graphics[{PointSize[.02], points, path}]];
```

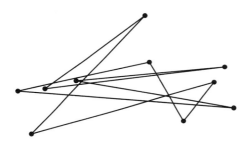

The second problem—the fact that our path is not simple—is geometric in nature. To find an algorithm that will insure that our path does not cross itself for *any* set of points in the plane, we will first pick a point from our set at random and call this the *base* point.

```
In[8]:= base = coords[[Random[Integer, {1, Length[coords]}]]]
```

```
Out[8]= {0.572902, 0.733683}
```

The path problem can be solved by first computing the counterclockwise (polar) angle between a horizontal line and each of the remaining points, using the base point as the vertex of the angle. Then, sorting the points according to this angle and connecting the points in this order will produce the desired result.

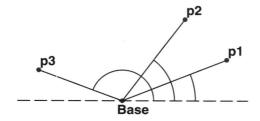

First we compute the angle between the points a and b. (The reader should verify the trigonometric analysis necessary to find this angle in the various cases. Note that we are computing the polar angle between two points and hence we need the ArcTan function.)

```
In[9]:= angle[a_List, b_List] := Apply[ArcTan, (b-a)]
```

We can use this function to compute the angle between our base point and each of the points in the list coords. (We need to make sure that we do not try to compute the angle between the base point and itself as this will evaluate to ArcTan[0, 0], which is undefined. This situation can be avoided by removing the base point from our list of coordinates when computing the angles.)

```
In[10]:= remain = Complement[coords, {base}];
```

```
In[11]:= Map[angle[base, #]&, remain]
Out[11]= {-2.46052, -2.18493, -2.35699, -2.2298, -1.49871,
           -1.30253, -0.911942, -0.706847, -0.944512}
```

Instead of computing the angles explicitly, we will just use the angle function as an ordering function on our list of coordinates. Sort [*list, rule*] will sort *list* according to *rule*, which is a two-argument predicate. We wish to sort coords according to our ordering function on the angles between each point and the base point. The following code accomplishes this:

```
In[12]:= s = Sort[remain, (angle[base, #1] <= angle[base, #2])&]
Out[12]= {{0.0783677, 0.332895}, {0.184384, 0.345779},
           {0.30547, 0.388389}, {0.132478, 0.109042},
           {0.590387, 0.49156}, {0.72443, 0.18245},
           {0.920595, 0.253071}, {0.842936, 0.384924},
           {0.884128, 0.467877}}
```

This is our list of coordinates sorted according to the polar angle between each point and the base point. In order to start and end with the base point, we Join three separate lists,

```
In[13]:= p = Join[{base}, s, {base}]
```

```
Out[13]= {{0.572902, 0.733683}, {0.0783677, 0.332895},
         {0.184384, 0.345779}, {0.30547, 0.388389},
         {0.132478, 0.109042}, {0.590387, 0.49156},
         {0.72443, 0.18245}, {0.920595, 0.253071},
         {0.842936, 0.384924}, {0.884128, 0.467877},
         {0.572902, 0.733683}}
```

and then create our path and display the graphic.

```
In[14]:= path = Line[p];
```

```
In[15]:= Show[Graphics[{PointSize[.02], points, path}]]
```

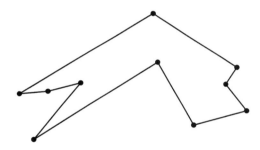

```
Out[15]= -Graphics-
```

If we collect the above commands into a program simpleClosedPath, then we can find such paths for arbitrary sets of coordinates.

```
In[16]:= simpleClosedPath[l_] :=
        Module[{points, base, angle, sorted, path},
           points = {PointSize[.02], RGBColor[1,0,0], Map[Point,l]};
           base = l[[ Random[Integer, {1, Length[l]}] ]];
           angle[a_, b_]:= Apply[ArcTan, (b - a)];
           sorted = Sort[Complement[l, {base}],
                       (angle[base, #1] <= angle[base, #2])&];
           path = Line[Join[{base}, sorted, {base}]];

           Show[Graphics[{{RGBColor[1,0,0], points}, path}]]]
```

Now we can create large sets of points and find the corresponding simple closed path readily:

```
In[17]:= data = Table[Random[], {25}, {2}];
```

```
In[18]:= simpleClosedPath[data]
```

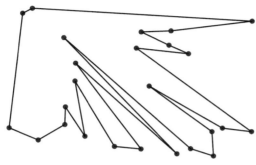

```
Out[18]= -Graphics-
```

```
In[19]:= data = Table[Random[], {100}, {2}];
```

```
In[20]:= simpleClosedPath[data]
```

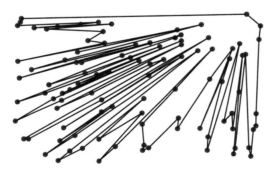

```
Out[20]= -Graphics-
```

10.4.2 | Drawing Trees

The trees drawn in Chapter 7 were drawn using a *Mathematica* program. We give a simpler version of the program here; the full version is developed in the exercises. Here, trees are drawn without their labels—with just a disk at each node—and, more importantly, the placement of nodes is not as good (aesthetically speaking). Still, it is a good example of using recursion to create a line-drawing.

When drawing trees, the central question is: How far should the children of a given node be separated? For example, in Figure 10.1, the separation of the children of node 2 is much greater than that of the children of node 1. That's because the *total width* of the trees below node 2 is so great that they require such a separation; or rather, the total width of the *right* side of the *left* subtree and the *left* side of the *right* subtree requires that separation. To illustrate this point, consider the trees in Figures 10.2(a) and 10.2(b). The subtrees of the root are the same, but in a different order; the result is that in Figure 10.2(a), the children of the root must be separated much more.

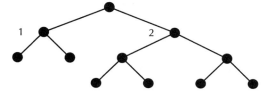

Figure 10.1: A tree with different separations.

(a)   (b)

Figure 10.2: Trees whose children have different separations.

Thus, to properly place subtrees, we need to know, for each one, its total width to the left and to the right of its root. Then, the two trees will be separated by an amount equal to the right width of the left subtree plus the left width of the right subtree, plus some arbitrary additional separation. This is illustrated in Figure 10.3. `lw1` represents the left width of the left subtree, `rw1` the right width of the left subtree, and `lw2` and `rw2` represent the corresponding widths for the right subtree. `minsep` is the additional separation always added between subtrees, and `sep` is the separation eventually computed for these two subtrees.

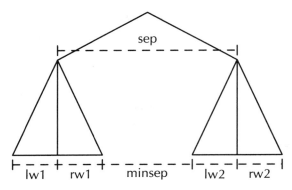

Figure 10.3: Calculation of the separation between children.

The function `placeTree` is given a binary tree (represented as in Section 7.6) and returns a list of three things:

- A *separation tree*—a tree having the same shape as the argument, labelled at each interior node with a number, the separation of that node's children.

- The *left width* of the tree—the distance it extends to the left from its root.

- The *right width* of the tree.

Now, computing `placeTree[{`*lab*, *lc*, *rc*`}]` is accomplished in these steps:

1. Recursively compute `placeTree[`*lc*`]` and `placeTree[`*rc*`]`; suppose the results are {*st1*, *lw1*, *rw1*} and {*st2*, *lw2*, *rw2*}, respectively.

2. The separation of *lc* and *rc* is equal to the right width of *lc* (*rw1*) plus the left width of *rc* (*lw2*), plus the additional separation. Call the total separation thus computed *sep*.

3. The left width of the total tree is *sep*/2 + *lw1*, and its right width is *sep*/2 + *rw2*.

This leads to the code:

```
placeTree[{_}] := {{ }, 0, 0}
placeTree[{_, lc_, rc_}] :=
   Module[{left = placeTree[lc], right = placeTree[rc],
           minsep = 1.0},
   Module[{sep = left[[3]] + right[[2]] + minsep},
          {{sep, left[[1]], right[[1]]},
           left[[2]] + sep/2, right[[3]] + sep/2}]]
```

Given a list {*st*, *lw*, *rw*} produced by `placeTree`, we no longer need *lw* or *rw* to draw the tree: the separation tree *st* suffices. Transforming *st* into a drawing is straightforward (the `Disk` primitive draws a filled circle with given center and radius).

```
drawSepTree[{}, lev_, xaxis_] := {Disk[{xaxis, lev}, 0.1]}
drawSepTree[{sep_, lc_, rc_}, lev_, xaxis_] :=
   Join[{Disk[{xaxis, lev}, 0.1],
         Line[{{xaxis, lev}, {xaxis-sep, lev-1}}],
         Line[{{xaxis, lev}, {xaxis+sep, lev-1}}]},
   drawSepTree[lc, lev-1, xaxis-sep],
   drawSepTree[rc, lev-1, xaxis+sep]]
```

Thus, to draw a tree `t`, enter,

```
placeTree[t];
drawSepTree[%[[1]], 0, 0];
Show[Graphics[%]]
```

or create a function to automate this process,

```
showTree[tree_]:=
   Show[Graphics[drawSepTree[placeTree[tree][[1]], 0, 0]]]
```

and a picture like the one in Figure 10.1 will appear. For example,

```
In[1]:= tree1 = {a, {b}, {a, {c, {e, {g}, {f}}, {d}}, {b}}}

Out[1]= {a, {b}, {a, {c, {e, {g}, {f}}, {d}}, {b}}}


In[2]:= showTree[tree1]
```

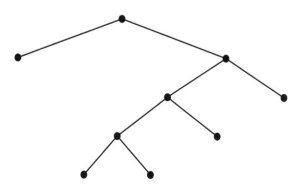

```
Out[2]= -Graphics-
```

Exercises

1. Although the program `simpleClosedPath` works well, there are conditions under which it will occasionally fail. Experiment by repeatedly computing `simpleClosedPath` for a set of 10 points until you see the failure. Determine the conditions that must be imposed on the selection of the base point for the program to work consistently.

2. Modify `simpleClosedPath` so that the point with the smallest x-coordinate of the list of data is chosen as the `base` point.

3. Modify `simpleClosedPath` so that the point that has the largest y-coordinate is chosen as the `base` point.

4. Given a set of coordinates in the plane and the simple closed polygon such as that produced by the `simpleClosedPath` algorithm in the text, write a program `outside[`*data,* `{x, y}]` to determine if an arbitrary point (x, y) is *inside* or *outside* of the polygon. Your program should take a list of coordinates in the plane (an $n \times 2$ list) and draw the polygon and the point, and `Print` a statement stating the solution. (*e.g.*, "The point (x, y) is outside of the polygon.")

5. Write a function `triangleArea` that computes the area of any triangle in the plane.

6. A polygon is called *convex* if any line connecting any two points inside the polygon lies completely inside the polygon. Most of the simple closed polygons we computed in this section are nonconvex. For a given set of n points, find those points which form a convex polygon that is a boundary for the entire point set. (The smallest such boundary is called the *convex hull* of the set of points.) That is, given a set of points in the plane,

write a function `convex` that outputs a graph such as the one at the top of the next page.

Exercises (cont.)

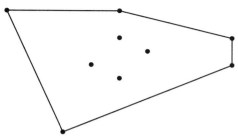

7. Write a function `pointInPolygonQ` that tests whether a given point is inside a specified polygon. For example, the origin is inside the polygon formed by joining the four unit vectors:

```
In[1]:= poinInPolygonQ[{0,0}, {{1,0},{0,1},{-1,0},{0,-1}}]

Out[1]= True
```

8. Another way of finding a simple closed path is to start with any closed path and progressively make it simpler by finding intersections and changing the path to avoid them. Prove that this process ends, and that it ends with a closed path. Write a program to implement this procedure and then compare the paths given by your function with those of `simpleClosedPath` given in the text.

9. The tree-drawing code we've presented is not the same code we used in drawing the trees in Chapter 7. The two trees drawn in Figure 10.4 show the difference: drawing (a) is the one produced by `placeTree`, and (b) is the one produced by the algorithm used in Chapter 7. That algorithm is due to Reingold and Tilford [Reingold and Tilford 1981], and basically what it does is just this: instead of basing the separation of subtrees on their total width, it does a level-by-level comparison, and separates them only as far as needed at any particular level.

Figure 10.4: Results from different tree-drawing algorithms.

Program this tree-drawing algorithm. There is one tricky part to it, which we'll leave you to discover for yourself, except to say this: your program should draw the tree shown in Figure 10.5 roughly as drawn there.

Exercises (cont.)

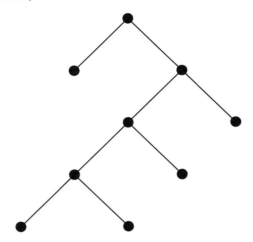

Figure 10.5: A tricky tree to draw.

10. Another difference between the code we've shown here and the code that was used in Chapter 7 is that the algorithm there was able to draw trees with labels at the nodes. Extend your algorithm from Exercise 9 to add labels; your trees should have strings as their labels. You need to take the width of the labels into account when computing the separation tree (this is a change to `placeTree`), and make sure the lines don't intersect the labels (this is a change to `drawSepTree`). Unfortunately, there is no way to compute the exact width of a text string as it will appear in a *Mathematica* graphic; just approximate using the number of characters in the label.

Solutions

1. There are a number of things that could go wrong with the algorithm of just choosing a base point randomly and then sorting according to the arctangent. The default branch cut for `ArcTangent` gives values between $-\pi/2$ and $\pi/2$. (The reader is encouraged to think about why this could occasionally cause the algorithm in the text to fail.) By choosing the base point so that it lies at some extreme of the diameter of the set of points, the polar angle algorithm given in the text will work consistently. If you choose the base point so that it is lowest and left-most, then all the angles will be in the range $(0, \pi]$.

```
In[1]:= simpleClosedPath[l_] :=
         Module[{points, base, angle, sorted, path},
            points = {PointSize[.02], RGBColor[1,0,0],
                       Map[Point,l]};
            base = Last[Sort[l,(#2[[2]] < #1[[2]])&]];
            angle[a_,b_]:= Apply[ArcTan, (b - a)];
            sorted = Sort[Complement[l,{base}],
                          (angle[base, #1] <= angle[base, #2])&];
            path = Line[Join[{base}, sorted, {base}]];
            Show[Graphics[{{RGBColor[1,0,0], points}, path}]]]
```

```
In[2]:= pts = Table[Random[], {20}, {2}];
```

```
In[3]:= simpleClosedPath[pts]
```

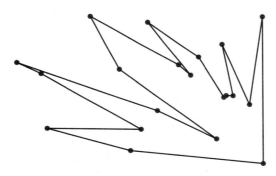

```
Out[3]= -Graphics-
```

3. A simple change to the program simpleClosedPath given in Exercise 1 chooses the base point with the largest *y*-coordinate.

```
In[1]:= simpleClosedPath[l_] :=
         Module[{points, base, angle, sorted, path},
            points = {PointSize[.02], RGBColor[1,0,0],
                       Map[Point,l]};
            base = Last[Sort[l,(#2[[2]] > #1[[2]])&]];
            angle[a_, b_]:= Apply[ArcTan, (b - a)];
            sorted = Sort[Complement[l, {base}],
                          (angle[base, #1] <= angle[base, #2])&];
            path = Line[Join[{base}, sorted, {base}]];
            Show[Graphics[{{RGBColor[1,0,0], points}, path}]]]
```

5. The area of a triangle is one-half the base times the altitude. For arbitrary points, the altitude requires a bit of computation that does not generalize.

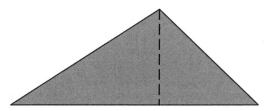

The magnitude of the cross product of two vectors gives the area of the parallelogram that they determine. Since the vectors we are working with are in two-dimensional space, we embed them in three-dimensional space in the plane $z = 0$ so that we can compute the cross product which only makes sense in three dimensions.

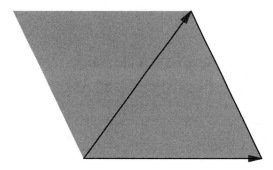

```
CrossProduct[{x2,y2}-{x1,y1}, {x3,y3}-{x1,y1}] /.
        {x_,y_} -> {x,y,0}
```

Here are the coordinates for a triangle.

```
In[1]:= a = {0,0};
        b = {5,0};
        c = {3,2};
```

And here is the computation for the cross product.

```
In[2]:= CrossProduct[(b-a),(c-a)] /. {x_,y_} -> {x,y,0}

Out[2]= {0, 0, 10}
```

So the given area is then just half the magnitude of the cross product.

```
In[3]:= Apply[Plus,%]/2

Out[3]= 5
```

Here is a function that computes the area of any triangle using the cross product.

```
In[4]:= triangleArea[v_List] := 1/2 Plus @@
            (CrossProduct[v[[2]]-v[[1]], v[[3]]-v[[1]]] /.
            {x_, y_} -> {x, y, 0})
```

This is done more simply using determinants and this method generalizes more easily to higher dimensions.

```
In[5]:= triangleArea[{v1_, v2_, v3_}] :=
            Det[{v1, v2, v3} /. {x_, y_} -> {x, y, 1}]/2


In[6]:= triangleArea[{a,b,c}]

Out[6]= 5
```

7. The key observation is that in computing the area of a triangle using the determinant formulation as in Exercise 5, the area will be a positive quantity if the points are given in counter-clockwise order, and will be negative if in clockwise order. So, for a given point p not on a line **ab**, the area of $\triangle abp$ will be positive (computed using determinants), if p is to the left of **ab**. Similarly, for each of the lines in a polygon, relative to the given point p. So, to perform the computation, we first partition the polygon into pairs of points, and then map the triangle area function (here called leftofQ) with the given point across each pair. If all such areas are greater than or equal to zero, then a value of True is returned.

```
In[1]:= pointInPolygonQ[p_, poly_]:= Module[{leftofQ},
            leftofQ[{v1_,v2_,v3_}]:= Det[{v1,v2,v3} /.
                    {x_,y_} -> {x,y,1}]/2 >= 0;
        Apply[And, Map[leftofQ[Join[{p},#]]&,
                Partition[poly /.
                    {a_,b__} :> {a,b,a},2,1]]
        ]]
```

Here are the coordinates for a quadrilateral and two distinct points.

```
In[2]:= quad = {{1,0},{0,1},{-1,0},{0,-1}};
```

```
In[3]:= p1 = {0,0};
```

```
In[4]:= p2 = {1,1};
```

```
In[5]:= Show[Graphics[{Line[quad /. {a_,b__}:>{a,b,a}],
           PointSize[.025], Point[p1], Point[p2]},
           AspectRatio->Automatic]]
```

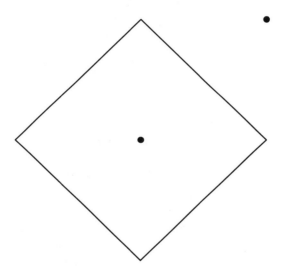

```
Out[5]= -Graphics-
```

Finally, here are the computations for these points and polygon.

```
In[6]:= pointInPolygonQ[p1, quad]
Out[6]= True
```

```
In[7]:= pointInPolygonQ[p2, quad]
Out[7]= False
```

9. RT (for Reingold-Tilford), replaces the `placeTree` function. In `placeTree`, the result was a separation tree plus two numbers: width of the left side of the tree and width of the right side of the tree. In RT, the result is instead a separation tree plus two lists, the first giving the width of the left side of each level of the tree, the second giving the corresponding widths on the right side. `sep` is calculated by adding the right widths of the left subtree to the left widths of the right subtree at each level, and taking the maximum separation. `drawSepTree` is unchanged.

```
RT[{_}] := {{}, {}, {}}
RT[{_, lc_, rc_}] :=
  With[{left = RT[lc],
        right = RT[rc],
        minsep = 2.0},
      With[{sep = (Max[0, Apply[Max, Map[Apply[Plus,#]&,
                  zip[left[[3]], right[[2]]]]]] + minsep) / 2},
          With[{newtree = {sep, left[[1]], right[[1]]},
                leftedge = Join[{sep}, extend[left[[2]],
                                right[[2]], sep]],
                rightedge = Join[{sep}, extend[right[[3]],
                                left[[3]], sep]]},
              {newtree, leftedge, rightedge}]]]]

drawTree[t_] := drawSepTree[RT[t][[1]], 0, 0]
```

The auxiliary functions are `zip` and `extend`. Given the *left* widths of each level of the *right* subtree, and the *right* widths of each level of the *left* subtree, the separation of the two subtrees is determined by adding those numbers at each level and taking the maximum. `zip` is used to join those two lists into a list of pairs; it facilitates this process:

```
zip[{}, _] := {}
zip[_, {}] := {}
zip[{x1_, y1___}, {x2_, y2___}] :=
   Join[{{x1, x2}}, zip[{y1}, {y2}]]
```

When the separation of a tree's subtrees is determined, the lists of left and right widths of the combined tree are computed from the corresponding lists for the subtrees. This is simple enough for the most part: The left widths of the tree are obtained, mainly, by taking the left widths of the left subtree and shifting them left; and similarly for the right widths. There is an exception, though: If the right subtree is taller than the left subtree, the left widths of the bottom part of the tree are obtained from the left widths of the bottom part of the right subtree. Combining the left widths of the two subtrees to create the list of left widths of the combined tree is done by `extend`:

```
extend[edges1_, edges2_, sep_] :=
    Join[edges1 + sep, Take[edges2,
                            Min[0, Length[edges1] -
                            Length[edges2]]] - sep]
```

The same reasoning applies to computing the right widths, and extend is also used for that.

Here is the "tricky" tree drawn in Figure 10.5.

```
In[1]:= t1 = {a, {b}, {a, {c, {e, {g}, {f}}, {d}}, {b}}}
```

```
In[2]:= Show[Graphics[drawTree[t1]],
            PlotRange -> All,
            AspectRatio -> Automatic]
```

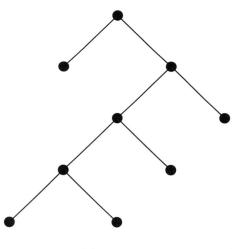

```
Out[2]= -Graphics-
```

10. RT is modified so that the left widths and right widths of each row take into account the width of the labels.

```
RT[{x_}] := {{}, {width[x]/2}, {width[x]/2}}
RT[{x_, lc_, rc_}] :=
   With[{left = RT[lc],
         right = RT[rc],
         minsep = 0.5},
      With[{sep = (Max[0, Apply[Max, Map[Apply[Plus, #]&,
                  zip[left[[3]], right[[2]]]]]] + minsep) / 2},
         With[{newtree = {sep, left[[1]], right[[1]]},
               leftedge = Join[{width[x]/2},
                              extend[left[[2]],right[[2]],sep]],
               rightedge = Join[{width[x]/2},
                              extend[right[[3]],left[[3]],sep]]},
            {newtree, leftedge, rightedge}]]]

width[t_] := StringLength[t]
```

Drawing the following tree using the new RT and the old drawSepTree will show the difference in the layout of the trees. However, since drawSepTree above only prints disks at each node, a new version of it is required.

```
In[1]:= t1 = {"a", {"abcdef"}, {"", {"abcdefghij"}, {"abc"}}}
```

The new version of drawSepTree draws the labels at each node instead of a disk. A complicating factor is that we can no longer just draw the lines from the center of the disk, since this would collide with the text. So the lines are now drawn in such a way as to leave a gap between the text and the line.

```
In[2]:= settext[lab_] := FontForm[lab, {"Helvetica", 10}]
```

```
In[3]:= drawSepTree[{lab_}, {}, lev_, xaxis_] :=
                {Text[settext[lab], {xaxis, lev}]}
```

```
In[4]:= drawSepTree[{lab_, lc_, rc_}, {sep_, ls_, rs_},
                    lev_, xaxis_] :=
           With[{h1 = If[lab == "", 0, .3],
                 h2 = If[lc[[1]] == "", 0, .3],
                 h3 = If[rc[[1]] == "", 0, .3]},
              Join[{Text[settext[lab], {xaxis, lev}],
                    Line[{{xaxis-sep*(h1/2), lev-h1/2},
                          {xaxis-sep+sep*(h2/2), lev-1+h2/2}}],
                    Line[{{xaxis+sep*(h1/2), lev-h1/2},
                          {xaxis+sep-sep*(h3/2), lev-1+h3/2}}]},
                 drawSepTree[lc, ls, lev-1, xaxis-sep],
                 drawSepTree[rc, rs, lev-1, xaxis+sep]]]
```

```
In[5]:= drawTree[t_] := drawSepTree[t, RT[t][[1]], 0, 0]
```

```
In[6]:= Show[Graphics[drawTree[t1]], PlotRange->All];
```

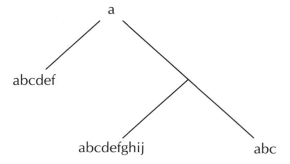

10.5 Sound

10.5.1 | The Sound of Mathematics

Our ears hear sound when the air around them compresses and expands the air near the eardrum. Depending upon how the eardrum vibrates, different signals are sent to the brain *via* the auditory nerves in the inner ear. These signals are then interpreted in the brain as various sounds. Musical tones compress and expand the air periodically according to sine waves. The human ear is able to hear these waves when the frequency is between 20 and 20,000 oscillations per second, or Hertz.

Recall that one oscillation of Sin[x] occurs between 0 and 2π.

```
In[1]:= Plot[Sin[x], {x,0,2Pi}]
```

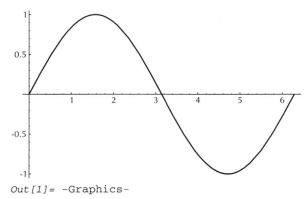

Out[1]= -Graphics-

Sin[4x] oscillates four times in the same interval.

In[2]:= **Plot[Sin[4x], {x,0,2Pi}]**

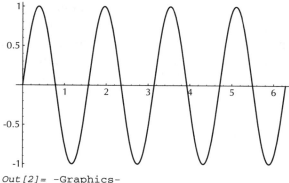

Out[2]= -Graphics-

Mathematica is able to take a function such as Sin and sample its amplitudes roughly 8000 times per second, and then send corresponding voltages to the speaker on your computer, if it has one, to produce the sound of the sine wave. The function that accomplishes this is Play which has the same syntax as the Plot command.

In[3]:= **?Play**

```
Play[f, {t, tmin, tmax}] plays a sound whose amplitude
    is given by f as a function of time t in seconds
    between tmin and tmax.
```

The function Sin[256 t] oscillates 256 times each 2π units, so if we want to *play* a function that oscillates 256 times per second, we want Sin[256 t (2Pi)]. This plays the function for two seconds.

In[4]:= **Play[Sin[256t (2Pi)], {t,0,2}]**

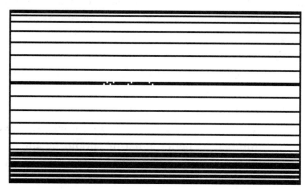

Out[4]= -Sound-

If your computer has sound capabilities, you should hear a C, one octave below middle C, played for two seconds. The graphic that *Mathematica* outputs with the Sound object is a somewhat primitive attempt to display the waveform. Since it does not contain very useful information, we will occasionally omit it from the display.

The Play function samples functions at a rate of about 8000 times per second, or hertz. This is good to keep in mind as anomalies can occur when playing a function whose periodicity is very close to the sample rate. Listen to the quite surprising result that follows (users will have to check the SampleRate on their computers and adjust the following code accordingly).

```
In[5]:= Options[Play, SampleRate]

Out[5]= {SampleRate -> 8192}


In[6]:= Play[Sin[8192 2Pi t], {t,0,3}];
```

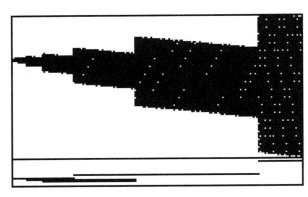

```
Out[6]= -Sound-
```

Although we would expect a tone at 8192 hertz, we get something quite different. You are encouraged to try other frequencies that are close to the sample rate on your computer and to think of the analogy with the graph of Cos[x] over the interval {0, 48Pi} that we gave earlier in this chapter on page 321.

Sounds that are generally thought to be pleasant to the human ear are modeled by periodic functions. Noise consists of random amplitudes. We can use these notions to find periodicity in sequences of numbers.

For example, recall that a rational number can be expressed as a finite or repeating decimal, whereas an irrational number cannot be so represented. If we were to "play" the digits of a rational number, its periodic nature should be apparent as a discernible tone. Playing the digits of an irrational number should result in noise.

The following displays the first 20 digits of the decimal expansion of 1/19.

```
In[7]:= RealDigits[N[1/19, 20]]
Out[7]= {{5, 2, 6, 3, 1, 5, 7, 8, 9, 4, 7, 3, 6, 8, 4, 2, 1,
         0, 5, 3}, -1}
```

The −1 at the end of the above list indicates the number of places to the left of the decimal point where the first nonzero digit occurs. Since the first digit of this real number is one place to the *right* of the decimal point, this is indicated with a negative number.

The periodic nature of this number is not apparent from such a short list. We can lengthen the list and pull off only the decimal digits as follows. We suppress the output using the semicolon.

```
In[8]:= digits = RealDigits[N[1/19, 1000]][[1]];
```

Now we can play this list of digits. ListPlay will play a sound where the amplitudes are given by the numbers in our list. (*Mathematica* scales the amplitudes to fit in a range that ListPlay can work with, and that is audible.)

```
In[9]:= ListPlay[digits]
```

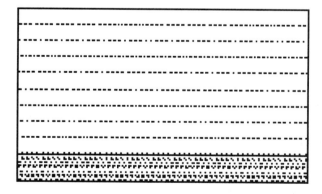

```
Out[9]= -Sound-
```

Clearly (from listening to the resulting tone), this sequence is periodic, whereas the following sequence of digits is not.

```
In[10]:= irratdigits = RealDigits[N[Pi, 1000]][[1]];
```

In[11]:= `ListPlay[irratdigits]`

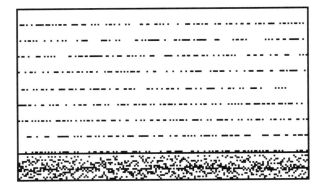

Out[11]= -Sound-

As the reader is probably well aware at this point, `Play` and `ListPlay` are audio analogues of `Plot` and `ListPlot`. This analogy will allow us to do *audio programming* in much the same way as we approached graphics programming earlier in this chapter. The next section contains a discussion of some very recent ideas in sound synthesis.

10.5.2 | White Music, Brownian Music, and Fractal Noise

White Noise, White Music

Imagine playing a recording of a certain sound at different speeds. Normally you would expect the character of the resulting sound to be quite different than the original. Speeding up a recording of your voice makes it sound cartoon-like, and if sped up fast enough, unintelligible. Slowing down a recording of the first few bars of Gershwin's *Rhapsody in Blue* would make the clarinet solo sound like a rumble.

There are some sounds though that sound roughly the same when played at different speeds. Benôit Mandelbrot of the IBM Thomas J. Watson Research Center described these sounds as "scaling noises." White noise is probably the most common example of a scaling noise. If you tuned your radio in between stations, recorded the noise and then played the recording at different speeds, you would hear roughly the same sound, although you would have to adjust the volume to get this effect.

Mandelbrot additionally characterized white noise as having zero "auto-correlation." This means that the fluctuations in such a sound at any moment are completely unrelated to any previous fluctuations.

In his book, *Fractal Music, Hypercards, and More . . .* Martin Gardner describes an algorithm for generating "white tunes", those having no correlation between notes

([Gardner 1992]). In this section we will implement his algorithms in *Mathematica* and compose such tunes. We will then see how to generate tunes that have varying degrees of correlation among the notes.

A simple "melody" with no correlation can be generated by randomly selecting notes from a scale. First we generate the frequencies of the twelve semitones from an equal-tempered C major scale. This is just a chromatic scale beginning with middle C.

```
In[12]:= cmajor = Table[N[261.62558 2^(j/12)], {j,0,11}]

Out[12]= {261.626, 277.183, 293.665, 311.127, 329.628, 349.228,
              369.994, 391.995, 415.305, 440., 466.164, 493.883}
```

We can play the first four notes of this scale.

```
In[13]:= notes4 = Take[cmajor, 4];
```

```
In[14]:= Do[Play[Sin[notes4[[j]] 2Pi t], {t,0,1}],
            {j,1,4}] //Timing
```

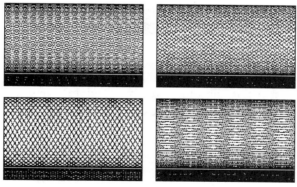

```
Out[14]= {45.7833 Second, Null}
```

The reader who executes the above code will certainly notice that it is terribly slow. Since we will be generating many sounds below, we will need to speed the execution of multiple sounds. The reason for the slowness has to do with how Play handles the functions on which it operates. Normally, Play will compile the function that appears as its argument, but it does not do this if what appears is only the name of a function defined elsewhere. notes4 was defined elsewhere, so it is not compiled. The following function playTones will speed the evaluation immensely. Note the time for execution of the same four notes as compared with the Do loop above. (We have made the playTones function Listable so that it will automatically map across lists of frequencies. Otherwise, we would have to manually Map it across such lists.)

```
In[15]:= SetAttributes[playTones, Listable]
```

```
In[16]:= playTones[freq_, time_]:=
            Play[Sin[2Pi t freq], {t, 0, time}]
```

```
In[17]:= playTones[notes4, 0.5]; //Timing
```

```
Out[17]= {3.05 Second, Null}
```

Now we can quickly generate the tune. Here we randomly generate twenty tones from the list cmajor (we have suppressed the display of the graphics images).

```
In[18]:= randomnotes =
            Table[cmajor[[Random[Integer, {1,12}]]], {20}]
```

```
In[19]:= playTones[randomnotes, 0.5]
```

```
Out[19]= -Sound-
```

A listener would be hard-pressed to find a pattern or any autocorrelation in this "tune" and the music is quite uninteresting as a result. Melodies generated using this scaling are referred to as $1/f^0$, where the 0 loosely refers to the level of correlation.

We leave as an exercise the writing of more sophisticated white melodies—one where the duration of each note varies randomly, and another where the likelihood of a note being chosen obeys a certain probability distribution.

Brownian Music

We now move in the other direction and generate melodies that are overly correlated. We will essentially perform a "random walk" through the C major scale. Music generated in such a way is called "Brownian" because it behaves much like the movement of particles suspended in liquid—Brownian motion.

Our melody will be constructed as follows: each note will be generated by randomly moving up or down a few semitones from the previous note. When a sequence gets to one end of the scale, we will simplify matters by having it wrap around to the other end.

We first create a function step that will randomly choose an integer from -2 to 2. These steps will determine how many semitones to move up or down.

```
In[1]:= step := Random[Integer, {-2, 2}]
```

Instead of alternating between choosing a step size and moving up and down the scale, we will first create a list of the steps in entirety. We will choose twenty steps corresponding to twenty notes.

```
In[2]:= s20 = Table[step, {20}]
Out[2]= {2, 1, 0, 2, -1, 2, -1, 2, 0, -2, -1, 0,
         2, 0, -1, 1, 1, 2, -1, -1}
```

This list will correspond to first moving two steps down, then one step up, then one step up, etc. So, starting (arbitrarily) with the sixth element of the list cmajor, the following gives the positions of the notes to play.

```
In[3]:= FoldList[Plus, 6, s20]
Out[3]= {6, 8, 9, 9, 11, 10, 12, 11, 13, 13, 11, 10,
         10, 12, 12, 11, 12, 13, 15, 14, 13}
```

There is one problem with this approach. If we get to the end of the list (twelfth position), and have to add two steps say, we would be stuck.

```
In[4]:= cmajor[[14]]
Part::partw: Part 14. of {261.626, 277.183,
   293.665, 311.127, 329.628, <<5>>, 466.164,
   493.883} does not exist.
Out[4]= {261.626, 277.183, 293.665, 311.127, 329.628, 349.228,
         369.994, 391.995, 415.305, 440., 466.164, 493.883}[[14]]
```

The way around this is to use modular arithmetic. This will have the effect of wrapping around to the opposite end of the list whenever you reach one boundary. Since the list cmajor is twelve elements long, we will use Mod 11 and add 1. This will give us positions 1 through 12, as opposed to 0 through 11 if we used Mod 12 alone.[4]

```
In[5]:= pos = Mod[FoldList[Plus, 4, s20], 11] + 1

Out[5]= {5, 7, 8, 8, 10, 9, 11, 10, 1, 1, 10, 9, 9, 11,
          11, 10, 11, 1, 3, 2, 1}
```

Finally, we create a list of those frequencies from cmajor at the positions given by the above list pos and then generate the tones (again suppressing the display of the graphics images).

```
In[6]:= brown = cmajor[[pos]]

Out[6]= {329.628, 369.994, 391.995, 391.995, 440., 415.305,
          466.164, 440., 261.626, 261.626, 440., 415.305,
          415.305, 466.164, 466.164, 440., 466.164, 261.626,
          293.665, 277.183, 261.626}
```

Here then, is a function for generating the tones from a brownian walk across the C major scale. This function is set up so that the default range of steps is -2 to 2 $(r\_:2)$.

```
In[7]:= brownMusic[n_Integer, r_:2] := Module[{cmajor,s},
          cmajor = Table[N[261.62558 2^(j/12)], {j,0,11}];
          s = Table[Random[Integer,{-r,r}], {n}];
          cmajor[[ Mod[FoldList[Plus,4,s],11]+1 ]]
          ]
```

This plays the tones.

```
In[8]:= playTones[brownMusic[20], 0.5]

Out[8]= -Sound-
```

This melody has a different character from the $1/f^0$ melody produced above. In fact, it is so over-correlated that it is often referred to as as $1/f^2$ music as a result of a computed spectral density of $1/f^2$. Although different in character from $1/f^0$ music, it is just as monotonous. The melody meanders up and down the scale aimlessly without

---

[4]Recall that Part[list, 0] gives the Head of the List.

any central theme. The exercises contain a discussion of $1/f$ music (or noise); that is, music that is moderately correlated. $1/f$ noise is quite widespread in nature and is intimately tied to areas of science that study fractal behavior. John Casti, in his book *Reality Rules: I, Picturing the World in Mathematics* gives the following characterization of $1/f$ *noise*: "If an electrical engineer were to compute the power spectrum (the squared magnitude of the Fourier transform) $f(x)$ of the relative frequency intervals x between successive notes in Bach's *Brandenburg Concerto*, it would be found that over a large range $f(x) = c/x$, where c is some constant. Thus Bach's music is characterized by the kind of 'noise' that engineers call $1/f$ *noise*." The interested reader should consult [Casti 1992] or [Mandelbrot 1982].

Exercises

1. Evaluate `Play[Sin[1000/x], {x,-2,2}]`. Explain the dynamics of the sound generated from this function.

2. Experiment with the `Play` function by creating arithmetic combinations of `Sin` functions. For example, you might try the following:

   ```
   Play[Sin[440 2Pi t] / Sin[660 2Pi t], {t,0,1}]
   ```

3. Create a tone that doubles in frequency each second.

4. A *square wave* consists of the addition of sine waves, each an odd multiple of a fundamental frequency; i.e., it consists of the sum of sine waves having frequencies f_0, $3f_0$, $5f_0$, $7f_0$, etc. Create a square wave with a fundamental frequency of 440 Hz. The more overtones you include, the 'squarer' the wave.

5. Create a square wave consisting of the sum of sine waves with frequencies f_0, $3f_0$, $5f_0$, $7f_0$, etc., and amplitudes 1, 1/3, 1/5, 1/7, etc. This is actually a truer square wave than that produced in the previous exercise.

6. Create a square wave consisting of overtones that are randomly out of phase. How does this wave differ from the previous two?

7. A *sawtooth wave* consists of the sum of both odd and even-numbered overtones (f_0, $2f_0$, $3f_0$, $4f_0$, etc. with amplitudes in the ratios 1, 1/2, 1/3, 1/4, etc.) Create a sawtooth wave and compare its tonal qualities to the square wave.

8. A wide variety of sounds can be generated using something called FM (frequency modulation) synthesis. The basic idea of FM synthesis is to use functions of the form,

$$a \sin (2\pi F_c, t + mod \sin(2\pi F_m t))$$

Exercises (cont.)

where a is the peak amplitude, F_c is the carrier frequency in Hz, *mod* is the modulation index, and F_m is the modulating frequency in Hz.

Determine what effect varying the parameters has on the resulting tones by creating a series of FM synthesized tones. First, create a function FM[Amp, Fc, mod, Fm, time] that implements the above formula and generates a tone using the Play function. Then you should try several examples to see what effect varying the parameters has on the resulting tones. For example, you can generate a tone with strong vibrato at a carrier frequency at middle A for one second, by evaluating FM[1, 440, 45, 5, 1].

9. Write a function pentatonic that generates $1/f^2$ music choosing notes from a five-tone scale. (A pentatonic scale can be played on a piano by beginning with C♯, and then playing only the black keys—C♯, E♭, F♯, A♭, C♯. The pentatonic scale is common to Chinese, Celtic, and Native American music.)

10. Modify the routine for generating $1/f^0$ music so that frequencies are chosen according to a specified probability distribution. For example, you might use the following distribution that indicates a note and its probability of being chosen: C – 5%, C♯ – 5%, D – 5%, E♭ – 10%, E – 10%, F – 10%, F♯ – 10%, G – 10%, A♭ – 10%, A – 10%, B♭ – 5%, B – 5%, C – 5%. (Hint: try the Which function.)

11. Modify the routine for generating $1/f^0$ music so that the *durations* of the notes obey $1/f^0$ scaling. Write a function tonesAndTimes that creates a two-dimensional list of frequencies and time durations. Consider using the function MapThread.

12. If you read musical notation, take a musical composition such as one of Bach's *Brandenburg Concertos* and write down a list of the frequency intervals x between successive notes. Then find a function that interpolates the power spectrum of these frequency intervals and determine if this function is of the form $f(x) = c/x$ for some constant c. (Hint: To get the power spectrum, you will need to square the magnitude of the Fourier transform: take Abs[Fourier[...]]^2 of your data.) Compute the power spectra of different types of music using this procedure.

13. Modify the routine for generating $1/f^2$ music so that the *durations* of the notes obey $1/f^2$ scaling.

14. The following series of exercises are designed to create $1/f$ music—music that is mildly correlated.

    (a) Write a function cmajor16 that extends cmajor to sixteen consecutive semitones.

Exercises (cont.)

(b) Write three functions red, green, and blue that simulate rolling 3 dice. The first note from cmajor16 is picked by rolling the dice and choosing the note in the position given by the sum (mod 16) + 1.

(c) To generate the next eight notes, think of the numbers 0 through 7 in binary. Let red correspond to the 1s digit, green to the 2s digit, and blue to the 4s digit. Starting from 0, and going to 1, only the 1s digit changes. So only the red die is retossed, the blue and green are left alone. This new sum (mod 16) of the red, green and blue is the next position from the list cmajor16. The third roll is obtained by noticing that in going from 1 to 2, both the 1s digit and the 2s digit change. Hence, reroll the red and green die, leaving the blue alone. The new sum of the three dice is the position of the next note. Continue in this fashion, rolling only those dice that correspond to digit changes when moving through the numbers 0–7, base 2. Finally, generate the tones corresponding to these frequencies.

(d) Extend the above algorithm to include four dice to produce sixteen notes from a 21 tone scale. If you have a sufficiently powerful computer with lots of memory and disk space, try ten dice to produce 1024 notes from a 55 tone scale.

Solutions

1. When x is close to -2, the frequency is quite low. As x increases, the fraction $1000/x$ gets larger, making the frequency of the sine function bigger. This in turn makes the tone much higher in pitch. As x approaches 0, the function is oscillating more and more, and at 0, the function can be thought of as oscillating infinitely often. In fact, it is oscillating so much that the sampling routine is not able to effectively compute amplitudes and hence we hear noise in this region.

```
Play[Sin[1000/x], {x, -2, 2}]
```

3. To generate a tone whose rate increases one octave per second, you need the sine of a function whose derivative doubles each second (frequency is a rate). That function is 2^t, so here is the command to produce the tone.

```
Play[Sin[2^t], {t,10,14}]
```

5. Here is a function that creates a square wave with decreasing amplitudes for higher overtones.

```
In[1]:= squarePlay[freq_, n_, time_] :=
            Play[Sum[Sin[freq i 2Pi t]/i, {i, 1, n, 2}],
            {t, 0, time}]
```

Here then, is an example of playing a square wave.

```
In[2]:= squarePlay[440, 7, .5]

Out[2]= -Sound-
```

7. This function creates a saw-tooth wave. The user specifies the fundamental frequency, the number of terms in the approximation, and the time duration of the tone.

```
In[1]:= sawPlay[freq_, n_, time_] :=
            Play[Sum[Sin[freq i 2Pi t]/i, {i, 1, n}],
            {t, 0, time}]
```

```
In[2]:= sawPlay[440, 6, 0.5]

Out[2]= -Sound-
```

Here is a true sawtooth with a fundamental frequency of 440Hz.

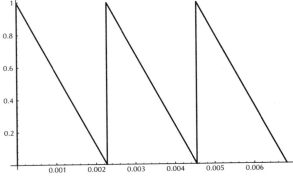

```
Out[3]= -Graphics-
```

```
In[4]:= Play[-440t-Floor[-440t], {t,0,.5}]

Out[4]= -Sound-
```

9. Here is a function that picks out frequencies from the pentatonic scale, using essentially brownian motion $1/f^2$ to select notes.

```
In[1]:= pentatonic[n_Integer, r_:2] := Module[{pscale, s},
            pscale = {277.183,311.13,369.99,415.30,466.16,554.37};
            s = Table[Random[Integer,{-r,r}], {n}];
            pscale[[ Mod[FoldList[Plus,3,s],4]+1 ]]
            ]
```

You could play a pentatonic "melody" as follows:

```
In[2]:= playTones[ pentatonic[24] ]

Out[2]= -Sound-
```

11. In this function, the notes are randomly chosen from the C major scale ($1/f^0$) and the durations are randomly chosen from the list that represents eighth notes, quarter notes, half notes, and whole notes (also $1/f^0$). playTones accepts two arguments, so MapThread threads corresponding notes and durations through playTones.

```
In[1]:= tonesAndTimes[n_]:= Module[{cmajor, notes, durs},
            cmajor = Table[N[261.62558 2^(j/12)], {j,0,11}];
            notes := Table[cmajor[[Random[Integer,{1,12}]]],{n}];
            durs := Table[1/2^(Random[Integer, {0,3}]), {n}];
            MapThread[playTones,{notes, durs}]]
```

13. Following the implementation in the text, we first create ten steps between -2 and 2 (you can alter the range of step movements). These steps will determine how to move up or down the list of tone durations (1/8, 1/4, 1/2, 1).

```
In[1]:= d10 = Table[Random[Integer,{-2,2}], {10}]

Out[1]= {0, -2, 1, 1, 0, 2, -1, -1, -1, 1}


In[2]:= Mod[FoldList[Plus,0,d10], 4] + 1

Out[2]= {1, 1, 3, 4, 1, 1, 3, 2, 1, 4, 1}
```

```
In[3]:= durs = {1/8,1/4,1/2,1}[[%]]
```

$$Out[3]= \{\frac{1}{8}, \frac{1}{8}, \frac{1}{2}, 1, \frac{1}{8}, \frac{1}{8}, \frac{1}{2}, \frac{1}{4}, \frac{1}{8}, 1, \frac{1}{8}\}$$

Here are some $1/f^2$ tones.

```
In[4]:= s10 = Table[Random[Integer,{-2,2}], {10}]

Out[4]= {0, -1, -2, 2, 2, 1, -2, 2, 0, 0}


In[5]:= pos = Mod[FoldList[Plus,0,s10],13]+1

Out[5]= {1, 1, 13, 11, 13, 2, 3, 1, 3, 3, 3}


In[6]:= tones = cmajor[[pos]]

Out[6]= {261.626, 261.626, 523.251, 466.164, 523.251, 277.183,
         293.665, 261.626, 293.665, 293.665, 293.665}


In[7]:= MapThread[playTones,{tones,durs}]

Out[7]= {-Sound-, -Sound-, -Sound-, -Sound-, -Sound-, -Sound-,
         -Sound-, -Sound-, -Sound-, -Sound-, -Sound-}
```

And finally, here is one function that puts this all together.

```
In[8]:= tonesAndTimes2[n_]:= Module[{cmajor, tones, durs,d,t},
          cmajor = Table[N[261.62558 2^(j/12)], {j,0,11}];
          d = Table[Random[Integer,{-2,2}], {n}];
          durs = {1/8,1/4,1/2,1}[[ Mod[FoldList[Plus,0,d], 4] + 1 ]];
          t = Table[Random[Integer,{-2,2}], {n}];
          tones = cmajor[[ Mod[FoldList[Plus,0,t],13]+1 ]];
          MapThread[playTones,{tones, durs}]]


In[9]:= tonesAndTimes2[3]

Out[9]= {-Sound-, -Sound-, -Sound-, -Sound-}
```

11 | Applications

The development of large-scale *Mathematica* programming projects is discussed and illustrated in this chapter. One of the most popular computer programs—the Game of Life—is used to illustrate functional and rule-based programming techniques to create programs that are both efficient and elegant. The random walk model is used to model random processes in physics, chemistry, biology, materials engineering, and economics. It is used here to demonstrate how *Mathematica* programs can be written, run, and analyzed both visually and numerically. A common problem in programming is reading and interpreting—parsing— a user's input. (For example, *Mathematica* does this each time you evaluate an expression.) The method of "recursive descent parsing" is illustrated by presenting the implementation of the language *PDL*, a language for describing simple pictures consisting of squares, circles, rectangles, and ovals.

11.1 The Random Walk

11.1.1 | Introduction

In the course of doing their work, scientists and engineers must perform a variety of different computing tasks. Thus, while each of the individual activities that can be carried out with *Mathematica* (symbolic manipulation, number crunching, graphics, programming, report writing) is useful, the ability to carry out all of these operations in a single, integrated computing environment makes *Mathematica* especially valuable.

The use of *Mathematica* to perform diverse computing tasks can be illustrated using the random walk model. This model, which can be envisioned by thinking of a person taking a succession of steps which are randomly oriented with respect to one another, is widely used to represent random processes in nature; physicists model the transport of molecules, biologists model the locomotion of organisms, engineers model heat conduction, and economists model the time behavior of financial markets, all with the random walk model.

In this section, we will develop a program for executing a random walk. Then we will run the program and create a visualization of the walk that is created. Finally, we will numerically analyze some average quantities obtained from running the program a number of times.

11.1.2 | The One-Dimensional Random Walk

The simplest random walk model consists of *n* steps of equal length, back-and-forth along a line. A step increment (or step) in the positive *x*-direction corresponds to a value of 1 and a step increment in the negative *x*-direction corresponds to a value of −1. A list of the successive step increments of an *n*-step random walk in one dimension is therefore a list of *n* randomly selected 1s and −1s. This list can be generated in many ways. We will use

```
Table[(-1)^Random[Integer], {n}]
```

For example, to generate a list of ten "steps", replace *n* by 10.

```
In[1]:= Table[(-1)^Random[Integer], {10}]
Out[1]= {1, -1, 1, 1, 1, -1, -1, 1, -1, 1}
```

A list generated in this manner can be used to generate a list of the $(n + 1)$ locations of a one-dimensional *n*-step walk which starts at the origin, using the folding operation.

```
In[2]:= FoldList[Plus, 0, %]
Out[2]= {0, 1, 0, 1, 2, 3, 2, 1, 2, 1, 2}
```

We can now write the program `Walk1D` to generate a list of the step locations of an *n*-step random walk, originating at the origin.

```
In[3]:= Walk1D[n_] :=
            FoldList[Plus, 0, Table[(-1)^Random[Integer], {n}]]
```

Here is a ten-step one-dimensional random walk using this `Walk1D` program.

```
In[4]:= Walk1D[10]
Out[4]= {0, 1, 0, 1, 2, 1, 0, -1, -2, -3, -2}
```

Note: A list of the step locations can also be generated without first creating a list of the step increments, using the nesting operation, but this method is a bit slower than the approach with `FoldList` above.

```
NestList[(# + (-1)^Random[Integer])&, 0, n]
```

11.1.3 | The Two-Dimensional Lattice Walk

The random walk model in two or more dimensions is more complicated than the random walk in one dimension. Although each step of a one-dimensional walk is at 0 degrees (forward) or 180 degrees (backward) with respect to the preceding step, in higher dimensions each step can take a range of orientations with respect to the previous step.

We'll consider a random walk on a lattice. Specifically, we'll look at a lattice walk on the two-dimensional square lattice. This walk consists of steps of uniform length, randomly taken in the North, East, South, or West direction. The list of the possible step increments in this walk is

```
{{0, 1}, {1, 0}, {0, -1}, {-1, 0}}
```

A list of n step increments can be created from this, using

```
In[5]:= {{0,1}, {1,0}, {0,-1}, {-1,0}}[[Table[Random[Integer,
                                                 {1, 4}], {n}] ]]
```

By analogy to the one-dimensional walk computation here is a program, called Walk2D, to generate a list of the step locations of an n-step lattice walk starting at the origin $\{0, 0\}$.

```
In[6]:= Walk2D[n_] :=
            FoldList[Plus, {0, 0},
                {{0,1}, {1,0}, {0,-1}, {-1,0}}[[
                Table[Random[Integer, {1,4}], {n}]]]]
```

A typical run of the Walk2D program is shown below for a value of n equal to 10.

```
In[7]:= Walk2D[10]
Out[7]= {{0, 0}, {0, 1}, {0, 0}, {0, 1}, {-1, 1}, {0, 1}, {0, 0},
            {-1, 0}, {-2, 0}, {-2, -1}, {-1, -1}}
```

11.1.4 | Visualizing The Two-Dimensional Lattice Walk

We can create a snapshot of the path of the lattice walk using the graphics primitive Line, to draw lines between successive points in the walk.

```
In[8]:= ShowWalk2D[coords_, opts___]:=
           Show[Graphics[Line[coords],
                 opts,
                 AspectRatio -> Automatic]]
```

Here we have set the value of the AspectRatio option to Automatic so that steps in the *x*- and *y*-directions have equal lengths. This option can be overwritten by specifying a different value in the list of options given by opts. Note the use of the triple blank in the definition of ShowWalk2D. The pattern opts___ matches any sequence (possibly empty) of rules which are used here to govern the display of the graphic by changing certain options to the Graphics function. It is important that opts appears *before* the option AspectRatio. This will allow you to override this (or any) option value. If *Mathematica* sees an option listed more than once in a list of options, it will only use the first such option. If opts had come at the end of this function, you would not be able to change the value for AspectRatio.

Here are a simple example of a two-dimensional lattice walk.

```
In[9]:= ShowWalk2D[ Walk2D[20] ]
```

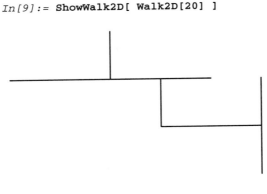

```
Out[9]= -Graphics-
```

As the graphics indicate, a lattice walk repeatedly revisits sites that have been previously visited in the course of its meandering. As a result, it is difficult, and usually impossible, to discern the history of the walk from a snapshot of the path. The best way to see the entire evolution process of the walk in an unobscured fashion is to create an animation.

Creating an animation of a random walk in *Mathematica* is straightforward. This can be explained using as an example a short lattice walk.

```
In[10]:= walk = Walk2D[10]
```

```
Out[10]= {{0, 0}, {1, 0}, {2, 0}, {2, -1}, {2, -2}, {3, -2},
          {3, -1}, {2, -1}, {1, -1}, {1, -2}, {0, -2}}
```

The animation consists of a sequence of graphics cells where the first cell shows the first step of the walk (consisting of a line drawn between the first two elements in walk for example) and each succeeding cell shows one more step than the previous cell. In general then, the mth cell is drawn using the Line function and the first $m + 1$ elements in walk. All of the graphics cells are drawn using

```
Map[(Show[Graphics[Line[Take[walk, #]]]])&,
    Range[2, Length[walk] ]]
```

In general, objects in a graphics cell are scaled to fill the monitor screen. Therefore, if we simply create cells, each containing a different number of steps of the walk using the above graphics command, steps in one cell will appear to be of a different length then the same steps in other cells. This will result in a jerky-looking animation.

We can make all of the step lengths in all of the graphics cells uniform by using the PlotRange option with the ordered pair of the minimum and maximum values of the components of the random walk in each direction, {{xmin, xmax}, {ymin, ymax}}. This quantity can be determined by separating the x- and y-components of the walk using Transpose and then mapping an anonymous function containing Min and Max onto it. For our example walk,

```
In[11]:= Map[{Min[#], Max[#]}&, Transpose[ walk ]]
```

```
Out[11]= {{0, 3}, {-2, 0}}
```

Here then, is the overall program for creating the animation.

```
In[12]:= AnimateWalk2D[coords_, opts___]:=
            Map[Show[Graphics[Line[Take[coords, #]]],
                     opts, AspectRatio -> Automatic,
                     PlotRange -> Map[{Min[#]-1, Max[#]+1}&,
                                      Transpose[coords]]]&,
            Range[2, Length[coords]]]
```

Note: We have added 1 to the maximum x and y values and subtracted 1 from the minimum x and y values in order to enhance the display by making the graphics a little smaller inside its bounding box.

While we can't see the random walk animation run in a book, we can look at the graphics cells in the animation by creating a graphics array. This is done for our example walk using

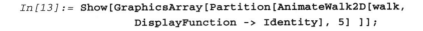

```
In[13]:= Show[GraphicsArray[Partition[AnimateWalk2D[walk,
              DisplayFunction -> Identity], 5] ]];
```

```
Out[13]= -GraphicsArray-
```

Note: The option `DisplayFunction -> Identity` is used to suppress the display of the individual graphics cells created by the `AnimateWalk` function (the GraphicsArray has its own `DisplayFunction` function option whose default value is `$DisplayFunction`) and the `Partition` function is used to specify the number of graphics in each row of the `GraphicsArray` picture.

11.1.5 | Numerical Analysis Of The Two-Dimensional Lattice Walk

In studying a random process, we are often interested in its mean, or average, properties. To obtain the mean value of a quantity, the quantity is computed a number of times, the values that are obtained are summed, and the result is divided by the number of computations. We'll look at two measures of the mean *size* of a two-dimensional lattice walk: the mean-square, end-to-end distance and the mean-square radius of gyration.

Mean-Square, End-To-End Distance

The square end-to-end distance r^2 of a two-dimensional lattice walk is given by $(x_f - x_i)^2 + (y_f - y_i)^2$ where $\{x_i, y_i\}$ and $\{x_f, y_f\}$ are the initial and final locations of the walk, respectively. Choosing the origin $\{0, 0\}$ as the starting point of the lattice walk simplifies the formula to $(x_f^2 + y_f^2)$.

The square end-to-end distance of a lattice walk starting at $\{0, 0\}$ and ending at $\{x_f, y_f\}$ is given by the following:

```
Apply[Plus, {xf, yf}^2]
```

We can write a program, called `SquareDistance`, using the above formula to compute r^2 for the two-dimensional lattice walk.

```
In[14]:= SquareDistance[walk_List] := Apply[Plus, Last[walk]^2]
```

A typical run of the `SquareDistance` program is shown below using our example lattice walk.

```
In[15]:= SquareDistance[walk]
```

```
Out[15]= 4
```

A program, called `MeanSquareDistance`, for computing the mean-square, end-to-end distance for m n-step lattice walks can be easily written, using the `Sum` function

```
In[16]:= MeanSquareDistance[n_Integer, m_Integer] :=
           Module[{walk2D},
              walk2D[s_] :=
                 FoldList[Plus, {0, 0},
                       {{0,1},{1,0},{0,-1},{-1,0}}[[
                    Table[Random[Integer,{1,4}],{s}]]]];
              N[Sum[Apply[Plus, Last[walk2D[n]]^2], {m}]/m]
              ]
```

A typical run of the `MeanSquareDistance` program is shown below for 25 ten-step walks.

```
In[17]:= MeanSquareDistance[10, 25]
```

```
Out[17]= 8.96
```

Mean-Square, Radius of Gyration

The mean square radius of gyration, $< R_g^2 >$, of a random walk is the sum of the squares of the distances of the step locations from the center of mass divided by the number of step locations, where the center of mass is the sum of the step locations divided by the number of step locations.

The computation of the center of mass and the sum of the squares of step distances from the center of mass is relatively straightforward.

For the list of the $(n + 1)$ locations of an n-step walk, which we'll call `locs`, the center of mass, which we'll call `cm`, is given by

```
cm = N[Apply[Plus, locs]/(n + 1)]
```

In the list $\{\{(x_0 - x_{cm})^2, (x_1 - x_{cm})^2, \ldots, (x_n - x_{cm})^2\}, \{(y_0 - y_{cm})^2, (y_1 - y_{cm})^2, \ldots, (y_n - y_{cm})^2\}\}$, x_j and y_j are the x- and y-coordinates of the jth step location, and x_{cm} and y_{cm} are the coordinates of the center of mass location. This list is given by

```
(Transpose[locs] - cm)^2
```

The sum of the squares of distances of the step locations from the center of mass is given by

```
Apply[Plus,  Flatten[(Transpose[locs] - cm)^2]]
```

The program for the mean-square-radius of gyration, called `MeanSquareRadiusGyration`, is written using the above expressions.

```
In[18]:= MeanSquareRadiusGyration[m_Integer, n_Integer] :=
      Module[{squareRadiusGyration},
          squareRadiusGyration[s_Integer] :=
          Module[{locs,cm,choices = {{1,0},{-1,0},{0,1},{0,-1}}},
              locs = FoldList[Plus,{0,0},
                  choices[[Table[ Random[Integer,{1,4}],{s}]]]]];
              cm = N[Apply[Plus, locs]/(s+1)];
              Apply[Plus, Flatten[(Transpose[locs]-cm)^2]]/(s+1)
              ];
          N[Sum[squareRadiusGyration[n], {m}]/m]
      ]
```

A typical run of the `MeanSquareRadiusGyration` program is shown below for 20 fifteen-step walks.

```
In[19]:= MeanSquareRadiusGyration[15, 20]

Out[19]= 3.12925
```

The Critical Exponent Of A Random Walk

Experimental and theoretical studies indicate that both $<r^2>$ and $<R_g^2>$ have a power law dependence on the number of steps in the walk, n. The power, ν, in the relationship $<r_2> = n^\nu$ is known as the *critical exponent* of the walk.

The critical exponent for a random walk can be determined by finding a fit of the form (coefficient $\cdot$ x) to a list of ordered pairs, $\{\text{Log}[<r^2>], \text{Log}[n]\}$. The computed value of the coefficient is the critical exponent. We will illustrate this with a specific example.

We first create a data set of the computed values of the ordered pairs $\{<r^2>,\ n\}$ for values of n from 10 to 90 in increments of 20, with a value of 75 being used to compute $<r^2>$.

```
In[20]:= data = Map[({MeanSquareDistance[#, 75], #})&,
            Range[10, 90, 20]]

Out[20]= {{10.3733, 10}, {28.5867, 30}, {45.0933, 50},
            {79.8667, 70}, {105.68, 90}}
```

Creating a list of ordered pairs, $\{\text{Log}[<r^2>],\ \text{Log}[n]\}$, from `data` and using it in the `Fit` function, we get

```
In[21]:= calculatedExponent = Fit[N[Log[data]], {x}, x]

Out[21]= 0.988123 x
```

In this example, the critical exponent is nearly one.

To see just how good the power law behavior of the random walk is, we can look at plots of the computed power law relationship and the data set together.

A plot of the computed power law relationship is created using

```
In[22]:= calculatedGraph = Plot[calculatedExponent, {x, 0, 90}]
```

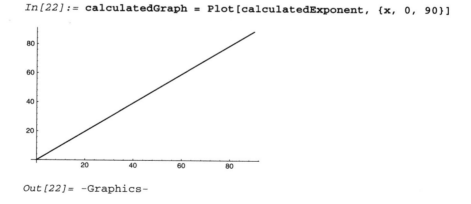

```
Out[22]= -Graphics-
```

A log-log plot of the data is created using the `LogLogListPlot` function defined in the *Mathematica* `Graphics` package.

```
In[23]:= Needs["Graphics`Graphics`"]
```

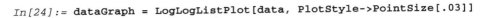

In[24]:= **dataGraph = LogLogListPlot[data, PlotStyle->PointSize[.03]]**

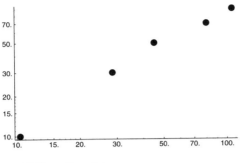

Out[24]= -Graphics-

The plots of the computed power law relationship and the data set can now be shown together.

In[25]:= **Show[dataGraph, calculatedGraph]**

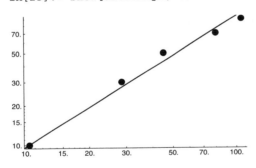

Out[25]= -Graphics-

Exercises

1. Write a program for computing the step locations of a three-dimensional walk on a cubic lattice. Create an animation of a ten-step cubic lattice walk.

2. The off-lattice walk in two dimensions consists of n steps of equal lengths, each step being randomly oriented with respect to preceding steps. In this model, the x- and y-components of each step increment are not independent because each step must have length one. This constraint can be satisfied by first randomly selecting an orientation angle θ, between 0 and 360 degrees, for each step, using

Exercises (cont.)

`Random[Real, {0, N[2 Pi]}]`, and then calculating the ordered pair of the *x*- and *y*-components of the step, with `{Cos[θ], Sin[θ]}`. A list of the step increments for the off-lattice, *n*-step walk is given by first generating a list of θ values and then mapping the anonymous function `{Cos[#],Sin[#]}&` onto the list.

```
OffLattice[n_] :=
    FoldList[Plus, {0.0 , 0.0},
        Map[{Cos[#], Sin[#]}&,
            Table[Random[Real, {0,N[2Pi]}], {n}]]]
```

Write a program for the step locations of an off-lattice walk in three dimensions.

3. Modify the function `AnimateWalk2D` by adding a red disk that moves to the "current position" in the walk. The viewer will then see this red disk moving along on the random walk as the animation is played.

Solutions

1.
```
Walk3D[n_] := FoldList[Plus, {0, 0, 0},
    {{-1,1,1}, {-1,-1,1}, {1,-1,1}, {1,1,1},
     {-1,1,-1},{-1,-1,-1},{1,-1,-1},{1,1,-1}}[[
    Table[Random[Integer, {1, 8}], {n}]]]]
```

3.
```
AnimateWalk2D[coords_, opts___]:=
        Map[Show[Graphics[{
                    {RGBColor[1,0,0], PointSize[.02],
                     Point[ coords[[#]] ]},
                    Line[Take[coords, #]]}],
            opts,
            AspectRatio -> Automatic,
            PlotRange -> Map[{Min[#]-.2, Max[#]+.2}&,
                                Transpose[coords]]]&,
                            Range[2, Length[coords]]]
```

11.2 The Game of Life

A cellular automaton is a system of discrete lattice sites, each of which has a value (usually an integer number) associated with it. The values of the sites change simultaneously, in each of a succession of discrete time steps, by applying rules that depend on the values of a site and the sites in its vicinity.

Cellular automata have been used to model various physical, chemical, biological, and social phenomena. [Gaylord and Wellin 1995]. In principle, any process that can be described by an algorithm or program can be modeled by a cellular automaton.

The Game of Life, created by the British mathematician John Conway, is the most well-known cellular automaton. It is the forerunner of so-called *artificial life* (or *a-life*) systems and it was the first program run on the first parallel processing computer. It has been estimated that more computer time has been spent (or wasted, depending on your point of view) running the Game of Life program than any other program.

We will show how to program the Game of Life in *Mathematica* so that it is optimized for efficiency (run speed).

The Game of Life is played on a two-dimensional square Boolean lattice where sites have values of either zero or one. A site with value one is said to be *alive* and a site with value zero is said to be *dead*. To illustrate the computations involved in the Game of Life program, we will use the following small lattice system:

```
In[1]:= (GameBoard = Table[Random[Integer], {4}, {4}])//TableForm

Out[1]//TableForm= 0   1   0   0
                   0   1   0   1
                   1   0   0   1
                   0   1   1   0
```

In order to update a site of GameBoard, the sum of the values of the sites in its neighborhood must be determined.

The neighborhood of a site in GameBoard consists of the site and the eight nearest neighbor sites lying North, Northeast, East, Southeast, South, Southwest, West, and Northwest of the site.

The neighborhood of a site located in the interior of the lattice is obvious. For example, the nearest neighbor sites to the {2, 3} site (which lies in the second row, third column of GameBoard) are the {1, 3}, {1, 4}, {2, 4}, {3, 4}, {3, 3}, {3, 2}, {2, 2}, and {1, 2} sites.

The neighborhood of a site lying on one of the borders of the lattice is less apparent. Employing what are known as periodic boundary conditions, some of the nearest neighbor sites are taken from the opposing borders. A non-corner site located in the first or last row (column) of the board has the corresponding site in the last or first row (column) as a nearest neighbor site, respectively, and a corner site has the two sites in the opposing corner as two of its nearest neighbor sites. For example, the nearest neighbor sites to the {2, 4} site (which lies in the second row, last column of GameBoard) are the {1, 4}, {1, 1}, {2, 1}, {3, 1}, {3,4}, {3, 3}, {2, 3}, and {1, 3} sites.

The sixteen neighborhoods of the sites in the lattice system can be generated in two steps.

An expanded matrix is created by first copying the first element in each row onto the end of the row and copying the last element in each row onto the front of the row, and then copying the first row onto the end of the list of rows and copying the last row onto the front of the list of rows. The following anonymous function can be used to perform this operation.

```
In[2]:= Wrap = Join[{Last[#]}, #, {First[#]}]&
Out[2]= Join[{Last[#1]}, #1, {First[#1]}] &
```

The application of the the Wrap function to GameBoard is shown below.

```
In[3]:= Wrap[Map[Wrap, GameBoard]]//TableForm
Out[3]//TableForm= 0   0   1   1   0   0
                   0   0   1   0   0   0
                   1   0   1   0   1   0
                   1   1   0   0   1   1
                   0   0   1   1   0   0
                   0   0   1   0   0   0
```

The expanded matrix created by applying the Wrap function to the lattice can be partitioned into overlapping three by three matrices to create a list of the neighborhoods of the sites in the lattice. In the case of the GameBoard lattice, we get

```
In[4]:= (Neighborhoods =
          Partition[Wrap[Map[Wrap, GameBoard]],
             {3, 3}, {1, 1}]) //MatrixForm
```

```
Out[4]//MatrixForm=  0 0 1    0 1 1    1 1 0    1 0 0
                     0 0 1    0 1 0    1 0 0    0 0 0
                     1 0 1    0 1 0    1 0 1    0 1 0

                     0 0 1    0 1 0    1 0 0    0 0 0
                     1 0 1    0 1 0    1 0 1    0 1 0
                     1 1 0    1 0 0    0 0 1    0 1 1

                     1 0 1    0 1 0    1 0 1    0 1 0
                     1 1 0    1 0 0    0 0 1    0 1 1
                     0 0 1    0 1 1    1 1 0    1 0 0

                     1 1 0    1 0 0    0 0 1    0 1 1
                     0 0 1    0 1 1    1 1 0    1 0 0
                     0 0 1    0 1 0    1 0 0    0 0 0
```

Given the neighborhoods of the sites on the lattice, we can determine whether a site is alive or dead and how many of its nearest neighbor sites are alive. These are the two quantities which appear in the rules used to update a site.

The three *life and death* rules for updating a site in the lattice are:

1. a living site (a site with value 1) with exactly two living nearest neighbor sites remains alive (its value is updated to 1).

2. any site (a site with value 0 or 1) with three living nearest neighbor sites stays alive or is born (its value is updated to 1).

3. any other site (a site with value 0 or 1) remains dead or dies (its value is updated to 0).

A conditional function which, given the neighborhood of a site, applies the appropriate rule is given below.

```
In[5]:= LiveOrDie[lis_] := Module[{Nghbrs},
        Nghbrs = Count[lis, 1, {2}] ;
        If[lis[[2, 2]] == 1 && Nghbrs == 4 || Nghbrs == 3, 1, 0]]
```

Applying the `LiveOrDie` function to the neighborhoods of `GameBoard` yields the updated `GameBoard`.

```
In[6]:= Map[LiveOrDie, Neighborhoods, {2}]//MatrixForm

Out[6]//MatrixForm=  0   1   0   0
                     0   1   0   1
                     0   0   0   1
                     0   1   1   1
```

Finally, the evolution of the lattice over *t* time steps, or until it stops changing, is carried out using the `FixedPointList` function. For the `GameBoard` example, we have

```
FixedPointList[Map[LiveOrDie,
                Partition[Wrap[Map[Wrap, #]],
                    {3, 3}, {1, 1}], {2}]&,
            GameBoard, t]
```

The code fragments developed above can be used to construct a program for the Game of Life. Indeed, this is done in the Life.m notebook that is distributed with *Mathematica*. However, while this program works, it is unduly slow. A much more efficient (faster running) program for the Game of Life can be developed by following some general *Mathematica* programming guidelines.

The most efficient way to program in *Mathematica* is to utilize the following approaches as much as possible:

- avoid looping

- minimize conditional branching

- manipulate data structures in their entirety

- employ built-in *Mathematica* functions

- use anonymous functions, higher-order functions, and nested function calls

- create look-up tables

The use of these principles is well illustrated in the Game of Life program we will now develop.

A matrix whose elements are the number of living, nearest neighbor sites to the corresponding sites in the Game of Life lattice can be computed directly from the lattice without having to first create the neighborhoods of the lattice, using the following function.

```
In[7]:= LivingNghbrs[mat_] :=
        Apply[Plus, Map[RotateRight[mat, #]&,
                {{-1,-1},{-1,0},{-1,1},{0,-1},
                {0,1},{1,-1},{1,0},{1,1}}]]
```

The LivingNghbrs function makes use of the fact that *Mathematica* adds lists by vector addition, adding the corresponding elements of the lists. Applying the function to the GameBoard example, we get

```
In[8]:= LivingNghbrs[GameBoard] //MatrixForm
```

```
Out[8]//MatrixForm=  4   3   5   2
                     5   2   4   2
                     4   4   5   3
                     4   3   3   3
```

Comparing this output with the Neighborhoods matrix created earlier, we can see that each element in LivingNghbrs[GameBoard] is the number of living nearest neighbor sites to the corresponding site in GameBoard.

We can write down site update rules, whose two arguments are the value of a site and the sum of the values of the nearest neighbor sites in its neighborhood. These rules are a direct translation of the life and death rules from words to code.

```
In[9]:= update[1, 2]  := 1

        update[_, 3]  := 1

        update[_, _]  := 0

        Attributes[update]  := Listable
```

The update rule is given the Listable attribute so when it is applied to a matrix of site values and also to a matrix of the number of living neighbors to these sites, a matrix is created whose elements are obtained by applying the update function to the corresponding elements of the two matrices. This behavior can be demonstrated using a general function, g, with the GameBoard and LivingNghbrs[GameBoard] matrices.

```
In[10]:= Attributes[g] = Listable;
```

```
In[11]:= g[GameBoard, LivingNghbrs[GameBoard]] //MatrixForm
```

```
Out[11]//MatrixForm=  g[0, 4]   g[1, 3]   g[0, 5]   g[0, 2]
                      g[0, 5]   g[1, 2]   g[0, 4]   g[1, 2]
                      g[1, 4]   g[0, 4]   g[0, 5]   g[1, 3]
                      g[0, 4]   g[1, 3]   g[1, 3]   g[0, 3]
```

Using the update rules with the GameBoard and LivingNghbrs matrices, and comparing the result obtained earlier by applying the LiveOrDie function to the Neighborhoods of GameBoard, we see that each site in the board has been correctly updated.

```
In[12]:= update[GameBoard, LivingNghbrs[GameBoard]] //MatrixForm
```

```
Out[12]//MatrixForm=  0   1   0   0
                      0   1   0   1
                      0   0   0   1
                      0   1   1   1
```

Note: While the three update rules overlap with one another, there is no confusion as to when each rule is used because *Mathematica* applies more specific rules before more general rules. Thus, while a site with value 1 and 2 nearest neighbor sites with value 1 will satisfy both the first and third rules, the first rule is used because it is the most specific applicable rule. Similarly, while a site having 3 nearest neighbor sites with value 1 will satisfy both the second and third rules, the second rule is used because it is the most specific applicable rule. The third rule is more general than the other two rules and hence is only used if neither of the other rules can be used.

The evolution of the lattice over t time steps can be carried out using an anonymous function

```
update[#, LivingNghbrs[#]]&
```

where # represents the lattice configuration, with the `FixedPointList` function. Using the `GameBoard` example and three time steps to illustrate this, gives

```
In[13]:= FixedPointList[update[#,LivingNghbrs[#]]&, GameBoard, 3]
```

```
Out[13]= {{{0, 1, 0, 0}, {0, 1, 0, 1}, {1, 0, 0, 1},
          {0, 1, 1, 0}}, {{0, 1, 0, 0}, {0, 1, 0, 1},
          {0, 0, 0, 1}, {0, 1, 1, 1}},
          {{0, 1, 0, 1}, {0, 0, 0, 0}, {0, 1, 0, 1},
          {0, 1, 0, 1}}, {{0, 0, 0, 0}, {0, 0, 0, 0},
          {0, 0, 0, 0}, {0, 1, 0, 1}}}
```

Taking the `Transpose` of this result in order to interchange rows and columns and facilitate a comparison with our previous results, we have

```
In[14]:= Map[Transpose, %] //MatrixForm
```

```
Out[14]//MatrixForm=  0   1   0   0
                      0   1   0   1
                      1   0   0   1
                      0   1   1   0

                      0   1   0   0
                      0   1   0   1
                      0   0   0   1
                      0   1   1   1

                      0   1   0   1
                      0   0   0   0
                      0   1   0   1
                      0   1   0   1

                      0   0   0   0
                      0   0   0   0
                      0   0   0   0
                      0   1   0   1
```

The code fragments given above are combined into the Game of Life program.

```
In[15]:= LifeGame[s_, t_] := Module[{initconfig,livingNghbrs,update},
            initconfig = Table[Random[Integer],{s},{s}];
            livingNghbrs[mat_]:=
               Apply[Plus,  Map[RotateRight[mat, #]&,
                              {{-1,-1},{-1,0},{-1,1},{0,-1},
                               {0,1},{1,-1},{1,0},{1,1}}]];
            update[1, 2] := 1;
            update[_, 3] := 1;
            update[_, _] := 0;
            Attributes[update] = Listable;
            FixedPointList[update[#,livingNghbrs[#]]&,initconfig,t]]
```

The input parameters, s and t, are respectively, the linear size of the lattice and the maximum number of time steps carried out.

Finally, the focus in playing the Game of Life is on identifying various patterns of 1s amongst the 0s, and observing their behaviors. This is more easily done using a graphical, rather than numerical, display. A program which can take the output of the LifeGame program and create an animation of the Game of Life is given below.

```
In[16]:= ShowLife[list_]:=
            Scan[(Show[Graphics[RasterArray[Reverse[list[[#]] /.
                           {1 -> RGBColor[1,0,0],
                            0 -> RGBColor[0,0,0]}]],
                   AspectRatio -> Automatic]])&,
               Range[Length[list]]]
```

Exercises

1. The Game of Life is most interesting to watch when persistent patterns, known as life-forms, occur during the evolution process.

One pattern that has been extensively studied is known as the glider which is defined by

```
In[1]:= glider[x_, y_] :=
          {{x, y}, {x+1, y}, {x+2, y}, {x+2, y+1}, {x+1, y+2}}
```

Modify the program for the Game of Life so that the lattice can be seeded with life forms and observe the behavior of a glider (it should appear, disappear, and then reappear in a shifted position every fifth generation). To better understand the use of the periodic boundary conditions, note what happens when a glider pattern moves beyond a border of the game board.

Solutions

1. We will use essentially the same function as before, but we will "overload" the function by providing a definition for the case when a third argument is provided.

```
In[1]:= LifeGame[s_, t_, lifeform_List]:=
          Module[{init=Table[0,{s},{s}], board,
                 livingNghbrs, update},
            board = ReplacePart[init, 1, lifeform];
            livingNghbrs[mat_]:=
              Apply[Plus, Map[RotateRight[mat, #]&,
                            {{-1,-1}, {-1,0}, {-1,1}, {0,-1},
                             {0,1}, {1,-1}, {1,0}, {1,1}}]
                   ];
            update[1, 2] := 1;
            update[_, 3] := 1;
            update[_, _] := 0;
            Attributes[update] = Listable;
            FixedPointList[update[#, livingNghbrs[#]]&,
                          board, t]]
```

If `LifeGame` is called with two arguments, then the definition given earlier will be applied (random initial game board). If `LifeGame` is called with three arguments, then this definition above will be matched.

Here is a game played on a 50 × 50 board, starting with a glider object initially at lattice site (20, 20), and played for ten generations.

```
In[2]:= lg50 = LifeGame[50, 10, glider[20,20]];
```

This game could then be animated by evaluating ShowLife[lg50].

11.3 Implementing Languages

The *Mathematica* programming language is just one example of a *computer language*. There are thousands of others, including C and FORTRAN for general-purpose programming, SQL for database queries, TEX and Postscript for typesetting, and on and on. The processing of these languages shares some basic methods which we will illustrate in this section by implementing a toy picture-description language, *PDL*.

PDL is used to describe pictures consisting of simple shapes either contained in or next to one another. An example of such a picture is shown in Figure 11.1; it is described by the picture specification

```
square (5)
  containing n/n (clear rectangle (5, 2)
            containing w/w (circle (2))
            containing c/c (circle (2))
            containing e/e (circle (2)))
  containing c/n (oval (3, 1)
        connecting sw/ne (square (1))
        connecting se/nw (square (1))
        connecting s/c (circle (1)))
```

The picture in Figure 11.1 contains one large and two small squares, one rectangle (but it is "clear," that is, invisible), four circles and an oval. The rectangle is contained in the square, and in turn contains three circles; the oval is contained in the square and has the two small squares and a circle connected to it. The numbers in parentheses give the sizes of the shapes, and the odd-looking notations like n/n and se/nw indicate how two shapes are connected.

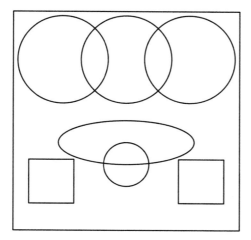

Figure 11.1: A picture produced by PDL.

For example, the n/n notation on the second line says that the top (or "north") of the rectangle is positioned next to the top of the square that contains it; the se/nw notation on the second to last line indicates that the upper-left ("northwest") point of the square is placed next to the lower-right ("southeast") point of the oval; on the last line, the south point of the oval connects to the center of the circle.

We will write a function PDL that will take such a description (as a character string) and convert it into *Mathematica* graphics primitives for display.

The first property all modern computer languages share is that their syntax can be formally defined. The formal definition guides the implementation in a direct and simple way. The formal definition is given as a *context-free grammar*, in which a set of rules, called *productions*, is used to define both the allowable picture specifications and the syntactic structure of those specifications. The formal syntax of *PDL* is given in Table 11.1.

In the *PDL* grammar, the names given in *slanted* font are called *variables*. The variables generate sets of strings; the legal picture specifications are all the strings generated by the variable *picture*. The items written in typewriter font appear literally in specifications. Aside from *integer* (which by definition generates all the integers) and *direction* (which by definition generates the strings n, e, s, w, c, ne, se, sw, nw), the variables generate strings in the following way: starting with a variable, replace it by the right-hand side of any rule in which it appears on the left-hand side; then continue to replace variables by the right-hand sides of rules for those variables (or, in the case of *integer* and *direction*, by any integer or direction), until a string without variables is obtained. (When production 2 or 8 is applied, the variable just disappears.)

1.	*picture*	⟶ *shape associations*
2.	*associations*	⟶
3.	*associations*	⟶ *connection associations*
4.	*associations*	⟶ *containment associations*
5.	*connection*	⟶ connecting *direction / direction (picture)*
6.	*containment*	⟶ containing *direction / direction (picture)*
7.	*shape*	⟶ *color primitive size*
8.	*color*	⟶
9.	*color*	⟶ clear
10.	*primitive*	⟶ square
11.	*primitive*	⟶ circle
12.	*primitive*	⟶ oval
13.	*primitive*	⟶ rectangle
14.	*size*	⟶ (*integer size2*
15.	*size2*	⟶)
16.	*size2*	⟶ , *integer*)

Table 11.1: Formal syntax of *PDL*

Consider the picture specification

```
square (20) containing c/w (oval (9, 18))
```

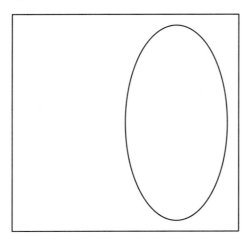

Figure 11.2: A simpler picture produced by *PDL*.

which produces the picture shown in Figure 11.2. It is generated from *picture* in this way (where we've indicated the number of the production being used in each case):

$$
\begin{array}{lll}
picture & \longrightarrow_1 & shape\ associations \\
 & \longrightarrow_7 & color\ primitive\ size\ associations \\
 & \longrightarrow_8 & primitive\ size\ associations \\
 & \longrightarrow_{10} & \texttt{square}\ size\ associations \\
 & \longrightarrow_{14} & \texttt{square}\ (\ integer\ size2\ associations \\
 & \longrightarrow & \texttt{square}\ (\ \texttt{20}\ size2\ associations \\
 & \longrightarrow_{15} & \texttt{square}\ (\ \texttt{20}\ )\ associations \\
 & \longrightarrow_4 & \texttt{square}\ (\ \texttt{20}\ )\ containment\ associations \\
 & \vdots &
\end{array}
$$

and so on.

A crucial observation is that the derivation of a string from a variable can be represented as a tree, called a *parse tree*. The derivation above corresponds to the following tree:

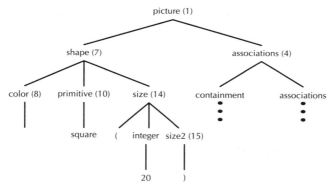

Figure 11.3: A parse tree.

Notice, by the way, that there is no need to include the variable at each node; the production number immediately determines the variable.

We will use the tree representation of the input—or a very similar representation, omitting uninteresting things like parentheses—extensively. The goal of *parsing* is to transform the sequence of characters in the input into a parse tree. Given that form, we can do the real work: finding the location of each shape and generating the *Mathematica* graphics primitives to draw the picture.

The parsing phase is divided into two steps, *lexical analysis* and *parsing*, and the remainder of the processing is also divided into two steps, computing information about each shape in the picture (especially, its location) and converting this information into graphics primitives. Thus, the function PDL is given by,

```
ShowPicture[p_] := Show[Graphics[p], AspectRatio->Automatic]
```

```
PDL[input_] :=
  ShowPicture[
    ConvertShapes[
      ComputeShapes[
        Parse[
          Lex[input]]]]]
```

For example,

```
PDL["square (20) containing c/w (oval (9, 18))"]
```

produces the graphic in Figure 11.2. We will, naturally, structure our discussion like-wise.

Before delving into programming details, we'll finish our brief "user's guide" begun earlier. As we've seen, shapes can be clear, in which case they are not drawn, or regular, in which case they are drawn in black. Each shape has a size (one integer for squares and circles, two for ovals and rectangles) with these meanings:

square:	length of a side
circle:	diameter
oval:	horizontal and vertical diameters
rectangle:	width and height

A shape can contain or be connected to any number of other shapes. (Since every shape has an explicit size, a "contained" shape is not necessarily completely contained.) The most complicated aspect of the language is determining where shapes go, depending upon the points at which they are connected. Each shape has a center and eight compass points. These are shown for each shape in Figure 11.4. When a shape is connected to or contained in another shape, the two directions given in the connecting or containing phrase match up. For example, Figure 11.5 shows the picture specified by `square (4) connected se/n (circle (2))`.

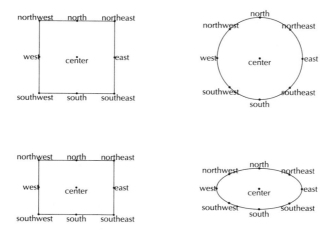

Figure 11.4: Compass points for the four types of shapes.

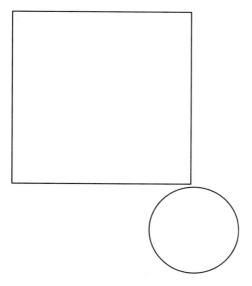

Figure 11.5: Picture produced by `square (4) connected se/n (circle (2))`

The top (north) of the circle is next to the lower-right (southeast) corner of the square. Similarly, Figure 11.6 shows an oval containing a rectangle.

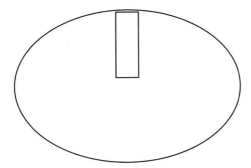

Figure 11.6: Picture produced by oval (10, 7) containing n/n (rectangle (1, 3))

The top of the oval touches the top of the rectangle. However, in both cases, the two figures don't exactly touch; rather, a gap of size 0.1 is left between them. The difference between connecting and containing is simply the direction of the gap. In

$$picture_1 \text{ connecting } d_1/d_2 \ (picture_2)$$

direction d_2 of $picture_2$ is placed at a point determined by finding the d_1 direction of $picture_1$ and then moving 0.1 units *away* from the center of $picture_1$. If connecting is replaced by containing, the correction is 0.1 units *toward* the center of $picture_1$.

Finally, the rules about correcting by 0.1 do not apply if either direction is c, for center. If the connecting directions are d/c or c/d, whatever d is, it is not adjusted by 0.1 in either direction. (Thus, in this case it doesn't matter whether $picture_2$ is connected or contained.) The reader is urged to try some examples using the code provided on the diskette before continuing.

To return to the programming of the *PDL* language processor, we'll start with a discussion of the syntactic analysis phase, consisting of lexical analysis (or *lexing*) and *parsing*. This division is conventional and appears in virtually all language processors.

Lexing is the process of dividing up the input (a character string) into significant syntactic units, called *tokens*. (Think of the entire picture specification as a sentence, the characters of the input as the letters, and the tokens as the words; lexing groups the letters into words, and parsing determines the syntactic structure of the sentence.) The function Lex is given a string and produces a list of symbols and numbers.

```
In[1]:= ex =  "square (20) containing c/w (oval (9, 18))"


In[2]:= Lex[ex]

Out[2]= {square, lparen, 20, rparen, containing, center, slash,
         west, lparen, oval, lparen, 9, comma, 18, rparen,
         rparen, eof}
```

```
In[3] := Map[Head, %]
```

Out[3]= {Symbol, Symbol, Integer, Symbol, Symbol, Symbol, Symbol,
          Symbol, Symbol, Symbol, Symbol, Integer, Symbol,
          Integer, Symbol, Symbol, Symbol}

In the lexed output, we have also replaced special characters like parentheses by symbols and we've added a final symbol, eof (a traditional name meaning "end of file").

Symbols are a little more convenient than strings for what we want to do. However, their use requires that we introduce a new operator for comparing symbols that we haven't needed until now, ===. The equality operator == works fine for numbers and strings, and for lists of same, but not for symbols:

```
In[4] := {a, b} == {a, b}
```

Out[4]= True

```
In[5] := {a, b} == {a, c}
```

Out[5]= {a, b} == {a, c}

== can tell when two lists of symbols are identical, but not when they are different. === compares symbols for identity:

```
In[6] := {a, b} === {a, b}
```

Out[6]= True

```
In[7] := {a, b} === {a, c}
```

Out[7]= False

All the code used by Lex is shown in Figure 11.3. The basic process is: find the first sequence of characters that form a token, say *t*, recursively lex the remaining characters, and join *t* to the result. Technicalities arise with the treatment of numbers and the desire to ignore blanks.

```
Needs["BaseConvert`"]

(* LEXICAL ANALYSIS *)

mainRules = {
  {"(", y___} -> {lparen, y},
  {")", y___} -> {rparen, y},
  {",", y___} -> {comma, y},
  {"/", y___} -> {slash, y},
  {"c", "o", "n", "n", "e", "c", "t", "i", "n", "g", y___}
      -> {connecting, y},
  {"c", "o", "n", "t", "a", "i", "n", "i", "n", "g", y___}
      -> {containing, y},
  {"s", "q", "u", "a", "r", "e", y___} -> {square, y},
  {"c", "i", "r", "c", "l", "e", y___} -> {circle, y},
  {"o", "v", "a", "l", y___} -> {oval, y},
  {"r", "e", "c", "t", "a", "n", "g", "l", "e", y___}
      -> {rectangle, y},
  {"c", "l", "e", "a", "r", y___} -> {clear, y},
  {"n", "e", y___} -> {northeast, y},
  {"s", "e", y___} -> {southeast, y},
  {"s", "w", y___} -> {southwest, y},
  {"n", "w", y___} -> {northwest, y},
  {"n", y___} -> {north, y},
  {"e", y___} -> {east, y},
  {"s", y___} -> {south, y},
  {"w", y___} -> {west, y},
  {"c", y___} -> {center, y}
}


convertDigits[L_] := Map[If[DigitQ[#], StringToInt[#], #]&, L]

numberRule =
  {{m_?NumberQ, n_?NumberQ, y___} -> {10m+n, y}}

removeBlanks = { {" ", y___} -> {y} }

Lex[input_] :=
   Module[{inp = Map[FromCharacterCode, ToCharacterCode[input]]},
     Lexaux[Join[convertDigits[inp], {eof}] //. removeBlanks]]

Lexaux[{eof}] := {eof}
Lexaux[input_] :=
   Module[{lexed = If[NumberQ[First[input]],
                     input //. numberRule,
                     input /. mainRules]},
         Join[{First[lexed]}, Lexaux[Rest[lexed] //. removeBlanks]]]
```

Program Listing 11.1: Code for lexing *PDL*

The first thing `Lex` does is "explode" the input string into a list of character codes. As we saw in section 7.6.2, we can do whatever we want with that list; however, this would involve looking up a lot of character codes, so a simpler approach is to convert each character code back to a string containing just that character. So, in `Lex`, `inp` is a list containing each of the characters in `input`. `convertDigits` is applied to change all digit characters to numbers (for example, the string `"4"` becomes the number 4), using `StringToInt` from the `BaseConvert` package. `eof` is added to the end of the list. As final preparation before calling `Lexaux`, the transformation rule `removeBlanks` is applied repeatedly (`//.`) to remove all leading blanks. Thus, the argument to `Lexaux` is a list of one-character strings, numbers, and a final `eof` symbol, with the first element non-blank. `Lexaux` repeatedly looks for characters that constitute a token at the start of the list and replaces those characters by the token; it does this either by a single use of a rule in `mainRules` or by repeated use of `numberRule`. It recursively lexes the rest of the list and returns. We have already shown the result for our running example.

`Parse` takes the list of tokens and, if it is a legal picture specification, returns its parse tree. The parser is the most interesting part of our language processor, as it shows a strong link between the grammar specification (page 390) and the program.

Our method here is called *top-down*, or *recursive descent*, *parsing*. The idea is to build the parse tree by starting with a variable and letting the input string guide us in adding nodes to the tree by telling us which production is applicable. For example, we are looking at the list of tokens

```
{square, lparen, 20, rparen, containing, center, slash, west,
     lparen, oval, lparen, 9, comma, 18, rparen, rparen, eof}
```

and we want to create a parse tree for this string from variable *picture*. The only production from *picture* is production 1, so we could just add it to the tree without even looking at the input. However, we'd also like to report any syntactic errors as soon as possible, so we will look at the first token in the input and see if it is legal at this point. It so happens that every string derivable from picture must begin with one of the words `square`, `circle`, `oval`, `rectangle`, or `clear`. If the first token is not one of these, we can report an error; if it is, we add *shape* and *associations* to the tree. We continue by trying to use variable *shape* to match part of the list of tokens. Again, there is only one production for *shape* (production 7), and after checking that `square` can be the first token in a string derivable from *shape*, we add production 7 to the tree. The first part of the right-hand side of production 7 is the variable *color*. We have a choice now, production 8 or 9, and we have to choose correctly. However, it is clear that `square` is not the first token in a string derived using production 9, so it must be production 8 and we fill that in. The next unfinished part of the tree is the node containing the variable *primitive*, which has four productions. A look at the input makes it immediately clear that only production 10 will work here, so we fill it in. Continuing in this way, we eventually get the tree shown on page 391 and use up all the input. The top-down parsing process is illustrated in the following series of parsing tree figures.

Parse Tree   **Input**

picture

Input: {square, lparen, 20, rparen, ...}

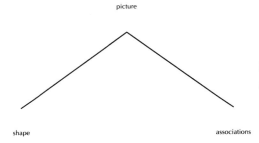

Input: {square, lparen, 20, rparen, ...}

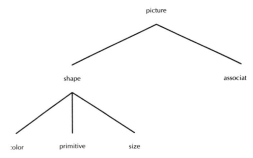

Input: {square, lparen, 20, rparen, ...}

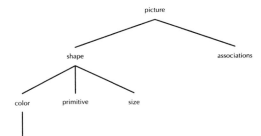

Input: {square, lparen, 20, rparen, ...}

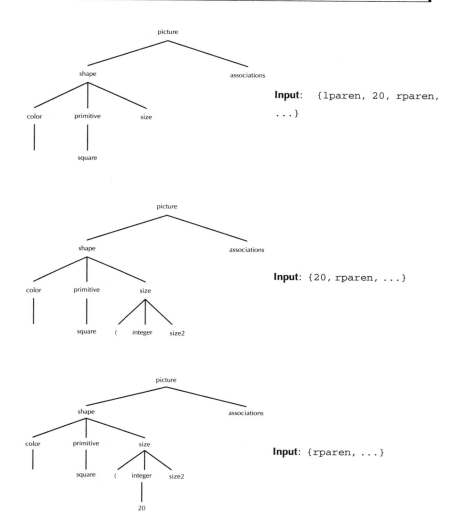

Input: {lparen, 20, rparen,
...}

Input: {20, rparen, ...}

Input: {rparen, ...}

Parse trees are represented as terms, using function symbols Prod1, Prod2, and so on. Only significant parts of the tree are retained, so that, for example, Prod5 has only three arguments, the two directions and the tree corresponding to the picture. Here is the parse tree for our running example; compare it to the tree on page 391:

```
In[8]:= Parse[Lex[ex]]

Out[8]= Prod1[Prod7[Prod8[], Prod10[], Prod14[5, Prod15[]],
              containing], Prod2[]]
```

The Parse function comes in three forms:

- `Parse[`*tokens_*`]` returns the parse tree corresponding to the list of tokens. This is the form we just used.

- `Parse[`*pns_*`,` *tokens_*`]`, where *pns* is a list of production numbers, derives a string matching part of the *tokens* using one of the productions in *pns*. It returns a pair containing the parse tree and the suffix of *tokens* not derived from the production.

- `Parse[`*pn_*`,` *tokens_*`]`, where *pn* is a production number, derives a prefix of *tokens* from production *pn* and, like the previous form, returns a parse tree and a suffix of *tokens*.

For the tokens in our example, we have:

```
In[9]:= Parse[7, Lex[ex]]

Out[9]= {Prod7[Prod8[], Prod10[], Prod14[20, Prod15[]]],
          {containing, center, slash, west, lparen, oval,
           lparen, 9, comma, 18, rparen, rparen, eof}}
```

In other words, the first four tokens were derived using production 7; the parse tree for production 7 and the remaining tokens are returned.

Parsing is quite simple. We are trying to generate a prefix of the input tokens from a given variable. We call the second form of `Parse`, passing a list of all the productions for that variable (the function `prodsFor`, shown in Figure 11.3 on page 391, gives us the list), and it looks at each one to see which might be usable given the first token (`matches`, also in Figure 11.3, tells whether a given production might apply for a given token). When it has found the correct production, it calls the third form of `Parse`, which uses that production to derive a prefix of the list of tokens.

In our example, `Parse[1, Lex[ex]]` first calls `Parse[{7}, Lex[ex]]`, since 7 is the only production for *shape*, which in turn calls `Parse[7, Lex[ex]]`, returning the pair shown above. It then calls `Parse[{2, 3, 4}, {containing, center, slash, ...}]`, 2, 3, and 4 being all the productions from *associations*, and containing, center, slash, ... being the tokens not matched by *shape*.

The first form of `Parse`, with one argument, is the one used by PDL. It starts the parsing off by attempting to derive the list of tokens from *picture*; if successful, it discards the `{eof}` and returns just the parse tree. The code for the three forms of `Parse` is given in the program listing below.

```
(* PARSING *)

prodsFor[picture] := {1}
prodsFor[associations] := {2, 3, 4}
prodsFor[connection] := {5}
prodsFor[containment] := {6}
prodsFor[shape] := {7}
prodsFor[color] := {8, 9}
prodsFor[primitive] := {10, 11, 12, 13}
prodsFor[size] := {14}
prodsFor[size2] := {15, 16}


matches[1, t_] := MemberQ[{clear, square, circle, oval, rectangle}, t]
matches[2, t_] := Not[MemberQ[{connecting, containing}, t]]
matches[3, t_] := MemberQ[{connecting}, t]
matches[4, t_] := MemberQ[{containing}, t]
matches[5, t_] := MemberQ[{connecting}, t]
matches[6, t_] := MemberQ[{containing}, t]
matches[7, t_] := MemberQ[{clear, square, circle, oval, rectangle}, t]
matches[8, t_] := MemberQ[{square, circle, oval, rectangle}, t]
matches[9, t_] := MemberQ[{clear}, t]
matches[10, t_] := MemberQ[{square}, t]
matches[11, t_] := MemberQ[{circle}, t]
matches[12, t_] := MemberQ[{oval}, t]
matches[13, t_] := MemberQ[{rectangle}, t]
matches[14, t_] := MemberQ[{lparen}, t]
matches[15, t_] := MemberQ[{rparen}, t]
matches[16, t_] := MemberQ[{comma}, t]


Parse[tokens_] := First[Parse[prodsFor[picture], tokens]]


Parse[{}, x_] :=
    (Print["Syntax error:  remaining input is ",
           Take[x, Min[Length[x], 10]], " ..."];
     Abort[])
Parse[{pn_, pns___}, tokens_] :=
    If[matches[pn, First[tokens]], (* if pn applies *)
       Parse[pn, tokens],          (* parse using it *)
       Parse[{pns}, tokens]]       (* else try other prod's *)


Parse[1, tokens_] :=
    Module[{part1 = Parse[prodsFor[shape], tokens]},
      Module[{part2 = Parse[prodsFor[associations], part1[[2]]]},
        {Prod1[part1[[1]], part2[[1]]], part2[[2]]}]]


Parse[2, tokens_] := {Prod2[], tokens}


Parse[3, tokens_] :=
    Module[{part1 = Parse[prodsFor[connection], tokens]},
      Module[{part2 = Parse[prodsFor[associations], part1[[2]]]},
        {Prod3[part1[[1]], part2[[1]]], part2[[2]]}]]
```

```
Parse[4, tokens_] :=
    Module[{part1 = Parse[prodsFor[containment], tokens]},
        Module[{part2 = Parse[prodsFor[associations], part1[[2]]]},
            {Prod4[part1[[1]], part2[[1]]], part2[[2]]}]]


Parse[5, tokens_] :=
    Module[{part1 = Parse[prodsFor[picture], Drop[tokens, 5]]},
        {Prod5[tokens[[2]], tokens[[4]], part1[[1]]], Rest[part1[[2]]]}]


Parse[6, tokens_] :=
    Module[{part1 = Parse[prodsFor[picture], Drop[tokens, 5]]},
        {Prod6[tokens[[2]], tokens[[4]], part1[[1]]], Rest[part1[[2]]]}]


Parse[7, tokens_] :=
    Module[{part1 = Parse[prodsFor[color], tokens]},
        Module[{part2 = Parse[prodsFor[primitive], part1[[2]]]},
            Module[{part3 = Parse[prodsFor[size], part2[[2]]]},
                {Prod7[part1[[1]], part2[[1]], part3[[1]]], part3[[2]]}]]]


Parse[8, tokens_] := {Prod8[], tokens}


Parse[9, tokens_] := {Prod9[], Rest[tokens]}


Parse[10, tokens_] := {Prod10[], Rest[tokens]}


Parse[11, tokens_] := {Prod11[], Rest[tokens]}


Parse[12, tokens_] := {Prod12[], Rest[tokens]}


Parse[13, tokens_] := {Prod13[], Rest[tokens]}


Parse[14, tokens_] :=
    Module[{part1 = Parse[prodsFor[size2], Drop[tokens, 2]]},
        {Prod14[tokens[[2]], part1[[1]]], part1[[2]]}]


Parse[15, tokens_] := {Prod15[], Rest[tokens]}


Parse[16, tokens_] := {Prod16[tokens[[2]]], Drop[tokens, 3]}
```

With the parse tree in hand, the remaining processing is a fairly routine matter of tree traversal, such as we used in sections 7.6 and 10.4.2. By computing the characteristics of each shape—its center, size, and compass points—we can compute the characteristics of the shapes it contains or is connected to. The coding has its occasional tricky moments, but is not basically very difficult.

Recall that there are two functions, ComputeShapes and ConvertShapes, in this part of the program. ComputeShapes does the tree traversal; ConvertShapes just converts the list of shapes to a list of *Mathematica* graphics. We take them in order.

ComputeShapes traverses the parse tree and produces a list of "shapes." The key point here is exactly what we mean by "shape." That is, how do we store the information about shapes that we mentioned above (center, size, compass points)? The structure is shown in the code at the top of Figure 11.3. A shape is represented by a 7–element list: its center (a pair of numbers), the primitive shape (a symbol), the color (a production, either Prod8 for a normal shape or Prod9 for a clear one), the distance from the center to the east compass point, the distance from the center to the north compass point, the angle of the northeast compass point (in radians), and the distance from the center to the northeast compass point. We've defined functions center, primitive, color, east, north, neangle, and nedist to extract these components from a shape.

```
(* SHAPE COMPUTATION - traverse parse tree, computing shape
   of every picture (Prod1 node).  A "shape" is a list of seven
   items: center (a pair of numbers), primitive, color, distance
   from center to north, distance from center to east, angle
   to northeast, distance to northeast
*)

center[{c_, ___}] := c
primitive[{_, p_, ___}] := p
color[{_, _, c_, ___}] := c
east[{_, _, _, e_, ___}] := e
north[{_, _, _, _, n_, ___}] := n
neangle[{_, _, _, _, _, a_, ___}] := a
nedist[{_, _, _, _, _, _, d_}] := d


angle[s_, north] := Pi/2
angle[s_, south] := -Pi/2
angle[s_, east] := 0
angle[s_, west] := Pi
angle[s_, northeast] := neangle[s]
angle[s_, southeast] := -neangle[s]
angle[s_, southwest] := Pi+neangle[s]
angle[s_, northwest] := Pi-neangle[s]
```

```
dist[s_, north] := north[s]
dist[s_, south] := north[s]
dist[s_, east] := east[s]
dist[s_, west] := east[s]
dist[s_, northeast] := nedist[s]
dist[s_, southeast] := nedist[s]
dist[s_, southwest] := nedist[s]
dist[s_, northwest] := nedist[s]


pointOf[s_, d_, delta_] :=
    center[s] + vector[angle[s, d], dist[s, d] + delta]
computeCenter[s_, p_, center] := p
computeCenter[s_, p_, d_] :=
    p + vector[Pi+angle[s, d], dist[s, d]]
vector[theta_, r_] := {r Cos[theta], r Sin[theta]}
```

We will need to compute points using angles and distances from the center of a shape. It is convenient to define the following functions, defined above in the program listing.

- angle[*shape, direction*] computes the angle from the center of *shape* to the given compass point. East is always 0, north always $\pi/2$, and so on, but the intermediate points depend upon the dimensions of the shape (at least for ovals and rectangles).

- dist[*shape, direction*] computes the distance from the center of a shape to the given compass point. Distances for intermediate points are all the same as the northeast distance.

- pointOf[*shape, direction, delta*] computes the compass point given by *direction* for *shape*, adjusted by *delta*. A positive *delta* moves the point away from the center of the shape, a negative *delta* towards it.

- computeCenter[*shape, point, direction*], where *shape* does not yet have a center, though it has all its other information, computes its center, given that the compass point named by *direction* is to be at *point*. For example, if s is a square with sides of length 10, computeCenter[s, {4, 2}, north] will return {4, -3}; if the square is centered at {4, -3}, its north point will be at {4, 2}.

The tree traversal is initiated by a call to the one-argument form of ComputeShapes, which is called with a Prod1 tree. It calls the three-argument form of ComputeShapes, which returns a list of shapes. N is applied to the list to evaluate all numerical formulas and all the "clear" shapes (Prod9) are removed. For our running example (Figure 11.2), we see the result in this session:

```
In[10]:= ComputeShapes[Parse[Lex[ex]]]

Out[10]= {{{0, 0}, square, Prod8[], 10., 10., 0.785398, 14.1421},
          {{4.5, 0}, oval, Prod8[], 4.5, 9., 1.10715, 7.11512}}
```

The main shape, centered at $(0, 0)$, is a 20×20 square (the 10s being the distance from the center to the side and the top). The 9×18 oval is centered at $(4.5, 0)$.

The three-argument form of ComputeShapes takes a tree given in the form Prod1[*shape, associations*] and computes the shape of *shape* and all the shapes in *associations*, returning a list. The second and third arguments are a point p and a direction d. First, *shape* is drawn with direction d at point p. This is done by calling computeShape; shapeInfo computes all the location-independent information, which is everything but the center, and the latter is filled in by a call to computeCenter, discussed above. Then the shapes in *associations* are drawn in positions computed with respect to *shape*. This is accomplished by calling computeAssociatedShapes, passing *associations* as the first argument and the shape computed for *shape* as the second. The *associations* parse tree (Prod2, Prod3, or Prod4) is traversed, and the shapes it contains are computed with respect to that second argument. The auxiliary function computePoint[*shape, dir1, dir2, relation*] computes the meeting point of *shape* with whatever shape it contains or connects to, given that *dir1* of *shape* is to meet *dir2* of the contained shape. The computation also depends upon whether the shape is contained or connected, as given by *relation*.

```
separation = .1

ComputeShapes[tree_] :=
        Select[N[computeShapes[tree, {0, 0}, center]],
               (color[#] =!= Prod9[])&]

computeShapes[Prod1[sh_, assoc_], p_, d_] :=
   Module[{s = computeShape[sh, p, d]},
      Module[{as = computeAssociatedShapes[assoc, s]},
         Join[{s}, as]]]


computeAssociatedShapes[Prod2[], _] := {}
computeAssociatedShapes[Prod3[Prod5[d1_, d2_, pic_], assoc_], s_] :=
   Module[{p = computePoint[s, d1, d2, connecting]},
      Module[{ss = computeShapes[pic, p, d2]},
         Join[ss, computeAssociatedShapes[assoc, s]]]]
computeAssociatedShapes[Prod4[Prod6[d1_, d2_, pic_], assoc_], s_] :=
   Module[{p = computePoint[s, d1, d2, containing]},
      Module[{ss = computeShapes[pic, p, d2]},
         Join[ss, computeAssociatedShapes[assoc, s]]]]
```

```
computePoint[s_, d1_, d2_, relation_] :=
  Which[d1===center, center[s],
        d2===center, pointOf[s, d1, 0],
        relation===connecting, pointOf[s, d1, separation],
        True, pointOf[s, d1, -separation]]

computeShape[s_, p_, d_] :=
  Module[{si = shapeInfo[s]},
    Join[{computeCenter[si, p, d]}, Rest[si]]]


shapeInfo[Prod7[color_, Prod10[], Prod14[i_, _]]] :=
  {0, square, color, i/2, i/2, Pi/4, i/Sqrt[2]}
shapeInfo[Prod7[color_, Prod11[], Prod14[i_, _]]] :=
  {0, circle, color, i/2, i/2, Pi/4, i/2}
ovalNE[a_, b_, theta_] :=
  a b Sqrt[(1 + Tan[theta]^2)/(a^2 + b^2 Tan[theta]^2)]
shapeInfo[Prod7[color_, Prod12[], Prod14[l_, Prod16[h_]]]] :=
  {0, oval, color, 1/2, h/2, ArcTan[h/l],
   ovalNE[h/2, 1/2, ArcTan[h/l]]}
shapeInfo[Prod7[color_, Prod13[], Prod14[l_, Prod16[h_]]]] :=
  {0, rectangle, color, 1/2, h/2, ArcTan[h/l], Sqrt[h^2 + l^2]/2}
```

Finally, the list of shapes is converted to a list of *Mathematica* graphics by mapping convertShape over the list. The *Mathematica* graphics primitives are well matched to our representation of shapes, making convertShape easy to write. The final output of our example is

```
In[11]:= ConvertShapes[ComputeShapes[Parse[Lex[ex]]]]

Out[11]= {Line[{{-10., -10.}, {-10., 10.}, {10., 10.}, {10., -10.},
              {-10., -10.}}], Circle[{4.5, 0}, {4.5, 9.}]}


(* OUTPUT - convert list of shapes to list of Mathematica
   graphics objects and draw
*)

makeRectangle[p_, l_, h_] :=
  Line[{p, p+{0,h}, p+{l,h}, p+{l,0}, p}]
convertShape[s_] /; MemberQ[{square, rectangle}, s[[2]]] :=
    makeRectangle[pointOf[s, southwest, 0], 2 east[s], 2 north[s]]
convertShape[s_] /; MemberQ[{circle, rectangle}, s[[2]]] :=
    Circle[center[s], north[s]]
convertShape[s_] /; s[[2]] === oval :=
    Circle[center[s], {east[s], north[s]}]

ConvertShapes[ss_] := Map[convertShape, ss]
```

12 | Contexts and Packages

In any interactive language with a multitude of built-in and user-defined functions, it can be difficult to keep all the names straight. It is for this reason that we introduced the `Module` construct. Contexts and packages are features that are included in *Mathematica* to help avoid name conflicts when entire files of value and function definitions are given. Packages are a mechanism that allows you to structure your programs so that they function like *Mathematica*'s built-in functions. This chapter contains a full discussion of contexts and gives two examples of constructing packages.

12.1 Introduction

When you begin a *Mathematica* session, the built-in functions are all available for you to use. There are, however, many more data and function definitions residing in files supplied with *Mathematica*. In principle, the only difference between those files and the ones you create is that those were written by professional programmers. There is another difference: the definitions in those files are placed in special structures called *packages*. Indeed, these files are often called "packages" instead of "files."

Packages are a *name localizing* construct, analogous to `Module`, but for entire files of definitions. Their purpose is to allow the programmer to define a collection of functions *for export*—for the clients of the package to use—and others *not for export*—auxiliary functions not to be used by clients. They allow the programmer to ensure that names used in the package will not conflict with names defined by the user at the top level or in other packages.

In this chapter, you will learn how to write your own packages. Much of the chapter is devoted to an explanation of a more primitive notion, that of *contexts*, which is a prerequisite to understanding packages. We then describe packages and give a simple example, showing the standard and accepted style for writing them.

12.2 Using Packages

Mathematica packages have been written for a great variety of problem domains. Many are provided with each version of *Mathematica* and are documented in the technical report [Boyland, Keiper and Martin 1992]; some are also discussed in the manual

[Wolfram 1991]. Below, we list some examples of some of the standard *Mathematica* packages. Note that package names always end with a back quote (`'`), and often have back quotes within them as well:

`Calculus'VectorAnalysis'`: This package provides a variety of variables and functions for doing calculus in various three-dimensional coordinate systems, *e.g.*, `SetCoordinates` to set the coordinate system (Cartesian, polar, etc.); `CrossProduct[`v_1`, `v_2`]` to compute cross products; `Curl[`f`]` to give the curl of the vector field f.

`Graphics'MultipleListPlot'`: Provides functions for superimposing several plots on the same graphic. `MultipleListPlot[`*list*$_1$`, `*list*$_2$`, ...]` is the main function in this package. It plots each *list*$_i$ as a separate plot on the same axes. Also provided is `MakeSymbol[`*symbolspecification*`]` which creates symbols to use in labeling the separate plots, plus a number of functions for specifying symbols.

12.2.1 | Loading Packages

Once you know which package you want to use, you can load it in one of two ways. For example, to load the package `Calculus'VectorAnalysis'`:

1. `<<Calculus'VectorAnalysis'` will read the file and evaluate each expression and definition as if it had been typed in. Actually, the argument of `<<` is a string, but the quotation marks can be omitted.

2. `Needs["Calculus'VectorAnalysis'"]` will read the package, just like `<<`, but only if it hasn't already been read.

Here is an example of using the `Calculus'VectorAnalysis'` package:

```
In[1]:= Needs["Calculus'VectorAnalysis'"]

In[2]:= CrossProduct[{1.0, 2.0, 3.0}, {-1.0, -2.0, -3.0}]
Out[2]= {0., 0., 0.}
```

12.2.2 | Finding Out What's in a Package

To use the *Mathematica* packages, you need to know what they provide. In fact, programmers always find that even remembering what's in their own packages isn't easy, if they haven't looked at them for a while.

If all you know is the name of the package and you want to know what it defines, load it, using *<<package`* or Needs [ "*package`* "] .

In[3]:= **Needs["Graphics`Graphics`"]**

Then type: Names ["*package`* * " ] or ?*package`* * for a list of the names defined in the package.

In[4]:= **Names["Graphics`Graphics`*"]**

Out[4]= {BarChart, BarEdges, BarEdgeStyle, BarGroupSpacing,
BarLabels, BarOrientation, BarSpacing, BarStyle,
BarValues, DisplayTogether, DisplayTogetherArray,
ErrorListPlot, GeneralizedBarChart, LabeledListPlot,
LinearLogListPlot, LinearLogPlot, LinearScale,
ListAndCurvePlot, LogGridMajor, LogGridMinor,
LogLinearListPlot, LogLinearPlot, LogListPlot,
LogLogListPlot, LogLogPlot, LogPlot, LogScale,
PercentileBarChart, PieChart, PieExploded,
PieLabels, PieLineStyle, PieStyle, PiScale,
PolarListPlot, PolarPlot, Scale, ScaledListPlot,
ScaledPlot, SkewGraphics, StackedBarChart,
TextListPlot, TransformGraphics, UnitScale}

Now you can use ? to get the usage message for any of those names.

In[5]:= **?LogLogPlot**

Out[5]= LogLogPlot[f, {x, xmin, xmax}] generates a plot of
Log[f] as a function of Log[x].

If, on the other hand, you forget the name of the package, you can look through [Boyland, Keiper and Martin 1992], or find out where the directory of packages is stored on your system, and browse through it. Alternatively, if you have *Mathematica* Version 2.2 or later, you can easily browse through the on-line Function Browser which will list all packages and names and usage messages of any functions defined in these packages.

12.3 Contexts

Every symbol you mention in *Mathematica*—whether to define it or use it—has a *full name* consisting of the symbol preceded by the *context* in which the name was first mentioned. When you first start your session, the *current context* is Global` (again note the back quote), and any symbol *sym* you mention now has full name Global`*sym*. A symbol can be given with its full name or in its regular, short form.

```
In[1]:= f[x_] := x + 1
```

```
In[2]:= Global`f[3]
Out[2]= 4
```

```
In[3]:= Global`f == f
Out[3]= True
```

Thus, a symbol is just an abbreviation for its full name. Symbols you define when your session begins have context Global`, and built-in operations have context System`.

```
In[4]:= System`Map[Global`f, {5, 7, 9}]
Out[4]= {6, 8, 10}
```

You can tell *Mathematica* to use a different context for any new symbols you mention by using the function Begin.

```
In[5]:= Begin["ContextA`"]
Out[5]= ContextA`
```

```
In[6]:= g[x_] := x + 2
```

```
In[7]:= g[3]
Out[7]= 5
```

```
In[8]:= ContextA`g[3]
Out[8]= 5
```

In this new context, the name g is an abbreviation for ContextA`g. Note that we can still refer to f, even though it wasn't defined in this context.

```
In[9]:= Map[g, {5, 7, 9}]
Out[9]= {7, 9, 11}

In[10]:= Map[Global`f, {5, 7, 9}]
Out[10]= {6, 8, 10}
```

After exiting the context using the End function, we may define a different g, having context Global`.

```
In[11]:= End[]
Out[11]= ContextA`

In[12]:= g[x_] := x + 3

In[13]:= g[3]
Out[13]= 6
```

We now have two definitions of g, or rather, one definition of Global`g and one of ContextA`g. Since our current context is Global`, when we just say g we get Global`g; but we can still refer to ContextA`g by its full name.

```
In[14]:= ContextA`g[3]
Out[14]= 5
```

The question arises: When you enter a symbol *sym*, how does *Mathematica* decide which version of *sym* to use? And how can you tell which one it has chosen?

To answer the second question first: The function Context gives the context of a symbol.

```
In[15]:= Context[g]
Out[15]= Global`

In[16]:= Context[Map]
Out[16]= System`
```

```
In[17]:= Context[ContextA`g]

Out[17]= ContextA`
```

or you can use ?:

```
In[18]:= ?g

Global`g
g[x_] := x + 3
```

How, then, does *Mathematica* decide which definition to use? It maintains two variables, $Context and $ContextPath. $Context contains a context (that is, a string giving the name of a context), which is the current context, and $ContextPath contains a list of contexts. *Mathematica* looks in $Context first, then in the contexts in $ContextPath (in the order in which they appear there); if it does not find the symbol at all, then it creates it in context $Context. Of course, none of this applies if you give the symbol's full name.

```
In[19]:= $Context

Out[19]= Global`
```

```
In[20]:= $ContextPath

Out[20]= {Global`, System`}
```

```
In[21]:= Begin["ContextA`"]

Out[21]= ContextA`
```

```
In[22]:= $Context

Out[22]= ContextA`
```

```
In[23]:= $ContextPath

Out[23]= {Global`, System`}
```

```
In[24]:= End[]

Out[24]= ContextA`
```

```
In[25]:= {$Context, $ContextPath}

Out[25]= {Global`, {Global`, System`}}
```

So the effect of entering a new context using `Begin` is simply to change the value of `$Context`; End changes it back. In either case, `$ContextPath` is not changed.

One final point about contexts: Contexts can be nested within contexts. That is, you can have context names like A`B`C`. To enter contexts like this, do the following:

```
In[26]:= Begin["A`"]          (* enter context A` *)

Out[26]= A`


In[27]:= Begin["`B`"]         (* enter context A`B` *)

Out[27]= A`B`


In[28]:= Begin["`C`"]         (* enter context A`B`C` *)

Out[28]= A`B`C`


In[29]:= End[]                (* back in context A`B` *)

Out[29]= A`B`C`


In[30]:= End[]                (* back in context A` *)

Out[30]= A`B`


In[31]:= End[]                (* back in context Global` *)

Out[31]= A`
```

Note the back quote *before* the context name in the second and third `Begin`'s. This is just a shorthand way of saying

```
In[32]:= Begin["A`"]

Out[32]= A`


In[33]:= Begin["A`B`"]

Out[33]= A`B`


In[34]:= Begin["A`B`C`"]

Out[34]= A`B`C`
```

Nested contexts are a way of managing the multiplicity of contexts. You will have noticed how the names of the standard packages we discussed earlier look just like nested contexts. In fact, package names are contexts. *Mathematica* organizes the standard packages into about ten major contexts (*e.g.*, `Calculus`` and `Graphics``), each with about ten nested contexts; it is just a way of keeping things organized. Most readers will recognize this as the idea behind hierarchical file systems. (In fact, when you load a package using `Needs` or `<<`, *Mathematica* translates the package name directly into a path name in the hierarchical file system on your computer.)

Summary

- Any name mentioned in a *Mathematica* session has a full name, containing a context and the short name.

- When using a name, you may give its full name. If you choose not to (as is customary), *Mathematica* will decide what the full name is, *i.e.*, what the context of the name is.

- Here is how *Mathematica* decides on the context:

  - First, it looks in the context given by the variable `$Context`.

  - Next, it looks in all the contexts given in the variable `$ContextPath`, in the order in which they appear there.

  - If those searches don't succeed, *Mathematica* assumes this is the first mention of the name, and so gives it the context `$Context`.

- `Begin["`*context*``"]` and `End[]` alter the value of `$Context` (but do not affect `$ContextPath`). Specifically, `Begin["`*context*``"]` sets `$Context` to *context*`, and `End[]` restores it to its prior contents before the `Begin`.

As of now, these functions are the only ways we know to alter the contents of these two variables. In the next section, we will see two other functions that change them in a subtly, but crucially, different way.

12.4 Packages

The idea of packages is to allow the definition of a collection of values and functions without having collisions with other definitions of those names. For example, if you load a package that defines functions `f` and `g`, and the definition of `g` contains a call to `f`, then `g` should always work—*i.e.*, call the `f` defined in the package—even if you've defined `f` separately in your session. Furthermore, packages can define their own auxiliary functions and constants that the user—or client—of the package will not ordinarily see at all.

All this is achieved using contexts, with two new functions:

- BeginPackage["*package*`"] sets $Context to *package*`, and $ContextPath to {*package*`, System`}.

- EndPackage[] resets both variables to their values prior to the evaluation of BeginPackage[], and then prepends *package*` to $ContextPath.

Thus, if you are in a *Mathematica* session, with current context Global`, and you read in a file containing,

```
BeginPackage["P`"]
  f[x_] := ...
  g[y_] := ...
EndPackage[]
```

then after it is read, the functions f and g, with full names P`f and P`g, will be defined, and the context P` will be in $ContextPath. If you don't have any other definitions of f, you can refer to it as just f; if you do, then use P`f; and similarly for the function g.

The precise definition of BeginPackage[*package*`] is important as it changes $ContextPath to {*package*`, System`}. Thus, all the names defined in the package will have context *package*`. In our example above, the f and g in the package are definitely P`f and P`g, regardless of any other definitions you may have given for them.

It is important to realize, too, that *Mathematica* determines the full name of any name *when it reads it in*. Thus, if g calls f, then the occurrence of f in the body of g becomes P`f when the package is loaded. g will always call this f, even if there is a different f defined in the context in which the call to g is made.

The BeginPackage function can be given multiple arguments. The second and subsequent arguments are the names of other packages that this one uses. They are treated as if they were arguments to the Needs function; that is, they are loaded if they haven't already been. Furthermore, they are included in $ContextPath *during the loading* of this package, so its functions can refer to their functions by their short names.

Summary

- BeginPackage["*package*`"] sets $Context to *package*`, and $ContextPath to {*package*`, System`}, so that any names subsequently mentioned—other than the names of built-in functions and constants—are defined in context *package*`.

- EndPackage[] resets $Context and $ContextPath to their prior values, except that *package`* is added to the front of $ContextPath.

12.5 Avoiding Name Collisions

You will sometimes read in a package P that defines a function f whose name you have already mentioned in your current session. It is very common, for example, to forget to load a package before calling one of its functions; mentioning the function's name creates a symbol in the current context. Then, if you try to make a call to f, *Mathematica* will assume you're talking about the f in the current context rather than the one defined in the package. Of course, you can always call the f from the package by using its full name, P`f.

If, however, you want to be able to call f by its short name, and forget the f you mentioned accidentally, use the function Remove[f]. This will make it seem that you had never mentioned the name f at all. Note that saying Clear[f] isn't enough: it removes the definition of f, if it had one, but doesn't "un-mention" it; only Remove does that.

There is a way to minimize this problem, if you have certain packages that you often use. The function

DeclarePackage["*package`*", "*name*₁", "*name*₂", ...}]

tells *Mathematica* that whenever you use one of the names *name*₁, *name*₂, ..., it should load *package* (if it hasn't already been loaded). It is a good practice to make a file containing a DeclarePackage for each package you frequently use, listing all the names of functions you use from that package. For example, if that file is called my-package.m then, whenever you start a *Mathematica* session, enter <<mypackage.m as your first input. Alternatively, you could put mypackage.m in the file init.m and *Mathematica* will automatically load it whenever you start a session.

12.6 The BaseConvert Package

The simple package BaseConvert will illustrate the above points, as well as some others not mentioned. BaseConvert is a package of functions to convert strings of digits to integers, possibly in a base other than 10. It defines the following names "for export," *i.e.*, for use by its clients:

- DefaultBase is initially set to 10; it is the base assumed in the conversion if no base is explicitly given.

- `StringToInt [`*str, b*`]` converts the string *str*, which must contain only base *b* digits, to a number. Given in one-argument form, it uses `DefaultBase`. The base must be a number between 2 and 10.

In addition, it defines the following auxiliary function not intended for clients to see, *i.e.*, not for export:

- `numeralToInt [`*numeral*`]` converts a single numeral to its numerical value.

Here is the definition:

```
BeginPackage["BaseConvert`"]

DefaultBase::usage = "Default base used by StringToInt."

StringToInt::usage = "StringToInt[str, b] converts str to
  an integer, assuming it is written in base b.
  StringToInt[str] is equivalent to StringToInt[str,
  DefaultBase]."

Begin["`Private`"]

DefaultBase = 10;
StringToInt[str_] := StringToInt[str, DefaultBase]
StringToInt[str_String, b_Integer] :=
  Which[Not[2 <= b <= 10],
        Print["StringToInt: second argument is
              inappropriate base."],
        Not[Apply[And,Map[(0<=numeralToInt[#]<b)&,
                      ToCharacterCode[str]]]],
        Print["StringToInt: first argument has
              incorrect digits."],
        True, (* arguments are legal *)
          Map[numeralToInt, ToCharacterCode[str]] .
          Table[b^i, {i,StringLength[str]-1,0,-1}]]

StringToInt[str_, b_] :=
  (If[Not[StringQ[str]],
     Print["StringToInt: ", str, " is not a string."]];
   If[Not[IntegerQ[b]],
     Print["StringToInt: ", b," is not an integer."]])


StringToInt[x___] :=
  Print["StringToInt: called with ", Length[{x}],
        " arguments; 1 or 2 arguments are expected."]
numeralToInt[numeral_] :=
  numeral - ToCharacterCode["0"][[1]]

End[]
EndPackage[]
```

The package begins by giving *usage messages* for every exported function. Usage messages will be explained below, but the important thing to note now is that the functions to be exported are *mentioned* here—*before* the subcontext `Private`' is entered—so that names `DefaultBase` and `StringToInt` have context `BaseConvert`'. `numeralToInt` is *not* mentioned here.

A new context `BaseConvert`'`Private`' is then begun *within* `BaseConvert`. All of the definitions of this package are given within this new context. When the definition of `DefaultBase` is read, the current context is `BaseConvert`'`Private`', and `$ContextPath` has the value `{BaseConvert`', `System`'}`. The result is that the two contexts `BaseConvert`'`DefaultBase` and `BaseConvert`'`StringToInt` are defined in the `System`' context. The context of `numeralToInt`, on the other hand, is `BaseConvert`'`Private`'.

After the `End[ ]` and `EndPackage[ ]` functions are evaluated, `$Context` and `$ContextPath` revert to whatever they were before, except that `BaseConvert`' is added to `$ContextPath`. Clients can refer to `DefaultBase` and `StringToInt` using their short names, but they can only refer to the auxiliary function `numeralToInt` by its full name. (The intent is to discourage clients from using `numeralToInt` at all, and doing so should definitely be avoided, since the author of the package may change or remove auxiliary definitions at any time.)

Thus, the package would be used like this:

```
In[1]:= Needs["BaseConvert'"]


In[2]:= StringToInt["123"]

Out[2]= 123


In[3]:= DefaultBase = 5

Out[3]= 5


In[4]:= StringToInt["123"]

Out[4]= 38


In[5]:= StringToInt["123", 12]

StringToInt: second argument is inappropriate base.
```

Above all, when you write functions important enough to put into a package, *you should strive to make them behave as much as possible like built-in functions.* One way we've done this is to capitalize exported names, as *Mathematica* does for all built-ins. More significant is that we've given them "usage messages." These are the little explanations you get by entering ? *name*. You can associate a usage message with any name by using the syntax shown above: *name*: :usage = *string*. (In this conventional style of writing package, they serve double duty: as documentation, and to introduce the name in the appropriate context.)

Another (even more important) way to make user-defined functions behave like built-in functions is to check the arguments to each function carefully and issue appropriate error messages. In this regard, we have been much more fastidious with StringToInt than with any of our previous functions. Naturally, it is always a good idea to have documentation and error-checking for any function you define, but putting it in a package carries the implication that it will become part of your permanent 'library' of completed code, so good style and good sense indicate that it should be made as *bullet-proof* as possible.

Summary

The standard format for a package P exporting functions f and g, which use auxiliary function h, is:

```
BeginPackage["P`"]
   f::usage = "f[...] ...explanation of f..."
   g::usage = "g[...] ...explanation of g..."
   Begin["`Private`"]
     f[args] := ...
     g[args] := ...
     h[args] := ...
   End[ ]
EndPackage[ ]
```

This package would normally be placed in a file called P.m and loaded using either <<P` or Needs["P`"]. (Another possibility, in cases where you know the name of the file, is <<P.m.)

It is conventional that functions being placed in packages be treated with special care, including complete error-checking of all arguments. Aside from having to load the package, future users of these functions should, as far as possible, see no difference between them and the built-ins.

12.7 The RandomWalks Package

12.7.1 | Adding Options and Defaults

When writing your own programs, it is often difficult to predict how a user will want to use some of your functions. Programmers usually try to provide a variety of ways to use their functions (allowing for different types of input, for example), or sometimes they write separate functions to handle special cases. The problem with having a separate function for each special case is that the user can soon become overloaded with the variety of functions to learn.

The use of options and defaults in your programs allows you to minimize the use of many parameters and function names for the user to remember. For example, the built-in function `FactorInteger` has an option `GaussianIntegers` which, when set to `True`, will factor a number over the gaussian integers.

```
In[1]:= FactorInteger[5, GaussianIntegers->True]

Out[1]= {{-I, 1}, {1 + 2 I, 1}, {2 + I, 1}}
```

The alternative to such an option would be to have a separate function, say `FactorGaussianInteger`, that the user would have to use. Since the main process here is factorization of numbers, it makes sense to have one function that covers various situations allowing for factorization over different domains by specifying different options.

On the other hand, polynomial factorization is a fundamentally different operation from integer factorization, and so a different function is used for that.

```
In[2]:= Factor[27x^5 + 81x^4 y + 9x^3 y^2 - 73x^2 y^3 +
            32x y^4 - 4y^5]

Out[2]= (3 x - y)^3 (x + 2 y)^2
```

In this section, we will show how to write options for your functions so that they behave like the built-in options in *Mathematica*. (In the next section, we will see how to incorporate the use of options into a full-fledged package.) We will create an option `LatticeWalk` for the `Walk2D` function from Chapter 11, that will generate a lattice walk when set to `True`, and will generate an off-lattice walk when set to `False`. For example, the syntax to create a 100-step off-lattice walk will be:

```
Walk2D[100, LatticeWalk->False]
```

The `LatticeWalk` option to the `Walk2D` function will be specified as a rule. The following definition both defines an option for the `Walk2D` function and specifies its default value (`True` in this case).

```
In[3]:= Options[Walk2D] = {LatticeWalk -> True}
```

If you were now to ask for information about the `Walk2D` function, you would see this new option listed.

```
In[4]:= ?Walk2D
```

```
Global`Walk2D
Options[Walk2D] = {LatticeWalk -> True}
```

We will use this option in the `Walk2D` function by branching to either a lattice walk or an off-lattice walk, depending upon the value of this new option. We will need to extract the value of this option inside the `Walk2D` function, and we can do this as follows (recall the definition, `Walk2D[n_, opts___] := ...`):

```
In[5]:= lattice = LatticeWalk /. {opts} /. Options[Walk2D];
```

From right to left, the values of the options to `Walk2D` are substituted into `opts`; then these (rules) are substituted to extract the value of `LatticeWalk`. This value is then assigned to the symbol `lattice`.

Here then is the full function `Walk2D`, using this option structure.

```
In[6]:= Walk2D[n_, opts___] := Module[{lattice},
           lattice = LatticeWalk /. {opts} /. Options[Walk2D];
           If[lattice == True,
             FoldList[Plus, {0,0},
               {{0,1},{1,0},{0,-1},{-1,0}}[[
               Table[Random[Integer,{1,4}], {n}]]]],
             FoldList[Plus, {0,0},
               Map[({Sin[#], Cos[#]})&,
               Table[Random[Real, {0, N[2Pi]}], {n}]]]
           ]]]
```

Notice that if `LatticeWalk` has a value of `True`, then the first branch of the `If` statement is followed, giving the lattice walk. If `LatticeWalk` has any other value (`False` for example), then the off-lattice definition is used.

This uses the default value of `LatticeWalk`.

```
In[7]:= Walk2D[4]
```

```
Out[7]= {{0, 0}, {1, 0}, {0, 0}, {1, 0}, {1, 1}}
```

This creates an off-lattice walk.

```
In[8]:= Walk2D[4, LatticeWalk -> False]
```

```
Out[8]= {{-0.78587, 0.618391}, {-0.90992, 0.414783},
          {0.489664, -0.871911}, {-0.95692, -0.290353}}
```

12.7.2 | The RandomWalks Package

In this section, we list the full RandomWalks package, including expressions for options and usage statements.

BeginPackage

First, we set the value of Context' which causes $ContextPath to be set to {RandomWalks', System'}.

```
In[9]:= BeginPackage["RandomWalks'"]

Out[9]= RandomWalks'
```

Importing Other Packages

You could import a package by using an optional argument to BeginPackage. In that case, you would have BeginPackage[Graphics'ArgColors'] above. The argument against this approach is that the package Graphics'ArgColors' will be left on the search path after the RandomWalks' package is read in. It is considered poor programming style to alter the user's environment by simply reading in a package—at least you should try to alter it as little as possible. There is another method of loading a package within a package, and that is to call Needs *after* the call to BeginPackage. Using this mechanism, the Graphics'ArgColors' context will not remain on the context path after the RandomWalks package is read in.

```
In[10]:= Needs["Graphics'ArgColors'"]
```

Usage statements

Defining usage messages for the functions in your packages creates symbols for the functions in the current context. Besides, it is good programming style on a number of counts. Making your functions behave much like the built-in functions will make it easier for users of your packages since they will expect usage messages and general functionality similar to that of *Mathematica*'s functions. It is also a good way for you to document what you have done. We would go so far as to suggest that you consider writing your usage messages *before* you write the function definitions in *Mathematica*. This will help you to clearly understand what it is you want your functions to do.

```
In[11]:= Walk1D::usage = "Walk1D[n] generates an n-step walk in one
            dimension. Steps are of unit length.";
```

In[12]:= `Walk2D::usage = "Walk2D[n] generates an n-step walk in two dimensions. The default behavior gives a lattice walk with steps in one of the four compass directions. Setting the option LatticeWalk->False will generate an off-lattice walk.";`

In[13]:= `Walk3D::usage = "Walk3D[n] generates an n-step walk in three dimensions. The walk is executed on a cubic lattice with each step in one of 8 lattice directions.";`

In[14]:= `LatticeWalk::usage = "LatticeWalk -> val is an option to Walk2D that determines whether the random walk will be a lattice walk or an off-lattice walk. Possible values are True and False.";`

In[15]:= `SquareDistance::usage = "SquareDistance[walk] computes the square of the end-to-end distance of a two-dimensional lattice walk. This can be thought of as the square of the diameter of the walk.";`

In[16]:= `MeanSquareDistance::usage = "MeanSquareDistance[n, m] computes the mean square end-to-end distance of m n-step lattice walks.";`

In[17]:= `MeanSquareRadiusGyration::usage = "MeanSquareRadiusGyration[m, n] computes the mean square radius of gyration of a random walk. This is the sum of the squares of the distances of the step locations from the center of mass, divided by the number of step locations, where the center of mass is the sum of the step locations divided by the number of step locations.";`

In[18]:= `ShowWalk2D::usage = "ShowWalk2D[walk] displays a two-dimensional random walk connecting each site with a line. Any Graphics options can be passed to ShowWalk2D. E.g., ShowWalk2D[walk, Background->GrayLevel[0]] to produce a black background.";`

In[19]:= **ShowPointWalk2D::usage = "ShowPointWalk2D[walk, opts]**
displays a two-dimensional walk with each site shown as
a point. Graphics options can be passed to ShowPointWalk2D.
E.g., ShowWalk2D[walk, Background->GrayLevel[0]] to produce
a black background.";

In[20]:= **ShowHistoryWalk2D::usage = "ShowHistoryWalk2D[walk, opts]**
displays a two-dimensional random walk with each site shown
as a point colored according to its history in the walk.
Graphics options can be passed to ShowHistoryWalk.";

In[21]:= **ShowWalk3D::usage = "ShowWalk3D[walk] displays a three-**
dimensional random walk connecting each site with a line.
Any Graphics3D options can be passed to ShowWalk3D. E.g.,
ShowWalk3D[walk, Boxed->False] to inhibit the display of
a bounding box.";

In[22]:= **ShowCubeWalk3D::usage = "ShowCubeWalk3D[walk] displays a**
three-dimensional random walk displaying each site as a cube.
Any Graphics3D options can be passed to ShowCubeWalk3D. E.g.,
ShowCubeWalk3D[walk, Boxed->False] to inhibit the display of
a bounding box.";

In[23]:= **AnimateWalk2D::usage = "AnimateWalk2D[walk, opts] creates**
an animation of a two-dimensional random walk. A red ball
will be seen to move to the current position in the walk
to aid in visualizing the animation.";

Options

In[24]:= **Options[Walk2D] = {LatticeWalk -> True}**

Out[24]= {LatticeWalk -> True}

Begin private context
The Begin command changes the current context without affecting the context path. By starting the argument `Private` with a context mark `, we change to a subcontext of the current context. This new subcontext is RandomWalks`Private`.

In[25]:= **Begin["`Private`"]**

Out[25]= RandomWalks`Private`

The Function Definitions

```
In[26]:= Walk1D[n_] :=
            FoldList[Plus, 0, Table[(-1)^Random[Integer],{n}]]


In[27]:= Walk2D[n_, opts___] := Module[{lattice},
            lattice = LatticeWalk /. {opts} /. Options[Walk2D];
            If[lattice == True,
               FoldList[Plus, {0,0},
                  {{0,1},{1,0},{0,-1},{-1,0}}[[
                  Table[Random[Integer,{1,4}], {n}]]]],
               FoldList[Plus, {0,0},
                  Map[({Sin[#], Cos[#]})&,
                  Table[Random[Real, {0, N[2Pi]}], {n}]]
             ]]]


In[28]:= Walk3D[n_] := FoldList[Plus, {0, 0, 0},
                {{-1,1,1},{-1,-1,1},{1,-1,1},{1,1,1},
                 {-1,1,-1},{-1,-1,-1},{1,-1,-1},{1,1,-1}}[[Table[
                      Random[Integer, {1, 8}], {n}]]]]


In[29]:= SquareDistance[walk_List] :=
              Apply[Plus, (Last[walk] - First[walk])^2]


In[30]:= MeanSquareDistance[n_Integer, m_Integer] :=
              N[Sum[SquareDistance[ Walk2D[n] ], {m}]/m]


In[31]:= MeanSquareRadiusGyration[m_Integer, n_Integer] :=
         Module[{},
            squareRadiusOfGyration[s_Integer] :=
            Module[{locs, cm, choices = {{1,0},{-1,0},{0,1},{0,-1}}},
               locs = FoldList[Plus,{0,0},
                        choices[[Table[Random[Integer,{1,4}],{s}]]]];
               cm = N[Apply[Plus, locs]/(s+1)];
               Apply[Plus, Flatten[(Transpose[locs] - cm)^2]]/(s+1)
            ];
            N[Sum[squareRadiusOfGyration[n], {m}]/m]
         ]
```

```
In[32]:= ShowWalk2D[coords_, opts___]:=
               Show[Graphics[Line[coords],
                    opts,
                    AspectRatio -> Automatic]]


In[33]:= ShowPointWalk2D[coords_, opts___]:=
             Module[{prwise, distances, c},
                 prwise = Map[({First[coords],#})&, coords];
                 distances = Map[N[Sqrt[Apply[Plus,
                                          (#[[1]]-#[[2]])^2]]]&,
                              prwise];
                 colors = Map[Hue, distances / Max[distances]];

                 Show[Graphics[Transpose[{colors, Map[Point,coords]}],
                       opts, AspectRatio -> Automatic]]]


In[34]:= ShowHistoryWalk2D[coords_, opts___]:=
             Module[{len = Length[coords],points,colors},
                 points = Map[Point, coords];
                 colors = Map[Hue, Range[len]/len];
                 Show[Graphics[{PointSize[.01],
                             Transpose[{colors,points}]}],
                       opts, AspectRatio -> Automatic]]]


In[35]:= ShowWalk3D[coords_, opts___]:=
               Show[Graphics3D[Line[coords],
                    opts, AspectRatio -> Automatic]]


In[36]:= ShowCubeWalk3D[walk_, opts___]:=
               Show[Graphics3D[Map[Cuboid, walk],
                    opts, AspectRatio -> Automatic]]


In[37]:= AnimateWalk2D[coords_, opts___]:=
               Scan[Show[Graphics[{
                       {RGBColor[1,0,0],PointSize[.025],
                        Point[ coords[[#]] ]},
                       Line[Take[coords, #]]}],
                   opts,AspectRatio -> Automatic,
                   PlotRange -> Map[{Min[#]-.2, Max[#]+.2}&,
                                     Transpose[coords]]]&,
               Range[2, Length[coords]]]
```

End private context

The End[] command closes the Begin[] and puts us back in the context RandomWalks'. Any symbols that were defined in the subcontext RandomWalks'Private' can no longer be accessed.

```
In[38]:= End[]

Out[38]= RandomWalks'Private'
```

End package

The EndPackage[] command puts us back in the context we were in prior to the BeginPackage[] command.

```
In[39]:= EndPackage[]
```

Examples

Starting with a new session, and making sure that the RandomWalks package is in a directory/folder where *Mathematica* can find it, this loads the package.

```
In[1]:= << RandomWalks'
```

This lists any options that are associated with the Walk2D function.

```
In[2]:= Options[Walk2D]

Out[2]= {LatticeWalk -> True}
```

Here is a list of all the names of symbols defined in the RandomWalks package.

```
In[3]:= Names["RandomWalks'*"]

Out[3]= {AnimateWalk2D, LatticeWalk, MeanSquareDistance,
         MeanSquareRadiusGyration, ShowCubeWalk3D,
         ShowHistoryWalk2D, ShowPointWalk2D, ShowWalk2D,
         ShowWalk3D, SquareDistance, Walk1D, Walk2D,
         Walk3D}
```

The usage messages we defined can be accessed in the usual manner.

```
In[4]:= ?ShowWalk2D
```

Out[4]= ShowWalk2D[walk] displays a two-dimensional random walk
         connecting each site with a line. Any Graphics
         options can be passed to ShowWalk2D. E.g.,
         ShowWalk2D[walk, Background->GrayLevel[0]] to
         produce a black background.

Finally, here are several graphics that demonstrate the usage and syntax of some of the functions defined in the RandomWalks package.

```
In[5]:= ShowWalk2D[ Walk2D[100] ]
```

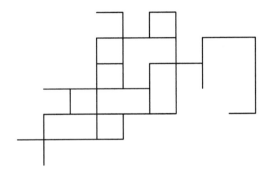

Out[5]= -Graphics-

```
In[6]:= ShowWalk2D[Walk2D[200, LatticeWalk -> False],
                   Frame -> True]
```

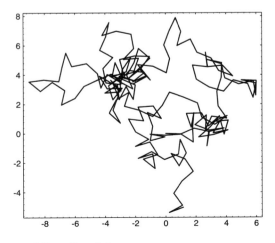

Out[6]= -Graphics-

In[7]:= **ShowCubeWalk3D[Walk3D[75]]**

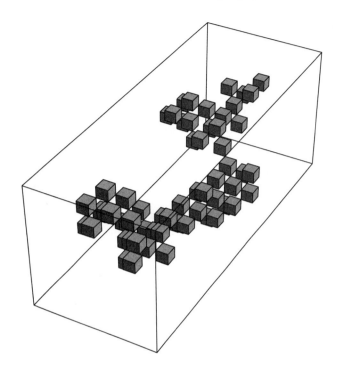

Out[7]= -Graphics3D-

References

M. Abell and J. Braselton. *Mathematica by Example.* Academic Press, Inc., Boston, 1992.

M. Abell and J. Braselton. *The Mathematica Handbook.* Academic Press, Inc., Boston, 1992.

M. Abell and J. Braselton. *Differential Equations with Mathematica.* Academic Press, Inc., Boston, 1993.

G. Baumann. *Mathematica in der theoretischen Physik.* Springer-Verlag, Heidelberg, 1993.

G. Baumann. *Mathematica in Theoretical Physics.* TELOS/Springer-Verlag, Santa Clara, 1995.

J. L. Bentley and J. H. Friedman. Algorithms for reporting and counting geometric intersections. Technical Report C-28, IEEE Transactions on Computing, 1979.

N. Blachman. *Mathematica, A Practical Approach.* Prentice-Hall, Englewood Cliffs, 1991.

N. Blachman. *Mathematica Quick Reference, Version 2.* Addison-Wesley Publishing Company, Redwood City, 1992.

P. Boyland, J. Keiper, and E. Martin. Guide to standard *Mathematica* packages. *Mathematica* technical report, Wolfram Research, Inc., Champaign, 1992.

B. Braden, D. Kurg, P. McCartney, and S. Wilkinson. *Discovering Calculus with Mathematica.* John Wiley & Sons, New York, 1991.

D. Brown, W. Davis, H. Porta, and J. Uhl. *CALCULUS&Mathematica.* Addison-Wesley Publishing Company, Reading, 1993.

D. C. M. Burbulla and C. T. J. Dodson. *Self-Tutor for Computer Calculus Using Mathematica.* Prentice-Hall, Englewood Cliffs, 1992.

R. L. Burden and J. D. Faires. *Numerical Analysis.* PWS-Kent Publishing Co., Boston, 4th edition, 1989.

J. Casti. *Reality Rules: I, Picturing the World in Mathematics—The Fundamentals.* John Wiley & Sons, Inc., New York, 1992.

J. Casti. *Reality Rules: II, Picturing the World in Mathematics—The Frontier.* John Wiley & Sons, Inc., New York, 1992.

R. E. Crandall. *Mathematica for the Sciences.* Addison-Wesley Publishing Company, Redwood City, 1991.

R. E. Crandall. Personal communication, March 1993.

R. E. Crandall. *Projects in Scientific Computation.* TELOS/Springer-Verlag, Santa Clara, 1994.

P. Crooke and J. Ratcliffe. *Guidebook to Calculus with Mathematica.* Wadsworth Publishing, Pacific Grove, 1991.

J. C. Culioli. *Introduction to Mathematica.* Ellipses, Paris, 1991.

J. H. Davenport, Y. Siret, and E. Tournier. *Computer Algebra, Systems and Algorithms for Algebraic Computation.* Academic Press, Inc., Boston, 1988.

W. Ellis and E. Lodi. *A Tutorial Introduction to Mathematica.* Brooks Cole Publishing Company, Pacific Grove, 1991.

J. Feagin. *Quantum Methods with Mathematica.* TELOS/Springer-Verlag Publishers, Santa Clara, 1993.

J. Finch and M. Lehman. *Exploring Calculus with Mathematica.* Addison-Wesley Publishing Company, Redwood City, 1992.

M. Gardner. *Fractal Music, Hypercards and More . . ..* W. H. Freeman and Company, New York, 1992.

M. Garey and D. Johnson. *Computers and Intractability: A Guide to the Theory of NP-Completeness.* W. H. Freeman, San Francisco, 1979.

R. J. Gaylord and P. R. Wellin. *Computer Simulations with Mathematica, Explorations in Complex Physical and Biological Systems.* TELOS/Springer-Verlag, Santa Clara, 1995.

O. Gloor, B. Amrhein, and R. Maeder. *Illustrated Mathematica.* TELOS/Springer-Verlag, Santa Clara, 1995.

M. Golin and R. Sedgewick. Analysis of a simple yet efficient convex hull algorithm. In *4th Annual Symposium on Computational Geometry.* ACM, 1988.

G. H. Golub and C. F. Van Loan. *Matrix Computations.* Johns Hopkins University Press, Baltimore, 1983.

R. L. Graham. An efficient algorithm for determining the convex hull of a finite planar set. *Information Processing Letters*, 1, 1972.

R. L. Graham, D. E. Knuth, and O. Patashnik. *Concrete Mathematics.* Addison-Wesley Publishing Company, Reading, 1989.

A. Gray. *Modern Differential Geometry of Curves and Surfaces.* CRC Press, New York, 1993.

A. Gray, M. Mezzino, and M. Pinsky. *Ordinary Differential Equations with Mathematica.* TELOS/Springer-Verlag, Santa Clara, 1995.

J. Gray. *Mastering Mathematica.* Academic Press, Inc., Cambridge MA, 1994.

T. Gray and J. Glynn. *Exploring Mathematics with Mathematica.* Addison-Wesley Publishing Company, Redwood City, 1991.

T. Gray and J. Glynn. *The Beginners Guide to Mathematica 2.* Addison-Wesley Publishing Company, Redwood City, 1992.

H. Gutowitz, editor. *Cellular Automata: Theory and Experiment.* The MIT Press, Cambridge, 1991.

A. Hayes. Sums of cubes of digits, driven to abstraction. *Mathematica in Education*, 1(4):3–11, 1992.

M. H. Hoft. *Laboratories for Calculus I Using Mathematica.* Addison-Wesley Publishing Company, Reading, 1992.

R. Honsberger. *Mathematical Gems II.* The Dolciani Mathematical Expositions, Number Two. The Mathematical Association of America, Providence, 1976.

D. Jacobson. Floating point in *Mathematica*. *The Mathematica Journal*, 2(3):42–46, 1992.

R. A. Jarvis. On the identification of the convex hull of a finite set of points in the plane. *Information Processing Letters*, 2, 1973.

S. G. Kang, G. W. Nam, and G. C. Jun. *Mathematica?!* Sungandang, Seoul, 1993.

S. Kaufmann. *Mathematica als Werkzeug, Eine Einfuhrung mit Anwendungsbeispielen*. Birkhauser Verlag, Basel, 1992.

S. Kaufmann. *Mathematica as a Tool*. Birkhauser, Basel, 1994.

V. Klee and S. Wagon. *Old and New Unsolved Problems in Plane Geometry and Number Theory*, volume 11 of *The Dolciani Mathematical Expositions*. The Mathematical Association of America, Providence, 1991.

D. E. Knuth. *The Art of Computer Programming*, volume 1, Fundamental Algorithms. Addison-Wesley Publishing Company, Reading, 2nd edition, 1973.

D. E. Knuth. *The Art of Computer Programming*, volume 2, Seminumerical Algorithms. Addison-Wesley Publishing Company, Reading, 2nd edition, 1981.

M. Kofler. *Mathematica. Einfuhrung und Leitfaden fur den Praktiker*. Addison-Wesley, Deutschland, Berlin, 1993.

S. Koike. *Mathematica: Introduction to Algebraic Computation*. Gijutsuhyooronsha, Tokyo, 1992.

R. R. Korfhage. *Discrete Computational Structures*. Academic Press, Inc., Orlando, 2nd edition, 1984.

J. C. Lagarias. The $3x + 1$ problem. *The American Mathematical Monthly*, 92:3–23, 1985.

E. Lawler, J. K. Lenstra, A. H. G. Rinnooy Kan, and D. B. Shmoys. *The Traveling Salesman Problem*. John Wiley & Sons, New York, 1985.

S. Lin. Computer solutions of the traveling salesman problem. *Bell System Technical Journal*, 44:2245–2269, 1965.

T. A. McMahon and J. T. Bonner. *On Size and Life*. Scientific American Books, Inc., New York, 1983.

R. E. Maeder. *Programming in Mathematica*. Addison-Wesley Publishing Company, Redwood City, 2nd edition, 1991.

R. E. Maeder. The design of the *Mathematica* programming language. *Dr. Dobbs Journal*, 17(4):86, 1992.

R. E. Maeder. *The Mathematica Programmer*. Academic Press, Inc., Cambridge MA, 1994.

B. Mandelbrot. *The Fractal Geometry of Nature*. W. H. Freeman and Company, New York, 1982.

Mathematica in Education and Research. TELOS/Springer-Verlag Publishers, Santa Clara.

The Mathematica Journal. Miller-Freeman Publishers, Inc., San Francisco.

M. V. Mathews. *The Technology of Computer Music*. MIT Press, Cambridge, 1969.

A. Nijenhuis and H. Wilf. *Combinatorial Algorithms*. Academic Press, Inc., New York, 2nd edition, 1978.

J. O'Rourke. *Computational Geometry in C*. Cambridge University Press, Cambridge, 1994.

E. Packel and S. Wagon. *Animating Calculus, Mathematica Notebooks for the Laboratory*. W.H. Freeman, New York, 1993.

J. R. Pierce. *The Science of Musical Sound*. W. H. Freeman and Company, New York, 1983.

L. K. Platzman and J. J. Bartholdi. Spacefilling curves and the planar traveling salesman problem. *Journal Assoc. for Computing Machinery*, 36:719–737, 1989.

W. Poundstone. *The Recursive Universe*. Oxford University Press, Oxford, 1985.

F. P. Preparata and M. I. Shamos. *Computational Geometry: An Introduction.* Springer-Verlag Publishing Co., New York, 1985.

E. M. Reingold and J. S. Tilford. Tidier drawings of trees. *IEEE Trans. Software Eng.*, SE-7:223–228, March 1981.

D. J. Rosenkrantz, R. E. Stearns, and P. M. Lewis. An analysis of several heuristics for the traveling salesman problem. *SIAM Journal of Computing*, 6:563–581, 1977.

T. D. Rossing. *The Science of Sound.* Addison-Wesley Publishing Company, Reading, 2nd edition, 1990.

B. Rust and W. R. Burrus. *Mathematical Programming and the Numerical Solution of Linear Equations.* American Elsevier Publishing Co., New York, 1972.

R. Sedgewick. *Algorithms.* Addison-Wesley Publishing Company, Reading, 2nd edition, 1988.

A. A. Sfeir. *Une Nouvelle Approche Du Calcul Scientifique.* Angkor Editeur and RITME Informatique, France, 1993.

M. I. Shamos and D. Hoey. Closest-point problems. In *16th Annual Symposium on Foundations of Computer Science.* IEEE, 1975.

R. Shepard. The analysis of proximities: Multidimensional scaling with an unknown distance factor. *Psychometrics*, 27:125–140, 1962.

R. D. Skeel and J. B. Keiper. *Elementary Numerical Computing with Mathematica.* McGraw-Hill, Inc., New York, 1993.

S. Skiena. *Implementing Discrete Mathematics: Combinatorics and Graph Theory with Mathematica.* Addison-Wesley Publishing Company, Redwood City, 1990.

D. E. Thomsen. Making music fractally. *Science News*, 117:187, 1980.

M. Trott. *The Mathematica Guidebook.* TELOS/Springer-Verlag, Santa Clara, 1995.

I. Vardi. *Computational Recreations in Mathematica.* Addison-Wesley Publishing Company, Redwood City, 1991.

I. Vardi. *Introduction to Symbolic Computation.* TELOS/Springer-Verlag, Santa Clara, 1995.

H. Varian, editor. *Economic and Financial Modeling with Mathematica.* TELOS/Springer-Verlag Publishers, Santa Clara, 1993.

R. F. Voss and J. Clarke. $1/f$ noise in music and speech. *Nature*, 258:317–318, 1975.

R. F. Voss and J. Clarke. $1/f$ noise in music. *The Journal of the Acoustical Society of America*, 63:258–263, 1978.

D. Vvedensky. *Partial Differential Equations with Mathematica.* Addison-Wesley Publishing Company, Redwood City, 1993.

S. Wagon. *Mathematica in Action.* W. H. Freeman and Company, New York, 1991.

G. H. Weiss. Random walks and their applications. *American Scientist*, 71:65–71, 1983.

T. Wickham-Jones. *Computer Graphics with Mathematica.* TELOS/Springer-Verlag Publishers, Santa Clara, 1994.

S. Wolfram. *Mathematica, A System for Doing Mathematics by Computer.* Addison-Wesley Publishing Company, Redwood City, 2nd edition, 1991.

S. Wolfram. *Mathematica Reference Guide.* Addison-Wesley Publishing Company, Redwood City, 1992.

Index

Since this field is fast-moving, we expect updates and changes to occur that might necessitate sending you the most current pertinent information by paper, electronic media, or both, regarding *An Introduction to Programming with Mathematica®, Second Edition*. Therefore, in order to not miss out on receiving your important update information, please fill out this card and return it to us promptly. Thank you.

Name: _____

Title: _____

Company: _____

Address: _____

City: _____ State: _____ Zip: _____

Country: _____ Phone: _____

E-mail: _____

Areas of Interest/Technical Expertise: _____

Comments on this Publication: _____

_____

_____

_____

☐ Please check this box to indicate that we may use your comments in our promotion and advertising for this publication.

Purchased from: _____
Date of Purchase: _____

☐ Please add me to your mailing list to receive updated information on *An Introduction to Programming with Mathematica®, Second Edition* and other TELOS publications.

☐ I have a   ☐ IBM compatible   ☐ Macintosh   ☐ UNIX   ☐ other

Designate specific model _____

THE
ELECTRONIC
LIBRARY
OF
SCIENCE

lay!

PLEASE TAPE HERE

FOLD HERE

‖‖‖‖

BUSINESS REPLY MAIL

FIRST CLASS MAIL PERMIT NO. 1314 SANTA CLARA, CA

POSTAGE WILL BE PAID BY ADDRESSEE

THE
ELECTRONIC
LIBRARY
OF
SCIENCE

3600 PRUNERIDGE AVE STE 200
SANTA CLARA CA 95051-9835